Energy Systems in Electrical Engineering

Series Editor

Muhammad H. Rashid, Florida Polytechnic University, Lakeland, USA

Energy Systems in Electrical Engineering is a unique series that aims to capture advances in electrical energy technology as well as advances electronic devices and systems used to control and capture other sources of energy. Electric power generated from alternate energy sources is getting increasing attention and supports for new initiatives and developments in order to meet the increased energy demands around the world. The availability of computer–based advanced control techniques along with the advancement in the high-power processing capabilities is opening new doors of opportunity for the development, applications and management of energy and electric power. This series aims to serve as a conduit for dissemination of knowledge based on advances in theory, techniques, and applications in electric energy systems. The Series accepts research monographs, introductory and advanced textbooks, professional books, reference works, and select conference proceedings. Areas of interest include, electrical and electronic aspects, applications, and needs of the following key areas:

- Biomass and Wastes Energy
- Carbon Management
- Costs and Marketing
- Diagnostics and Protections
- Distributed Energy Systems
- Distribution System Control and Communication
- Electric Vehicles and Tractions Applications
- Electromechanical Energy Conversion
- Energy Conversion Systems
- Energy Costs and Monitoring
- Energy Economics
- Energy Efficiency
- Energy and Environment
- Energy Management, and Monitoring
- Energy Policy
- Energy Security
- Energy Storage and Transportation
- Energy Sustainability
- Fuel Cells
- Geothermal Energy
- Hydrogen, Methanol and Ethanol Energy
- Hydropower and Technology
- Intelligent Control of Power and Energy Systems
- Nuclear Energy and Technology
- Ocean Energy
- Power and Energy Conversions and Processing
- Power Electronics and Power Systems
- Renewable Energy Technologies
- Simulation and Modeling for Energy Systems
- Superconducting for Energy Applications
- Tidal Energy
- Transport Energy

The books of this series are reviewed in a single blind peer review process by the main editor and/or the advisory board.

Kumari Namrata · R. P. Saini · D. P. Kothari

Wind and Solar Energy Systems

Springer

Kumari Namrata
Department of Electrical Engineering
National Institute of Technology
Jamshedpur
Jamshedpur, Jharkhand, India

R. P. Saini
Department of Hydro and Renewable
Energy
Indian Institute of Technology Roorkee
Roorkee, Uttarakhand, India

D. P. Kothari
Visvesvaraya National Institute
of Technology
Nagpur, Maharashtra, India

ISSN 2199-8582 ISSN 2199-8590 (electronic)
Energy Systems in Electrical Engineering
ISBN 978-981-99-9709-1 ISBN 978-981-99-9710-7 (eBook)
https://doi.org/10.1007/978-981-99-9710-7

© The Editor(s) (if applicable) and The Author(s), under exclusive license to Springer Nature
Singapore Pte Ltd. 2024

This work is subject to copyright. All rights are solely and exclusively licensed by the Publisher, whether
the whole or part of the material is concerned, specifically the rights of translation, reprinting, reuse
of illustrations, recitation, broadcasting, reproduction on microfilms or in any other physical way, and
transmission or information storage and retrieval, electronic adaptation, computer software, or by similar
or dissimilar methodology now known or hereafter developed.
The use of general descriptive names, registered names, trademarks, service marks, etc. in this publication
does not imply, even in the absence of a specific statement, that such names are exempt from the relevant
protective laws and regulations and therefore free for general use.
The publisher, the authors and the editors are safe to assume that the advice and information in this book
are believed to be true and accurate at the date of publication. Neither the publisher nor the authors or
the editors give a warranty, expressed or implied, with respect to the material contained herein or for any
errors or omissions that may have been made. The publisher remains neutral with regard to jurisdictional
claims in published maps and institutional affiliations.

This Springer imprint is published by the registered company Springer Nature Singapore Pte Ltd.
The registered company address is: 152 Beach Road, #21-01/04 Gateway East, Singapore 189721,
Singapore

Paper in this product is recyclable.

*In a heartfelt tribute this book is dedicated to my dearest parents (**Dr. Bimal Kumar and Mrs. Nirmala Kumar**), whose unwavering guidance, faith, and unconditional love have shaped me into the person I am today.*

and

*To my husband **Prof. (Dr.) Om Shankar**, whose boundless love, strength, motivation, and devotion continue to inspire me every day.*

Preface

In an era where the world faces pressing challenges related to energy consumption and environmental sustainability, the exploration and utilization of renewable energy sources have become critical. Among these sources, solar and wind energies stand out as two of the most promising and rapidly growing alternatives to traditional fossil fuels. The book you hold in your hands, *Wind and Solar Energy Systems*, serves as a comprehensive guide and a source of inspiration for anyone interested in understanding, harnessing, and advocating for these clean, abundant, and endlessly renewable resources.

As we enter the twenty-first century, the need for sustainable energy solutions has never been more urgent. The overreliance on fossil fuels, coupled with the environmental consequences of their extraction and combustion, has led to detrimental effects on our planet. Climate change, air pollution, and resource depletion are just a few of the challenges we face, necessitating a shift towards greener alternatives. Solar and wind energy, with their immense potential, offer us a beacon of hope in our quest for a cleaner and more sustainable future.

This book is a testament to the remarkable progress that has been made in the field of renewable energy. Its pages delve into the science, technology, and practical applications of solar and wind energy, providing readers with a comprehensive understanding of these systems. Whether you are a student, an engineer, a policymaker, or simply a concerned citizen, this book offers valuable insights and practical knowledge that can empower you to contribute to the transition to renewable energy sources.

Solar energy, harnessed by capturing the power of the sun's rays, has experienced a tremendous leap forward in recent years. The advancements in solar photovoltaic (PV) technology have made it more affordable, efficient, and accessible than ever before. From rooftop installations on residential buildings to large-scale solar farms, solar energy is transforming the way we generate electricity. The book explores the fundamentals of solar energy, its conversion mechanisms, and the latest advancements in PV technology. It also delves into the integration of solar power into the existing energy grid and the potential for off-grid applications in remote areas.

Wind energy, derived from the natural power of the wind, has also emerged as a leading renewable energy source. Wind turbines, with their majestic presence on the landscape, capture the kinetic energy of the wind and convert it into electricity. The book provides an in-depth exploration of wind energy, covering topics such as wind turbine design, wind resource assessment, and the integration of wind power into the electrical grid. It also examines the environmental impacts of wind energy and the measures taken to mitigate potential challenges.

Beyond the technical aspects, *Wind and Solar Energy Systems* highlights the economic, social, and environmental benefits of transitioning to renewable energy systems. It sheds light on the job opportunities, economic growth, and energy independence that can be achieved through a renewable energy revolution. Furthermore, it underscores the significance of renewable energy in mitigating climate change, reducing carbon emissions, and preserving our planet for future generations.

The authors of this book, experts in the field of renewable energy, have skillfully synthesized a vast body of knowledge into a coherent and accessible resource. Their passion for the subject matter is evident in the depth of research, the clarity of explanations, and the inclusion of realworld case studies and examples. Whether you are seeking to expand your understanding of renewable energy or looking for guidance in implementing solar or wind projects, this book will undoubtedly become an invaluable companion on your journey.

In conclusion, *Wind and Solar Energy Systems* serves as a beacon of hope in the quest for a sustainable and cleaner future. By exploring the science, technology, and practical applications of solar and wind energy, this book equips readers with the knowledge and inspiration to contribute to the renewable energy revolution. Let us embark on this transformative journey together and harness the power of the sun and wind to create a brighter and more sustainable world for generations to come.

Jamshedpur, India   Dr. Kumari Namrata
Roorkee, India   Prof. (Dr.) R. P. Saini
Nagpur, India   Prof. (Dr.) D. P. Kothari

Acknowledgements

It is a great pleasure for me to express my respect and deep sense of gratitude to my Ph.D. supervisor Prof. S.P. Sharma, Department of Mechanical Engineering, National Institute of Technology, Jamshedpur for his wisdom, vision, expertise, guidance, enthusiastic involvement and persistent encouragement during the planning and development of this book.

Writing a book is possible only if the right ambience exists both at the workplace and at home. NIT Jamshedpur has been my workplace for many years. I have grown up with the institute, worked for its cause and its good, and savoured the pleasure of having wonderful colleagues and students.

I am highly obliged to the honourable Director, National Institute of Technology, Jamshedpur for providing all the facilities, help and encouragement for carrying out this work.

I would like to express my gratitude to my research scholar Mr. Mantosh Kumar, whose dedication and patience greatly contributed to the completion of this book. I also extend my thanks to Dr. Askhit Samadhiya whose support and guidance were unconditional in initiating this work. Additionally, I am grateful to Dr. Nishant Kumar for his valuable contributions from various perspectives.

I would also like to thank my department and person who have directly or indirectly supported me for this achievement.

I express my gratitude to my brothers Er. Kumar Dhiraj, Dr. Kumar Gaurav and Dr. Kumar Vaibhav for their continuous encouragement and support. I also appreciate the patience, support,love and concern of my daughter Ms. Harshita Shankar and sons Harsh Shankar and Harshit Shankar for the completing this book.

I would like to thank my parents for raising me in a way to believe that I can achieve anything in life with hard work and dedication.

I would like to take a moment to express my deepest gratitude to Springer Nature for publishing my book. Working with Priya Vyas, Senior Editor and Silky Abhay Sinha has been an incredible experience, and I will be always obliged to them for their constant support throughout the journey of this book. I look forward to the continued success of ("*Wind and Solar Systems*") with your esteemed publishing house.

Above all, I would like to express my gratitude to all mighty for showering the blessings as the supreme source of knowledge and wisdom in this world.

Introduction

Solar and wind energy are at the forefront of the global transition towards a sustainable and clean energy future. As the world grapples with the challenges posed by climate change and the need to reduce greenhouse gas emissions, renewable energy sources have emerged as the key to mitigating environmental impacts while meeting the ever-increasing demand for power. Among these renewable sources, solar and wind energy stand out for their immense potential and versatility.

This book, *Wind and Solar Energy Systems*, delves into the fundamental concepts, technical aspects, and practical applications of solar and wind energy systems. It provides a comprehensive overview of these two renewable energy sources, exploring their availability, characteristics, environmental impacts, and the methods employed to harness their power. The book is divided into eight chapters, each addressing a specific aspect of solar and wind energy.

Chapter 1 serves as an introduction, examining the diverse energy resources available to us, their unique characteristics, and the environmental implications associated with their extraction and use. It sets the stage for a deeper exploration of solar and wind energy, highlighting the importance of these renewable alternatives in a world seeking sustainable solutions.

Chapter 2 delves into the fundamentals of solar energy, elucidating the principles behind capturing sunlight and converting it into usable energy. From the basic concepts of photovoltaic technology to the intricacies of solar radiation and its measurement, this chapter lays the groundwork for understanding the potential and limitations of solar energy systems.

Building upon the foundation laid in the previous chapter, Chap. 3 introduces the reader to photovoltaic solar energy. It explores the various types of solar cells, their materials, and the mechanisms by which they convert sunlight into electricity. The chapter also delves into the design, operation, and performance evaluation of photovoltaic systems, providing a comprehensive understanding of their applications and potential for future development.

Shifting the focus to wind energy, Chap. 4 presents an overview of this dynamic renewable resource. It explores the basic principles of wind power generation, discussing the factors that influence wind patterns, wind turbine technology, and the environmental considerations associated with wind energy projects. The chapter aims to equip readers with the knowledge necessary to comprehend and appreciate the immense power harnessed from the wind.

Chapter 5 builds upon the foundations laid in the previous chapter by delving into the intricacies of wind energy conversion systems. It explores different types of wind turbines, their components, and the principles of operation. From horizontal-axis and vertical-axis turbines to offshore wind farms, this chapter explores the diverse approaches to capturing wind energy and converting it into electricity.

As the integration of solar and wind energy systems into existing power grids becomes increasingly critical, Chap. 6 focuses on grid integration techniques. It examines the challenges and opportunities associated with the large-scale deployment of solar and wind energy, discussing grid stability, power quality, and the role of advanced control systems in ensuring a seamless integration of renewable energy into the existing infrastructure.

Chapter 7 explores solar collectors and thermal conversion, showcasing the applications of solar energy beyond electricity generation. It explores the principles of solar thermal technology, the various types of solar collectors, and their applications in heating, cooling, and industrial processes. The chapter highlights the versatility of solar energy and its potential to revolutionize a wide range of sectors.

Finally, Chap. 8 presents an in-depth exploration of solar ponds, a lesser-known but promising technology for harnessing solar energy. It elucidates the principles behind solar pond design, operation, and the conversion of solar energy into heat. The chapter explores the applications of solar ponds in heating, desalination, and other industrial processes, underscoring their potential as a cost-effective and sustainable energy solution.

Wind and Solar Energy Systems aims to be a comprehensive guide for students, researchers, and professionals seeking to deepen their understanding of the principles, technologies, and applications of solar and wind energy. It provides a holistic perspective on these renewable energy sources, addressing not only their technical aspects but also the environmental and socio-economic implications associated with their adoption.

The chapters in this book are carefully structured to provide a progressive learning experience, starting from the fundamentals and gradually delving into more advanced topics. Each chapter is written in a clear and accessible manner, with a focus on presenting complex concepts in an understandable way, making it suitable for readers with varying levels of technical expertise.

Throughout the book, real-world examples, case studies, and practical applications are incorporated to enhance the reader's understanding and demonstrate the diverse range of possibilities offered by solar and wind energy. The intention is to inspire and encourage readers to explore innovative solutions and contribute to the ongoing energy transition.

It is important to note that this book does not claim to be an exhaustive compendium on solar and wind energy. Rather, it aims to provide a solid foundation for further exploration and learning. The rapidly evolving nature of renewable energy technologies necessitates continuous research and development, and this book is intended to serve as a stepping stone for those eager to delve deeper into the field.

The authors of this book bring together their expertise and knowledge in the fields of renewable energy, engineering, and environmental science. Their collective experience enables them to provide a comprehensive and multidisciplinary perspective on solar and wind energy, ensuring that readers gain a holistic understanding of the subject matter.

In closing, *Wind and Solar Energy Systems* is an invitation to embark on a journey through the fascinating world of renewable energy. It is our hope that this book will empower readers to embrace sustainable energy solutions, foster innovation, and contribute to a greener and more sustainable future.

Kumari Namrata
R. P. Saini
D. P. Kothari

Contents

1 Energy Resources: Availability, Characteristics, and Environmental Impacts 1
- 1.1 Introduction 1
  - 1.1.1 Human Beings, Oil Crises, and the Energy Revolution 1
  - 1.1.2 Oil Crisis 1973 2
  - 1.1.3 Energy Evolution 3
- 1.2 Different Forms and Conversion of Energy 4
- 1.3 Global Energy Outlook—History and Current Scenario 8
- 1.4 Energy Scenario—India 16
- 1.5 Classification of Energy Resources 19
- 1.6 Advantages and Disadvantages of Conventional Energy Sources 21
- 1.7 Future of Energy Renewables 22
- 1.8 Origin of Renewable Energy Resources 23
  - 1.8.1 Solar Energy 23
  - 1.8.2 Wind Energy 24
  - 1.8.3 Geothermal Energy 25
  - 1.8.4 Ocean Thermal Energy Conversion (OTEC) 26
  - 1.8.5 Wave Energy 27
  - 1.8.6 Tidal Energy 28
  - 1.8.7 Biomass Energy 29
- 1.9 Impact of Renewables on Energy Sector 30
- 1.10 Renewable Energy: Global Statistics and Analysis 32
- 1.11 Renewable Energy Regulations and Policies 34
- 1.12 Environmental Aspects of Energy Sources 41
- References 49

2 Fundamentals of Solar Energy ... 53

2.1	Introduction ...	54
2.2	The Sun ...	55
2.3	The Solar Constant ...	56
2.4	Variation of Spectral Distribution and Extra-Terrestrial Radiation ...	58
	2.4.1 Spectral Distribution ...	58
	2.4.2 Variation of Extra-Terrestrial Radiation ...	59
2.5	Solar Irradiance Falling at the Earth's Surface ...	61
2.6	Solar Radiation Geometry ...	65
2.7	Sun's Apparent Motion ...	69
	2.7.1 Day Length, Sunrise, and Sunset ...	69
	2.7.2 Local Apparent Time ...	71
	2.7.3 Why Solar Radiation Data is Needed? ...	73
2.8	Measurement of Solar Radiation ...	74
2.9	Calculating the Sun Radiation Availability ...	78
	2.9.1 Monthly Average Daily Global Radiation ...	78
	2.9.2 Monthly Average Daily Diffuse Radiation ...	83
	2.9.3 Monthly Average Hourly Global Radiation ...	85
	2.9.4 Monthly Average Hourly Diffuse Radiation ...	86
	2.9.5 Hourly Global Beam and Diffuse Radiation Under Clear Skies ...	88
2.10	Solar Radiation on Tilted Surfaces ...	90
	References ...	99

3 Introduction to Photovoltaic Solar Energy ... 101

3.1	Fundamentals of Photovoltaic ...	101
	3.1.1 Semiconductor Materials ...	103
	3.1.2 Photon Energy ...	105
	3.1.3 A P–N Junction ...	107
	3.1.4 Photovoltaic Effect ...	109
	3.1.5 Photovoltaic Cell Materials ...	109
3.2	Types of Photovoltaic Cells and Efficiency ...	110
	3.2.1 Amorphous PV Cells ...	110
	3.2.2 Monocrystalline PV Cells ...	110
	3.2.3 Polycrystalline PV Cells ...	112
3.3	Analytical Model of the Solar Cell ...	113
	3.3.1 Analysis of the Single-Diode Model ...	113
	3.3.2 Analysis of the Two-Diode Model ...	114
3.4	Electrical Parameters of Solar Cell ...	115
	3.4.1 Current in a Short Circuit ...	116
	3.4.2 Voltage on an Open Circuit ...	116
	3.4.3 Fill Factor ...	116
	3.4.4 Maximum Power ...	117
	3.4.5 Solar Cell Efficiency ...	117

3.5	Electrical Characteristics of PV Cells	117
	3.5.1 PV Cell I-V Characteristics	118
	3.5.2 P–V Characteristics of PV Cell	118
3.6	Maximum Power Point Tracking (MPPT)	119
3.7	Effect of Parameters and Atmospheric Conditions on PV Cell Characteristics	120
3.8	Photovoltaic Modules and Array	120
	3.8.1 Theory and Construction	122
	3.8.2 Packing Factor of PV Module	123
	3.8.3 Efficiency of PV Module	123
3.9	Overview of Photovoltaic System Applications	124
3.10	Overview of Photovoltaic-Based Power System	124
	3.10.1 Standalone Photovoltaic System	124
	3.10.2 Grid-Connected Photovoltaic System	125
3.11	Power Converter Topologies for PV-Based Power System	126
	3.11.1 DC/DC Converters	126
	3.11.2 DC/AC Converters	127
3.12	Control of Photovoltaic-Based Power Systems	127
	3.12.1 Maximum Power Point Tracking (MPPT) Control	128
	3.12.2 DC/DC Converter Control	128
	3.12.3 DC/AC Inverter Control	129
References		134

4 Introduction to Wind Energy 135

4.1	Wind—The Resource	135
	4.1.1 The Nature of Wind	135
	4.1.2 Geographical Variations in the Wind	137
4.2	Worldwide Status of Wind Power	139
	4.2.1 Global Wind Power Statistics	139
	4.2.2 Indian Wind Power Statistics	143
	4.2.3 Environmental Aspects	149
4.3	Wind Energy Basics	151
	4.3.1 Power	151
	4.3.2 Air Density	153
	4.3.3 Swept Area	154
	4.3.4 Cube of Wind Speed	155
	4.3.5 Tower Height Effect	156
4.4	Analysis of Wind Data	158
	4.4.1 Average Wind Speed	158
	4.4.2 Wind Speed Distribution	159
	4.4.3 Wind Data Statistical Analysis	160
4.5	Overview of Wind Turbines and Its Components	162
	4.5.1 Introduction	162
	4.5.2 Classification of Wind Turbines	165
	4.5.3 Aerodynamics of Rotor	168

4.5.4	Transmission System	170
4.5.5	Generator	173
4.5.6	Power Electronics Interface	175
4.5.7	Control System	178

4.6 Power Coefficients and Characteristics ... 184
- 4.6.1 Introduction ... 184
- 4.6.2 Tip Speed Ratio ... 185
- 4.6.3 Wind Power Coefficient and Betz's Law ... 186
- 4.6.4 Power Coefficient Versus Tip Speed Ratio Curve ... 188

References ... 195

5 Wind Energy Conversion System ... 197

5.1 Introduction ... 197

5.2 Overview of Wind Turbine Topologies ... 198
- 5.2.1 Wind Turbine Architectures ... 199
- 5.2.2 Fixed and Variable-Speed Wind Turbines ... 200
- 5.2.3 Horizontal and Vertical Axis Wind Turbine ... 203
- 5.2.4 Stall and Pitch Aerodynamic Power Control ... 205

5.3 Generators for Wind Turbines ... 206

5.4 Power Electronics in Wind Energy ... 210
- 5.4.1 Soft Starters ... 210
- 5.4.2 Capacitor Bank ... 220
- 5.4.3 Rectifiers and Inverters ... 223
- 5.4.4 Frequency Converters ... 262
- 5.4.5 Maximum Power Point Tracking Control and Converter Control ... 263

References ... 271

6 Grid Integration Techniques in Solar and Wind-Based Energy Systems ... 273

6.1 Introduction ... 273
- 6.1.1 Integration of Small-Scale Generation into Grids ... 273
- 6.1.2 Large-Scale Generation Integration into Grids ... 278

6.2 Integration Issues Related to Wind and Solar Power ... 283
- 6.2.1 Consumer Requirements ... 283
- 6.2.2 Requirement for Wind Farm and Solar Farm Operators ... 285
- 6.2.3 The Integration Issues ... 289

6.3 Grid Requirements for Solar-Based Energy Systems ... 304
- 6.3.1 Power Quality Requirements ... 304
- 6.3.2 Response to Abnormal Grid Conditions ... 305
- 6.3.3 Anti-Islanding Requirements ... 306

6.4 Grid Requirements for Wind-Based Energy System ... 310
- 6.4.1 Voltage and Frequency Variation Under Normal Operation ... 311
- 6.4.2 Active and Reactive Power Control ... 311

Contents   xix

| | 6.4.3 | Behaviour Under Grid Disturbances | 317 |

6.4.3   Behaviour Under Grid Disturbances   317
6.4.4   Harmonic Requirements for Grid-Connected Wind Power System   319
6.5   Solar and Wind-Based Hybrid Renewable Energy Systems   320
    6.5.1   Hybrid Energy Systems   321
    6.5.2   Hybrid Energy System Characteristics   322
    6.5.3   Technology Used in Hybrid Energy   323
    6.5.4   Strategy for Implementation   330
    6.5.5   Constraints   333
    6.5.6   Issues and Challenges   334
    6.5.7   Applications   334
    6.5.8   Hybrid System Economics   335
References   339

7   Solar Collectors and Thermal Conversion   341
7.1   The Solar Option   341
    7.1.1   Low-Temperature Systems   343
    7.1.2   Medium-Temperature Systems   346
    7.1.3   High-Temperature Systems   348
7.2   Solar Collectors and Thermal Conversion   351
    7.2.1   Devices for Thermal Collection and Storage   353
7.3   Solar Concentrating Collectors   359
    7.3.1   Introduction   359
    7.3.2   Definitions   359
    7.3.3   Methods of Classification   360
    7.3.4   Types of Concentrating Collectors   360
    7.3.5   Thermal Analysis of Concentrating Collectors   362
7.4   Flat-Plate Collectors with Plane Reflectors   363
7.5   Cylindrical Parabolic Collector   364
    7.5.1   Description   364
    7.5.2   Orientation and Tracking Modes   366
    7.5.3   Performance Analysis   370
    7.5.4   Correlations Between the Overall Loss Coefficient and Heat Transfer   373
    7.5.5   A Numerical Example   379
    7.5.6   Parametric Study of Collector Performance (Kelkar 1982)   385
7.6   Compound Parabolic Collector (CPC)   391
    7.6.1   Geometry   391
    7.6.2   Tracking Requirements   394
    7.6.3   Performance Analysis   397
7.7   Paraboloid Disc Collector   402

7.8	Central Receiver Collector		403
	7.8.1	Heliostats	404
	7.8.2	Receiver	405
	7.8.3	Analysis	409
References			417

8 Solar Pond 421

8.1	Introduction	421
8.2	Working Principle of Solar Pond	422
8.3	Description of Solar Pond	424
8.4	Performance Analysis	425
	8.4.1 Transmissivity Based on Reflection–Refraction at the Air–Water Interface	426
	8.4.2 Transmissivity Based on Absorption	427
	8.4.3 Temperature Distribution and Collection Efficiency	430
References		439

About the Authors

Dr. Kumari Namrata is an accomplished academician and researcher with a rich background in Electrical Engineering. She earned her Ph.D. from the esteemed National Institute of Technology Jamshedpur, India, in 2017, complementing her academic journey that began with an M.Tech. in Power Systems in 2001 from the same institution. Her foundation in the field was laid with a B.Tech. in Electrical Engineering from the National Institute of Technology, Patna, in 1998. Currently holding the position of Associate Professor in the Department of Electrical Engineering at NIT Jamshedpur, she brings nearly 23 years of extensive experience in academia and research. Her academic contributions are substantiated by over 75 published research papers in esteemed journals and international conferences, guided five Ph.D. scholars, and 100 M.Tech. students. She has also authored two books and served as the editor of Springer's distinguished work, *Smart Energy and Advancement in Power Technologies*. Her scholarly pursuits are centred around vital areas such as solar power generation and conversion, solar radiation estimation, renewable energy-based system modelling and simulation, microgrid operation and control, hybrid energy systems, and renewable power system energy management. She is a dedicated **Board Member of IEEE Women in Systems Engineering** and a lifetime member of the Indian Society for Technical Education. She has been honoured with the distinguished **"Young Observer Award 2023"** for her unwavering commitment and remarkable impact on higher education.

Dr. R. P. Saini is currently a professor at the Department of Hydro and Renewable Energy Centre, Indian Institute of Technology Roorkee, Uttarakhand. He obtained his B.E. (Mechanical) from the University of Mysore, M.E. (Mechanical) from the Indian Institute of Technology Roorkee in 1989 and Ph.D. from Indian Institute of Technology Roorkee in 1996. His major areas of research interest include small hydropower-hydroturbine design and performance testing. Hydrokinetic energy, silt erosion in hydroturbines, optimal selection of SHP equipment, cost optimization of SHP schemes, solar energy-solar thermal energy utilization-performance enhancement of solar air heaters, integrated renewable energy systems, modelling of renewable energy systems, modelling of hybrid energy systems, etc. He has published more than 300 papers in reputed international journals. He has also pioneered and successfully completed 50 research and development projects. Multiple honours and awards are on the name of Dr. R. P. Saini. The Dr. A. P. J. Abdul Kalam Award is one of them.

Dr. D. P. Kothari Ex-Director IIT Delhi, VC, VIT Vellore, and Principal, VRCE Nagpur has published and presented 840 research papers, guided 57 Ph.D. scholars and 68 M.Tech. students, and authored 77 books in various allied areas. He has delivered several keynote addresses, 150 plus webinars. His research area is energy systems.

Chapter 1
Energy Resources: Availability, Characteristics, and Environmental Impacts

Abstract The field of energy resources is undergoing a transformative shift driven by factors such as an environmental concerns, technological advancements, and geopolitical considerations. This chapter provides a comprehensive overview of various energy resources, their characteristics, availability, and environmental impacts.

Beginning with an introduction to the historical context of energy revolutions and the 1973 oil crisis, it deals with the different forms of energy and their conversion processes. It then presents a detailed analysis of the global energy outlook, highlighting historical trends and the current scenario, with a specific focus on India's energy scenario.

Keywords Energy Resources · Environmental Impacts · Energy Revolution · Oil Crisis · Conversion of Energy

1.1 Introduction

1.1.1 Human Beings, Oil Crises, and the Energy Revolution

Human beings have needed and utilized energy at an increasing rate for their survival and well-being since they appeared on earth a few million years ago. The energy was primarily needed by early human beings in the form of food. They gained this by eating the plants or animals they had hunted. Their energy needs to be increased once they subsequently found fire because they started using wood and other biomass to fulfil their requirements for cooking meals and warming themself. As time went on, humans began to prepare land for agriculture. They domesticated animals and taught them to labour for them, which gave the use of energy a new dimension. When the need for energy increased, humans started to use the wind to propel ships and power windmills as well as the force of flowing water to turn water wheels. It would not be incorrect to claim that up until this point, humans were solely reliant on renewable energy sources to meet all of their energy demands, whether directly or indirectly. Several changes were made as a result of the Industrial Revolution, which

© The Author(s), under exclusive license to Springer Nature Singapore Pte Ltd. 2024
K. Namrata et al., *Wind and Solar Energy Systems*, Energy Systems in Electrical Engineering, https://doi.org/10.1007/978-981-99-9710-7_1

1

started in AD 1700 with the invention of the steam engine. For the first time, humans started utilizing coal as a major source of energy. Oil and natural gas started to be utilized extensively a short while after the internal combustion engine was developed (AD 1870). Energy was now readily available in a concentrated form as the era of fossil fuels, which used non-renewable sources, had begun. Fossil fuels and the development of heat engines made energy portable and gave man the much-needed mobility flexibility. For the first time, man was no longer constrained to a particular location, such as a swiftly flowing stream for powering a water wheel or a windy hill for powering a windmill. The invention of electricity and the growth of central power plants that either used fossil fuels or water power increased this flexibility. After the Second World War, nuclear energy became a viable energy source. Almost 50 years after the first major nuclear power plant was put into operation, several nations are already relying on nuclear energy for a sizable portion of their energy needs (Sukhatme and Nayak 1996). So nowadays, every nation uses a range of sources to meet its energy needs. Commercial and non-commercial sources can be broadly separated into these two groups. Wood, animal waste and agricultural waste are examples of non-commercial sources, whereas fossil fuels (coal, oil, and natural gas), hydroelectricity, nuclear power, and wind power are examples of commercial sources. In a developed country like the USA, most of the country's energy requirements are met by commercial sources, whereas in a developing country like India, both commercial and non-commercial sources are employed to a roughly comparable degree.

1.1.2 Oil Crisis 1973

The age of reliable and affordable oil came to an end in 1973. Organization of Petrol Exporting Countries (OPEC) disbanded in October of that year. This led to escalating price increases for different commercial energy sources, which further fuelled worldwide inflation. It is now clear those fossil fuel resources are swiftly running out and that the era of fossil fuels is progressively coming to an end. When it comes to oil and natural gas, this is especially true. All nation's governments took this issue very seriously, and for the first time, the need to find alternate energy sources became apparent. Alternative energy sources were given careful consideration, and significant resources were set aside to develop them. As a result, 1973 is remembered as the first "oil shock" year (Khan 2024). In the same decade, the globe experienced one more "oil shock" in 1979, which sharpened the focus on alternative energy sources. By the end of 1980, the cost of crude oil had increased 19 times from where it had been ten years earlier. In order to provide some indicators of the available reserves, it will be helpful to first look at the rates of consumption of the various energy sources.

1.1.3 Energy Evolution

Humans have been harnessing energy for thousands of years, starting with the use of fire for warmth and cooking. Over time, humans have developed increasingly sophisticated methods for harnessing energy, from the invention of the wheel to the creation of complex power grids. The evolution of word energy sources can be traced through key historical periods, from the earliest civilizations to the modern day (https://www.unep.org/resources/emissions-gap-report-2022).

- Early Energy Sources: Fire and Muscle Power

  The earliest human energy sources were simple and direct. Fire was the primary source of energy for warmth, light, and cooking. This was followed by the use of muscle power, as humans began to domesticate animals and use them for transportation, agriculture, and other tasks. The invention of the wheel around 3500 BCE allowed humans to harness the power of animals more effectively, enabling the creation of more complex societies and economies.

- Ancient Civilizations: Water and Wind Power

  Ancient China, Greece, and Rome all made significant contributions to the creation of energy sources. The Greeks and Romans used water wheels to power mills for grinding grain and other materials. The Chinese developed windmills for pumping water and grinding grain. These early water and wind power technologies were limited in their scope and efficiency, but they laid the foundation for future developments.

- The Industrial Revolution: Steam Power

  An important turning point in the development of energy sources occurred during the Industrial Revolution in the eighteenth and nineteenth centuries. James Watt's discovery of the steam engine in 1775 revolutionized the manner in which people utilize energy. Steam engines powered factories, trains, and ships, allowing for greater productivity and transportation. The development of coal mining also played a critical role in the Industrial Revolution, providing fuel for steam engines and other machinery.

- The Twentieth Century: Fossil Fuels and Nuclear Power

  The twentieth century saw significant changes in the evolution of energy sources. Fossil fuels became the primary energy source for transportation, heating, and the production of electricity as a result of the discovery and exploitation of oil and natural gas. A new form of energy was made possible with the advent of nuclear energy in the middle of the twentieth century, but it also brought up issues with safety and the environment.

- Renewable Energy: Solar, Wind, and Hydropower

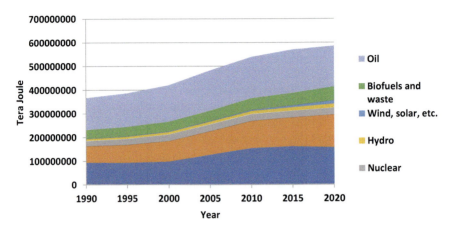

Fig. 1.1 Energy sources evolution (in Tera joule). (*Source* IEA–International Energy Agency) (Energy Agency 2021)

In the last few centuries, the concern has shifted towards renewable energy sources that are sustainable and less damaging to the environment. Solar power, which harnesses the energy of the sun through photovoltaic panels, has become increasingly popular for residential and commercial use. Another growing form of energy is wind power, which makes use of turbines to harness wind energy, particularly in regions with stable wind patterns. Hydropower, which uses the energy of moving water to generate electricity, has been used for centuries but is now being developed on a larger scale.

- The Future of Energy Sources

As technology advances and concerns about climate change and sustainability grow, the evolution of energy sources is likely to continue. New developments in battery storage and smart grids may make it easier to harness and store renewable energy sources. Advances in fusion energy may provide a new source of clean and sustainable energy. Whatever the future holds, the evolution of energy sources is a critical part of human history and will continue to shape our societies and economies for years to come. The evolution of the energy supply of different sources has been shown in Fig. 1.1 (Energy Agency 2021).

1.2 Different Forms and Conversion of Energy

The only currency that is accepted everywhere is energy. The conversion of ever-increasing volumes of ever-more-concentrated and more-variable kinds of energy has been essential to the evolution of human society. The conversion matrix of different forms of energy has been shown in Table 1.1 (Smil 1994).

1.2 Different Forms and Conversion of Energy

Table 1.1 Energy conversion of different sources (Smil 1994)

From / To	Electromagnetic	Chemical	Nuclear	Thermal	Kinetic	Electrical
Electromagnetic		Chemiluminescence	Nuclear bombs	Thermal radiation	Accelerating charges	Electromagnetic radiation
Chemical	Photosynthesis	Chemical processing		Boiling	Dissociation by radiolysis	Electrolysis
Nuclear	Gamma neutron reactions					
Thermal	Solar absorption	Combustion	Fission/Fusion	Heat exchange	Friction	Resistance heating
Kinetic	Radiometers	Metabolism	Radioactivity/ Nuclear bombs	Thermal expansion/ internal combustion	Gears	Electric motors
Electrical	Solar cells	Fuel cells/batteries	Nuclear batteries	Thermo electricity	Electricity generators	

Energy is necessary for human existence in all known forms. This fact makes it impossible to rank them in order of importance. The planetary and universal fluxes of energy and their local or regional expressions have shaped and limited much of history. Gravitational energy controls the fundamental properties of the universe, directing countless galaxies and star systems similar to our solar system. Moreover, gravity holds the atmosphere of our planet and keeps it orbiting the sun at the ideal distance, making the earth habitable.

The sun uses nuclear energy to power itself, as do all stars in the universe. Electromagnetic (solar or radiant) energy, the by-product of those thermonuclear reactions, travels to earth. A third of this enormous flow of energy is reflected by clouds and surfaces, and it has a wide wavelength spectrum, including visible light. The globe emits nearly all of the remaining energy after it has been transformed into thermal energy by the oceans, land, and atmosphere. Geothermal energy comes from both the radioactive decay of materials and the initial gravitational accretion of the earth's planetary mass. Grand tectonic processes, which continuously reorganize the oceans and continents and result in volcanic eruptions and earthquakes, are driven by these flows.

Photosynthesis only converts a very small portion of light energy into fresh chemical energy stored in plants. The unbreakable foundation for all higher life is offered by these stores. A mammal's physiological functions, including maintaining a steady body temperature, are all maintained by an animated metabolism that reorganizes nutrients into developing tissues. Moreover, digestion produces kinetic (mechanical) energy for moving muscles. Animals use their muscles to convert food into energy for things like reproduction, escape, and defence, but these actions are restricted by the size of their bodies and the accessibility of food.

By using tools and utilizing energy outside of their bodies, humans can transcend these physical boundaries. These extra somatic energies, unlocked by the human mind, are utilized for a wide range of activities; they act as both potent prime movers and fuels that emit heat during combustion. For people to transform this energy for their purposes, two requirements must be met. First, there must be sufficient amounts of the necessary energy flows (wind, water), potentials (animals, biomass, fossil, or nuclear fuels), or both. In order to capture or lease these flows and potentials in meaningful forms, humans must first execute the necessary activities or apply the controls. Energy supply triggers are influenced by information flow and a wide range of artefacts.

These inventions have included anything from straightforward levers and hammerstones to sophisticated fuel-burning engines and nuclear fission-capable reactors. It is simple to summarize in general qualitative terms the fundamental evolutionary and historical sequence of these advancements. Humans need nourishment, just like any other non-photosynthesizing organism. Their most fundamental energy requirement is this. Early hominids' methods of foraging and scavenging for food were remarkably similar to those of their ape relatives (Whiten and Widdowson 1992). Even though other primates are capable of manufacturing simple tools, only hominids have actively pursued this possibility.

1.2 Different Forms and Conversion of Energy

People now have several mechanical advantages thanks to tools when it comes to providing food, housing, and clothing. Humanity's ability to control fire significantly increased its range of settlement and further separated humans from other animals (Goudsblom 1992). Later, improved tools allowed for the domestication of animals, the creation of intricate muscle-powered machines, and the conversion of a tiny portion of the enormous kinetic energy of wind and water into usable mechanical power.

The useable power that could be controlled by humans was substantially increased by these new prime movers, but for a very long time, their practical use was constrained by the nature and size of the captured flows. This was most obviously the case with sailing. The fundamental patterns of atmospheric and oceanic circulation are governed by solar energy inputs. Location and the interaction of land and sea masses shape dominant wind flows and persistent ocean currents. Even after nearly three centuries of sailing in the Pacific, these great flows guided European transatlantic trips in the late fifteenth century to the Caribbean and kept the Spaniards from discovering Hawaii.

The ability to convert plant chemical energy into thermal energy was made possible by the advent of controlled combustion in fireplaces, stoves, and furnaces. In addition to using it directly in homes, society started using this heat for industrial applications. They were able to use it to treat and finish countless goods, including bricks, metals, and other materials. All of these conventional methods of producing heat were made more prevalent and effective through the burning of fossil fuels. The conversion of thermal energy from the combustion of fossil fuels to mechanical energy was made possible by a number of important discoveries. These inventions began with steam and internal combustion engines and then came gas turbines and rockets. Fossil fuels, together with the kinetic energies of wind and water, have all been used to produce electrical energy since the end of the nineteenth century in all modernizing societies. Since the 1950s, a growing number of countries have also been harnessing nuclear energy produced by the fissioning of heavy atoms to generate power. Electricity and fossil fuels have spawned a brand-new, high-energy society that has quickly taken over the entire world.

What is much more difficult than outlining this grand sequence is to set these developments in a broader perspective. This attempt requires a number of approaches straightforward and others more complex. It is relatively easy to evaluate the advantages and drawbacks of the more potent prime movers and the better fuels. It is much more difficult to identify the factors that promote or impede innovation—that is the processes that enable a society to take the intellectual and technical steps needed to unlock great energy potentials. As these changes take place, they have important consequences for farming, industry, transport, settlement patterns, warfare and the earth's environment. An appraisal of these impacts is equally complex. But no serious attempt to address these matters could be solely qualitative. Quantitative accounts are essential in order to appreciate not only the magnitude and the limits of human achievements but also their consequences. Naturally, their comprehension requires knowledge of basic concepts and measures.

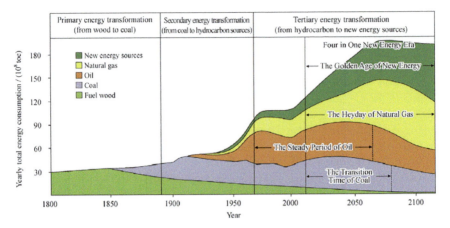

Fig. 1.2 Trends and forecasts of global energy consumption in a tonne of oil equivalent (toe) (https://www.iea.org/reports/key-world-energy-statistics-2021). (*Source* IEA) (Key World Energy Statistics 2021)

1.3 Global Energy Outlook—History and Current Scenario

Every energy crisis has historical precedents, and the current market stress is being compared to the worst energy disruptions in recent memory, most notably the oil shocks of the 1970s. Both then and now, there were significant geopolitical forces driving price increases, which resulted in high inflation and negative economic effects. The crises then, as now, exposed some fundamental linkages and fragilities in the energy system. As they do now, high prices produced powerful economic incentives to act, and those motivations were boosted by ideas about the security of the economy and the energy supply.

But, compared to other energy crises, the current one is substantially bigger and more complicated. Oil was the source of the shocks in the 1970s, and reducing dependency on oil, particularly oil imports, was the problem facing policymakers at the time. In contrast, the current energy crisis spans several areas, including natural gas, oil, coal, power, food security, and the environment. As a result, the solutions are also comprehensive. In the end, it will be necessary to change the character of the energy system itself in addition to diversifying away from a single energy commodity, all the while keeping energy services accessible and safe. The trends and forecasts of global energy consumption of various energy sources have been shown in Fig. 1.2 (Key World Energy Statistics 2021).

> **Key Finding Highlights (Energy Agency 2021)**
>
> - **The COVID-19 pandemic continues to impact global energy demand.** The third wave of the pandemic is prolonging restrictions on movement and

1.3 Global Energy Outlook—History and Current Scenario

continues to subdue global energy demand. But stimulus packages and vaccine rollouts provide a beacon of hope. Global economic output is expected to rebound by 6% in 2021, pushing the global GDP more than 2% higher than 2019 levels.

- **Emerging markets are driving energy demand back above 2019 levels.** Global energy demand is set to increase by 4.6% in 2021, more than offsetting the 4% contraction in 2020 and pushing demand 0.5% above 2019 levels. Almost 70% of the projected increase in global energy demand is in emerging markets and developing economies, where demand is set to rise to 3.4% above 2019 levels. Energy use in advanced economies is on course to be 3% below pre-COVID levels.
- **Global energy-related CO_2 emissions are heading for their second-largest annual increase ever.** Demand for all fossil fuels is set to grow significantly in 2021. Coal demand alone is projected to increase by 60% more than all renewables combined, underpinning a rise in emissions of almost 5%, or 1500 metric tonnes (Mt). This expected increase would reverse 80% of the drop in 2020, with emissions ending up just 1.2% (or 400 Mt) below 2019 emissions levels.
- **Sluggish demand for transport oil is mitigating the rebound in emissions.** Despite an expected annual increase of 6.2% in 2021, global oil demand is set to remain around 3% below 2019 levels. Oil use for road transport is not projected to reach pre-COVID levels until the end of 2021. Oil use for aviation is projected to remain 20% below 2019 levels even in December 2021, with annual demand more than 30% lower than in 2019. A full return to pre-crisis oil demand levels would have pushed up CO_2 emissions a further 1.5%, putting them well above 2019 levels.
- **Global coal demand in 2021 is set to exceed 2019 levels and approach its 2014 peak.** Coal demand is on course to rise 4.5% in 2021, with more than 80% of the growth concentrated in Asia. China alone is projected to account for over 50% of global growth. Coal demand in the United States and the European Union is also rebounding but is still set to remain well below pre-crisis levels. The power sector accounted for only 50% of the drop in coal-related emissions in 2020. However, the rapid increase in coal-fired generation in Asia means the power sector is expected to account for 80% of the rebound in 2021.
- **Among fossil fuels, natural gas is on course for the biggest rise relative to 2019 levels.** Natural gas demand is set to grow by 3.2% in 2021, propelled by increasing demand in Asia, the Middle East and the Russian Federation ("Russia"). This is expected to put global demand more than 1% above 2019 levels. In the United States—the world's largest natural gas market—the annual increase in demand is set to amount to less than 20% of the 20 billion cubic metres (bcm) decline in 2020, squeezed by the continued growth of renewables and rising natural gas prices. Nearly three-quarters of the global demand growth

in 2021 is from the industry and buildings sectors, while electricity generation from natural gas remains below 2019 levels.
- **Electricity demand is heading for its fastest growth in more than 10 years.** Electricity demand is due to increase by 4.5% in 2021, or over 1000 TWh. This is almost five times greater than the decline in 2020, cementing electricity's share in final energy demand above 20%. Almost 80% of the projected increase in demand in 2021 is in emerging markets and developing economies, with the People's Republic of China ("China") alone accounting for half of global growth. Demand in advanced economies remains below 2019 levels. The energy demand, GDP, and CO_2 emission scenario have been shown in Fig. 1.3 (Energy Agency 2021).

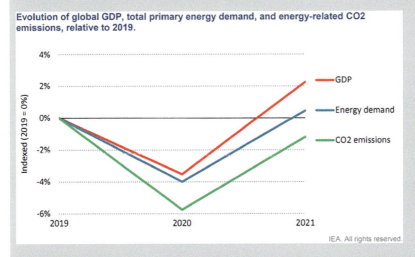

Fig. 1.3 Global GDP evolution, total energy demand, and energy-related CO_2 (Energy Agency 2021)

- **Renewables remain the success story of the COVID-19 era.** Demand for renewables grew by 3% in 2020 and is set to increase across all key sectors—power, heating, industry, and transport—in 2021. The power sector leads the way, with its demand for renewables on course to expand by more than 8%, to reach 8300 TWh, the largest year-on-year growth on record in absolute terms.
- **Renewables are set to provide more than half of the increase in global electricity supply in 2021.** Solar PV and wind are expected to contribute two-thirds of renewables' growth. The share of renewables in electricity generation is projected to increase to almost 30% in 2021, their highest share since the beginning of the Industrial Revolution and up from less than 27% in 2019. Wind is on track to record the largest increase in renewable generation, growing by 275 terawatt hours (TWh), or around 17%, from 2020. Solar PV electricity

1.3 Global Energy Outlook—History and Current Scenario

generation is expected to rise by 145 TWh, or almost 18%, and to approach 1000 TWh in 2021.

- **China alone is likely to account for almost half the global increase in renewable electricity generation.** It is followed by the United States, the European Union and India. China is expected to generate over 900 TWh from solar PV and wind in 2021, the European Union around 580 TWh, and the United States 550 TWh. Together, they represent almost three-quarters of global solar PV and wind output.

As shown in Fig. 1.4 (Energy Agency 2021), the primary energy demand has declined in the USA and European Union with respect to other countries. An exponential increase in clean energy investment is necessary to get net zero emissions. More than tripling present levels, the net zero emission (NZE) will have spent USD 4 trillion yearly in sustainable energy by 2030. Even while it will be challenging to get such a sizable investment, it will open up a tremendous number of market opportunities for manufacturers of equipment, services, developers, and engineering, procurement, and construction companies along the whole clean energy supply chain. Until 2050, the combined overall market for wind generators, photovoltaic cells, rechargeable batteries fuel cells and electrolysers in the NZE is estimated to be worth USD 27 trillion. The estimated market for clean energy technological devices in 2050 is led by batteries, accounting for about 60% of the total as shown in Fig. 1.5 (IEA 2021). The new energy economy, which will have 3 terawatt-hours (TWh) of installed battery storage and more than 3 billion electric vehicles (EVs) on the road, depends on batteries. Also, they become the only market for a number of necessary minerals, including lithium, nickel, and cobalt (IEA 2021).

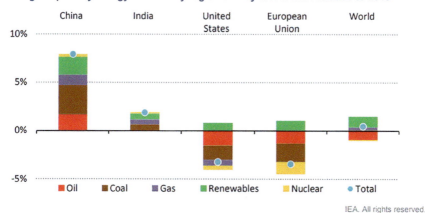

Fig. 1.4 Primary energy demand by region and by fuel (Energy Agency 2021)

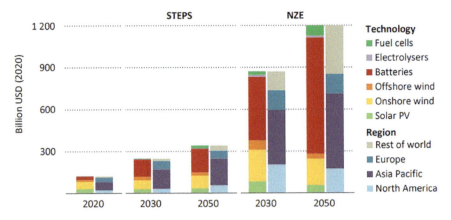

Fig. 1.5 Estimated market size for selected clean energy technologies by technology and region, 2020–2050 (IEA 2021)

Electricity Energy Demand

- **In 2020**

Almost 1% less power was consumed worldwide in 2020, with lockdowns restricting business and industrial activities throughout the first half of the year. At times, demand was 20–30% lower than during the hours before the shutdown. After adjusting for weather fluctuations, China's demand decreased by more than 10% in February compared to the same times in 2019. The second-largest user of energy in the world after China, the United States, had a reduction of about the same size in May when orders for stay-at-home were at their highest. Weekly demand decreased from March to April in Spain and Italy by more than 25%, and in Germany, France, and the United Kingdom by more than 15%. Similar trends were observed in India, where demand fell by more than 20% over a number of weeks from mid-March to the end of April. Demand decreased by around 8% in May in Japan and Korea, where COVID-19 instances were lower than in Europe and the US. Advanced economies made a comeback in the second half of 2020, although overall they continued to perform below levels from 2019. Certain developing nations and emerging markets had rapid development in the last quarter of the year, particularly China and India, which saw year-over-year growth rates of more than 8% and 6%, respectively (Energy Agency 2021).

- **In 2021**

With the economy likely to pick up and key developing economies like China experiencing strong expansion, the demand for electricity is projected to rise by 4.5% in 2021. In developed economies, COVID-19 immunization efforts should make it possible to gradually eliminate restrictions between spring and autumn. Demand should increase by 2.5%, which should be enough to keep it within 1% of 2019 levels. Demand in the United States is anticipated to rise by around 2% as a result of the economy's stimulus and the colder weather in the first half of 2021. With this rise, demand ought to be within 1.6% of 2019 levels. Germany, France, Italy, and Spain are predicted to continue to consume less than they did in 2019, with a gain of roughly 3% in 2021 failing to entirely offset decreases of 4 to 6% in 2020. The situation is comparable in Japan, where demand is predicted to increase by only 1% from 2020 levels, which is far from enough to stop the 4% decrease in 2020. The growth trajectory of demand in emerging and developing nations has not changed since it started in the second half of 2020. The anticipated robust economic rebound for China and India will quicken this trend. Electricity demand is anticipated to increase by around 8% in both China and India in comparison to 2020, with 2021 GDP growth in both nations estimated to be 9% and 12%, respectively. For China, the expected rise comes on top of 2020 growth, pushing demand in 2021 about 12% above 2019 levels. Southeast Asian nations are also likely to witness a significant return to growth, with demand growing 5% in 2021, pushing overall demand 3% above 2019 levels.

The electricity demand by region from year 2019 to 2021 has been shown in Fig. 1.6 (Energy Agency 2021).

Electricity Supply

- **In 2020**

A drop in electricity consumption worldwide and historic growth of renewable energy sources, driven by wind and solar PV, which climbed by 12 and 23%, respectively, in 2020, put fossil fuel-fired and nuclear energy plants in a difficult position. More than 3% less demand from non-renewable sources was generated. With a decrease of 440 TWh, coal suffered the most among all the power sources in 2020.

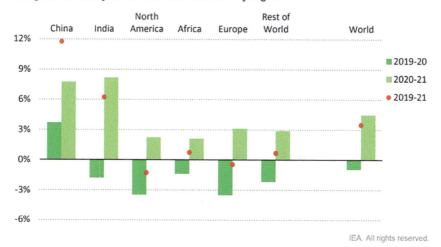

Fig. 1.6 Electricity demand by region in 2020 and 2021 (Energy Agency 2021)

The generation from coal decreased by 4.4%, which was the highest absolute reduction and the largest relative decline in the previous fifty years. The United States alone was responsible for roughly half of the net reduction globally, driven by low petrol costs. An extra 23% of the drop was attributed to the European Union, however, this decline was mainly mitigated by improvements in the output of renewable energy. Compared to coal-fired power plants, gas-fired power plants had production decreases in 2020, albeit only by 1.6%. Due to competitive prices, particularly in the middle of the year, gas was less affected. While gas-fired generation rose by 2% in the US in 2020, coal-fired generation fell by a startling 20%, or 210 TWh. With a drop of 4.4%, oil resumed the unbroken worldwide fall that began in 2012 (Energy Agency 2021).

- **In 2021**

Current trends indicate that 2021 will mark the generation of energy based on renewable source's 20th straight year of increase. A little more than half of the growth in electricity supply in 2021 is anticipated to come from increased production of renewable energy. Coal and gas-fired power plants fill the remaining increase in energy demand after nuclear power output is anticipated to increase by about 2%. With their production anticipated to rise by 480 TWh, coal-fired power plants are predicted to contribute the majority of the growth in energy generation from fossil fuels. Just little (+1%) does natural gas profit from the demand to increase gas prices. In the United States, where coal-fired generation fell by roughly 20% in 2020, we anticipate that nearly half of this loss will be made up in 2021 when several regions of the nation unwind their coal-to-gas switching policies. As a result, gas-fired power reduces by roughly 80 TWh in 2021 in the United States (https://gwec.net/global-wind-energy-outlook-2000-gigawatts-2030/; https://www.eia.gov/outlooks/aeo/). China is expected to account for well over half of the increase in coal-fired energy output in 2021. While making up nearly 45% of the increased worldwide renewable output in 2021, about half of China's 8% growth in energy supply will come from fossil fuels, driving boosting coal generation in China by 330 TWh (or 7%) compared to 2019 levels. 70% of the new power demand in India, where the absolute demand growth is anticipated to be second only to China, will be met by thermal generating in 2021, nearly exclusively using coal.

The change in electricity generation in the years 2020 and 2021 has been represented in Fig. 1.7 (Energy Agency 2021).

Analysis of the Asian Pacific Scenario

Due to auctions in India, feed-in tariffs in Japan, and new laws in the Association of Southeast Asian Nations (ASEAN) nations, solar PV has become the most widely implemented renewable energy technology in the region. The renewable energy capacity additions of ASEAN countries and capacity addition by technology have been shown in Fig. 1.8 (Energy Agency 2021).

Under the first scenario, it is anticipated that between 2021 and 2026, the Asia Pacific region (apart from China) will add over 330 GW of additional renewable energy capacity, a roughly 70% increase. Solar PV has the highest deployment rate

1.3 Global Energy Outlook—History and Current Scenario

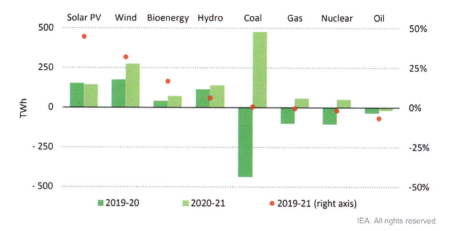

Fig. 1.7 Change in electricity generations in 2020 and 2021 (Energy Agency 2021)

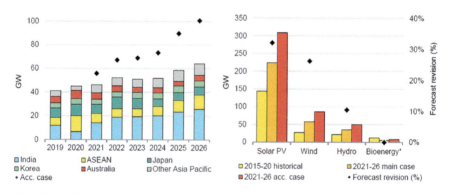

Fig. 1.8 Asia Pacific renewable capacity addition by country and by technology (Energy Agency 2021)

(68%) followed by wind (18%) and hydropower 11%. Annual additions in 2026 will be 42% greater than in 2020, according to the estimate, which is up 27% from last year. India will contribute roughly 40% of the growth in the region's renewable capacity during the forecast timeframe due to the continuation of auctions and better dispersed PV incentives. The ASEAN countries, led by Vietnam, are anticipated to continue to advance as a result of the implementation of a new auction and Feed Tariff (FIT) scheme, with the Philippines and Indonesia emerging as important development partners (Ali et al. 2017).

Under the accelerated scenario, renewable energy growth in Asia Pacific is almost 40% higher than in the base case. Increased investment in grid development is required in Australia, ASEAN nations, and India, as well as faster and more efficient implementation of new support policies (especially in Japan and ASEAN nations), the easing of permitting regulations, the introduction of more bankable power of purchase agreements (PPAs) (in ASEAN nations), and the resolution of land acquisition issues (in India and ASEAN countries).

1.4 Energy Scenario—India

India has a considerable presence on the global energy scene. Its increasing population—soon to be the largest in the world—and a period of rapid economic progress are to blame for the more than fourfold increase in energy use since 2000. In 2019, almost all houses had access to electricity, indicating that more than 900 million individuals had a connection in less than 20 years (https://mopng.gov.in/en).

India's continuing urbanization and industrialization will put a great deal of strain on the nation's energy sector and authorities. Energy use per capita is far below half the global average, and there are substantial differences in service quality and energy use across states and between rural and urban areas. The two primary problems for Indian clients with energy are its availability and price. The Indian Power Sector has seen a tremendous transformation in the 75 years since independence, starting with ensuring that everyone had access to power and ending up leading the energy transition. The milestones achieved till now in the area of the power sector have been shown in Fig. 1.9.

The COVID-19 pandemic has interfered with India's energy usage because of lockdowns and other associated limitations. Our most recent projections indicate

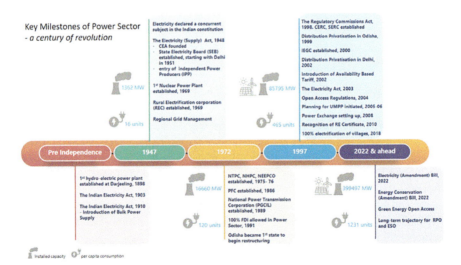

Fig. 1.9 Milestones achieved in the Indian power sector (https://www.linkedin.com/pulse/azaadi-se-aaj-tak-narrative-indian-power-sector-since-srinivasan/)

1.4 Energy Scenario—India

that the nation's energy consumption will probably decrease by about 5% in 2020, with the biggest declines occurring in the usage of coal and oil. Together with existing financial strains, the pandemic also had an effect on energy investment, which is anticipated to have fallen by 15% by 2020, especially among India's power distribution businesses. The speed at which the virus's spread can be stopped, together with the implementation of recovery plans and regulatory measures, will determine how long the effects remain.

Almost 80% of India's power needs are satisfied by coal, oil, and solid biomass combined. Coal, which still dominates the energy mix, has fostered the growth of industries and the generation of electricity. The consumption and imports of oil have grown along with the number of automobiles on the road. Although it no longer makes up as much of the energy mix, cooking fuel produced primarily of wood fuel is still often used in biomass. Despite recent successes in expanding liquid petroleum gas (LPG) penetration in rural areas, 660 million Indians have not transitioned to modern, eco-friendly cooking fuels or technology.

Natural gas and the more recent sources of clean energy, which have started to gain pace, were the two energy sources that were least affected by the COVID-19 outbreak in 2020. Solar PV, in particular, has had a sharp growth; it has vast resource potential, aspirational aims, and strong legislative support and has quickly risen to the position of the least cheap option for new power generation.

Analysis of the Indian Scenario

India's use of clean energy is increasing; however, there are still obstacles that must be solved in order to meet commitments. The renewable energy capacity addition up to 2026 and the future targets for the year 2030 have been shown in Fig. 1.10 (Energy Agency 2021).

Between 2021 and 2026, India is expecting for addition of 121 GW of clean energy, an increase of 86% over present capacity, making it the third developing market in the world leaving China and the US behind. The most extensively used technology is solar photovoltaic (74%) with onshore wind (16%) and hydropower.

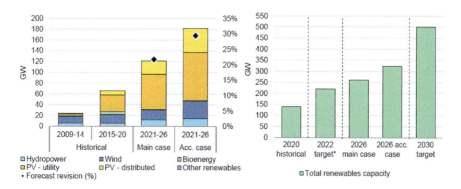

Fig. 1.10 Renewable capacity and targets for India (Energy Agency 2021)

The improved high prediction for India is mostly due to faster capacity additions after the forecast period after a fall in 2020 that have overcome severe limitations (https://coal.gov.in/en/public-information/reports/annual-reports/annual-report-2021-22).

The state of India's distribution companies (DISCOMs) continues to be a serious problem, delaying the signing of PPAs with auction winners and increasing the risk that certain projects would be delayed or abandoned. Further growth is being driven by developments in the agricultural distributed PV scheme (KUSUM) and notable oversubscription in prior PPA auctions. Renewable energy installations decreased by 44% in 2020 as a result of COVID-19's restricted mobility and supply chain problems.

In order to meet its ambitious goal of 175 GW of renewable energy by 2022, India reiterated its dedication to sustainable power while at the same period signing an agreement for a record amount of solar and wind energy (apart from large hydropower). In addition to 500 GW of total non-fossil capacity, 50% more renewable energy generation (up from 22% in 2020), and net-zero carbon emissions by 2070, India unveiled enhanced 2030 objectives at COP26 (The UN Climate Change Conference in Glasgow). These initiatives confirm the intention of energy transitions.

After falling by 20% from 2019 to 2020, India's solar photovoltaic potential additions are expected to rise steadily during the planned timeframe. Low costs and ongoing policy support are expected to facilitate deployment. The renewable energy capacity awarded in the year 2018–2021 via auctions and DISCOM's overdue payment to generators in the year 2019–2021 has been shown in Fig. 1.11 (Energy Agency 2021). The business sector may benefit from new recommendations for more humane criteria for commercial installations and an enhancement in the highest photovoltaic output suitable for utility grids from 10 to 500 kW. However, there are still a number of significant obstacles to overcome. DISCOMs are cautious about assisting quicker adoption of distributed PV because they are concerned about losing money from lower energy sales owing to self-consumption and paying greater grid expenditures. Because of expensive transaction charges, a lack of particular financial products, and the difficulties in obtaining a credit rating, small enterprises and households still have few investment choices. Public knowledge is very poor, resulting in a cautious deployment despite the strong economic appeal of many Indian states. Meanwhile, the KUSUM initiative, which promotes the growth of solar PV for agricultural customers in rural areas, seeks to put roughly 31 GW of PV capabilities online by 2024. Nevertheless, due to financial and implementation challenges, we project that only around 10 GW of this PV capacity will be operational between 2021 and 2026.

India's deployment of renewable energy between 2021 and 2026 will be increased by 50% in the accelerated scenario, putting it on track to achieve the government's 2030 goals. If the DISCOMs receive financial incentives, improved payment plans that fairly split grid expenses, and inventive business models that provide them advantages, the adoption of distributed PV in the accelerated scenario is around 82% higher. To enhance customer demand, new financial products and education efforts will be required. If the financial and implementation issues with the KUSUM programme are rectified, the initiative has upside potential. Utility-scale photovoltaic (PV) and

1.5 Classification of Energy Resources

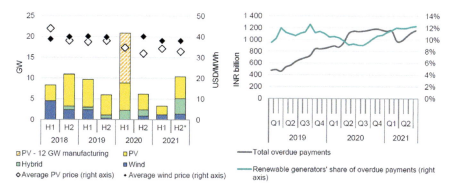

Fig. 1.11 Renewable capacities awarded and overdue payments for India (Energy Agency 2021)

wind deployments might rise over the baseline period by 37% and 70%, respectively. In the expedited situation, it is also assumed that DISCOMs will spend more quickly since they will sign PPAs with auction winners more quickly. The hastened situation necessitates increased collaboration from the federal and state governments in locating ideal areas for wind and solar PV as well as in supplying grid connections. It also speeds up the construction of large-scale projects, particularly in onshore wind.

1.5 Classification of Energy Resources

Energy sources can be categorized as follows (Khan 2024):

1. **Based on Energy Utilization**

   (a) Primary Resources: Before being changed or transformed by people, these resources already existed in nature. A few examples of primary energy sources are crude oil, coal, wind, sunshine, flowing rivers, vegetation, uranium, and so on. They are referred to as raw energy resources since they are typically only available in raw form. In most cases, this type of energy cannot be consumed directly; instead, it must be found, explored, extracted, processed, and transformed into the form needed by the consumer. In order to make the resource usable by a user, some measures are taken to make it accessible. The energy yield ratio of a power extraction method is described as follows:

   $$\text{Energy yield ratio} = \text{Energy received from raw energy source} / \text{Energy spent to get the raw energy source}$$

   Only resources with a moderately high energy yield ratio are seen to be worth exploring.

(b) Intermediate Resources: They serve as energy carriers and are created by transforming primary energy through one or more processes.

(c) Secondary Resources: Secondary or useful energy is the type of energy that is ultimately delivered to a consumer for use, such as chemical energy (in the form of hydrogen or fossil fuels), electrical energy, and thermal energy (in the form of steam or hot water). Certain energy sources, including electricity and hydrogen, can be classified as intermediate and secondary resources.

2. **Based on Conventional Utilization**

(a) The term "traditional energy resources" refers to sources of energy that have been traditionally utilized for many years and were widely used around the time of the 1973 oil crisis, such as nuclear power, fossil fuels, and hydropower.

(b) Clean energy sources like wind, solar, and biomass are power supplies that are highly suitable for widespread usage as a result of the 1973 oil crisis.

3. **Based on Future Accessibility**

(a) Non-renewable resources include fossil fuels, uranium, and other materials that have a limited supply and cannot be replaced after being used up.

(b) Renewable resources include sun, wind, biomass, ocean (thermal, tidal, and wave), geothermal, hydro, etc. that are continually replenished by nature and whose supply is unaffected by the pace of their consumption.

4. **Based on Market Usage**

(a) Commercial Energy Resource: This category includes the secondary forms of usable energy, such as gasoline, diesel, gas, and electricity, which are required for commercial activity. The economy of a country is based on its ability to convert natural raw power into commercial power.

(b) Non-commercial Energy: Non-commercial resources include things like wood, animal dung cake, and agricultural leftovers that are directly used to generate energy from nature without first going via a commercial outlet.

5. **Based on Origin**

(a) Nuclear energy
(b) Hydroenergy
(c) Fossil fuels energy
(d) Wind energy
(e) Ocean wave energy
(f) Solar energy
(g) Geothermal energy
(h) Biomass energy
(i) Ocean thermal energy
(j) Tidal energy.

1.6 Advantages and Disadvantages of Conventional Energy Sources

The following benefits and drawbacks are associated with the use of conventional energy source use (Khan 2024).

Advantages

1. Cost: Currently, these are less expensive than unconventional sources. The current estimated costs of electrical energy from various sources are as follows: Rs. 4.50 per kWh from gas, Rs. 3.97 per kWh from coal, Rs. 16.0 per kWh from diesel, Rs. 2.50–4.00 per kWh from hydropower, and Rs. 2–3.50 per kWh from nuclear power. (As of 2018–19 data available from IEA).
2. Security: The availability of energy may be guaranteed for a specific duration by storing a specific quantity due to storage's simplicity and convenience.
3. Convenience: These sources are particularly easy to employ because the technology for their conversion is widely accessible.

Disadvantages

1. Pollutants are produced by burning fossil fuels. The main pollutants produced while using these sources are heat, CO, CO_2, NO_x, and SO_x particle matter. In spite of the fact that these pollutants harm the environment, are hazardous to human health, and result in several other issues, CO_2 is the primary driver of global warming (Hafner and Luciani 2022; Editorial Board 2009).
2. The petrochemical, chemical, pharmaceutical, and paint industries all use coal as a key raw resource. Conserving coal for future needs is preferable from a long-term perspective.
3. Nuclear plant safety is a controversial issue. The following are the main issues with nuclear energy:

   a. The waste produced by nuclear power plants contains dangerously high radioactivity quotients; they persist over the permissible limit for a very long time, posing a risk to human health. Its proper disposal is a difficult undertaking but is necessary to stop radioactive contamination. In order to prevent radioactive waste from getting into the wrong hands, it must also be preserved for a significant duration (till its radiation value reduces to a safe range).
   b. Radioactive material leaking from the reactor is a possible case (like it occurred in Chernobyl, former USSR, 1986).
   c. Uranium resource, for which the technology presently exists, has limited availability.
   d. Using nuclear energy requires sophisticated technologies. The technology needed to use nuclear energy is only available in a few countries.

1.7 Future of Energy Renewables

Energy obtained from renewable natural resources, like solar, biomass, hydropower, geothermal, and wind, may be refilled or renewed. Several nations have set ambitious goals for switching to renewable energy, which is becoming more acknowledged as an essential part of the global energy mix. In this section, we will explore the future of renewable energy, including its potential, challenges, and opportunities (Demirbaş 2006).

- Potential of Renewable Energy

Several analysts agree that renewable energy will be essential in supplying the world's future energy requirements because of its vast potential. The IEA observes that clean energy sources might generate 56% of the world's power by 2035, up from 26% in 2018. By 2030, the IEA predicts that solar and wind energy will be the least expensive sources of power in the majority of nations.

Scalability is one of the main benefits of renewable energy. Several renewable energy sources can be utilized to generate power on a large scale. For example, large wind parks and solar parks can generate significant amounts of electricity, which can be distributed to homes, businesses, and industries.

Renewable energy sources are also becoming increasingly efficient and cost-effective. Advances in technology have made it possible to produce renewable energy at lower costs than traditional fossil fuels. As a result, renewable energy is becoming more attractive to investors and consumers, and it is driving significant growth in the renewable energy industry.

- Challenges Facing Renewable Energy

Despite its potential, renewable energy faces several challenges that must be overcome for it to become a dominant energy source. The inconsistent nature of clean energy sources is one of the biggest obstacles. Solar and wind energy are reliant on the weather, which may be unstable and change throughout the day. This means that renewable energy sources cannot always provide a steady supply of energy, which can be a problem for industries and households that require a constant supply of energy.

The lack of infrastructure to support renewable energy is another issue. Large-scale renewable energy production and distribution need major investments in new transmission and distribution infrastructure. These systems must have the ability to move electricity from rural regions with a wealth of renewable energy sources to metropolitan areas where it is required. Building new infrastructure can be costly and time-consuming, and it requires significant coordination between governments, utilities, and private companies.

Renewable energy also faces regulatory challenges, particularly in countries where fossil fuel industries are well-established. Governments frequently offer tax incentives and subsidies to the fossil fuel sector, making it challenging for renewable energy to compete better. In addition, regulatory frameworks may not be designed to support renewable energy, which can create barriers to entry for new renewable energy projects.

- Opportunities for Renewable Energy

Despite these challenges, there are many opportunities for renewable energy to become a dominant energy source in the future. One of the primary opportunities is the growing demand for clean energy. As awareness of the environmental impact of fossil fuels increases, more consumers and businesses are looking for alternative sources of energy that have a lower carbon footprint. This demand for clean energy is driving significant growth in the renewable energy industry, and it is creating new opportunities for investors and entrepreneurs.

Another opportunity for renewable energy is the development of new technologies. Advances in energy storage, such as batteries and pumped hydro, are making it possible to store renewable energy for use when it is needed. This can help to address the intermittency challenge and make renewable energy more reliable. Moreover, improvements in smart grid technology are allowing utilities to better control energy distribution, which may boost the dependability and effectiveness of non-conventional energy systems.

Renewable power also offers opportunities for job creation and economic growth. The renewable energy industry is labour-intensive and requires a diverse range of skills.

1.8 Origin of Renewable Energy Resources

1.8.1 Solar Energy

The term "solar energy" refers to all the energy that the sun produces. The sun undergoes nuclear fusion, which generates solar energy. The protons of striking hydrogen bonds combine to generate a helium nucleus during fusion in the sun's core. This method generates such a large amount of energy due to a proton-proton (PP) chain reaction. At its nucleus, the sun absorbs 620 million tonnes (Mt. of hydrogen each second. Several planets with sizes similar to our sun experience the PP chain reaction, which supplies them with constant energy and heat. These stars have a Kelvin temperature of around 4 million degrees (nearly 4 million degrees Celsius).

Stars that are around 1.3 times the size of the sun create energy through the CNO cycle. Similar to the carbon–nitrogen-oxygen (CNO) cycle, which uses carbon, nitrogen, and oxygen to create helium from hydrogen. Nowadays, less than 2% of the sun's energy is produced via the CNO cycle. Nuclear fusion releases massive quantities of energy in waves and particles via the PP chain reaction or CNO cycle. Solar energy is continuously travelling away from the sun throughout the solar system. The sun's energy warms the planet, causes wind and weather, and keeps living organisms alive.

The sun emits electromagnetic radiation (EMR) that contains heat, light, and energy. There are many different frequencies and lengths of waves in the electromagnetic spectrum. The wave's frequency indicates how frequently it will repeat over a specific period. High-frequency waves recur often over a short period and have very tiny wavelengths. In contrast, low-frequency waves have significantly larger wavelengths. The bulk of electromagnetic radiation is invisible to humans. The sun emits gamma rays, X-rays, and ultraviolet light at its greatest frequencies (UV rays).

The troposphere almost entirely absorbs the most dangerous UV rays. Sunburn can result from the atmosphere's weaker UV radiation. Infrared energy from the sun also creates waves, although they have a considerably lower frequency. The majority of the heat from the sun is produced by infrared radiation. The region between infrared and ultraviolet light is known as the visible spectrum and contains all the colours we perceive on earth. The wavelengths closest to UV are in violet, while the longest wavelengths are in red (closest to infrared).

For usage in structures like homes, workplaces, schools, and hospitals, solar power is a clean energy resource that may be directly gathered by a number of methods. Solar energy technology includes things like concentrated solar energy, solar architecture and photovoltaic cells and panels. In order to collect solar energy and transform it into useful energy, many techniques can be used. The strategies make utilization of both passive and active solar energy. Active solar systems actively convert solar energy into a different kind of energy using machinery or electricity, most typically heat or power. With passive solar technology, no further tools are needed. Instead, they take advantage of the local climate to heat buildings in the winter and reflect heat from the sun in the summer. Photovoltaic is probably the most common way to get energy from the sun right now. The most typical part of a photovoltaic array is a solar panel, which is composed of hundreds of solar cells (Turney and Fthenakis 2011).

1.8.2 Wind Energy

Engineers and scientists are exploiting the kinetic energy of the wind to produce power since anything that moves includes kinetic energy. Wind turbines, also referred to as wind generators or wind turbines, are devices that harness the power of the wind to produce electricity. The blades of the turbine, which are attached to a rotor, are blown over by the wind. A generator is then spun by the rotor to produce electricity.

Vertical-axis wind turbines (VAWTs) and horizontal-axis wind turbines (HAWTs) are the two types of wind turbines. HAWTs are the most common type of wind turbine. A common feature of them is the presence of two or three long, thin blades that resemble an aeroplane propeller. The blades are oriented to the face directly towards the wind. VAWTs have curved blades that are shorter and wider than electric mixer beaters. A house can be powered by 100 kilowatts of energy produced by small, independent wind turbines.

In water pumping stations and other locations, small wind turbines are deployed. A bit larger wind turbines have rotor blades that are around 40 m (130 feet) long and are mounted on towers that may reach 80 m in height (260 feet). The power output of these turbines is 1.8 megawatts. On top of towers that are 240 m (787 feet) high, you may find wind turbines with rotor blades that are longer than 162 m (531 feet). The output power range of these massive turbines is 4.8 to 9.5 megawatts.

Energy can be used straight away, connected to the power grid, or stored for later use depending on how it is created. The US Department of Energy is working with the National Laboratories to develop and improve technologies like batteries and pumped-storage hydropower as a means of storing surplus wind energy. Companies like General Electric place batteries next to wind turbines so that power generated by the wind may be promptly stored. The US Geological Survey estimates that there are 57,000 wind turbines in the country, including both onshore and offshore ones.

A grouping of wind turbines that are either standalone structures or are clustered together is known as a wind farm. A single turbine might be able to supply a single home with all the electricity it needs, but a wind farm can power thousands of homes. Wind farms are often located on top of hills or in other windy regions to make use of natural breezes. The world's largest offshore wind farm is the Walney Extension. The location of this wind farm is in the Irish Sea, roughly 19 km (11 miles) west of the northwest coast of England. The Walney Extension is a massive wind farm that, at 149 square kilometres (56 square miles), is bigger than both the city of San Francisco, California and the island of Manhattan in New York. Due to the grid of 87 turbines, which is 195 m (640 feet) tall, these offshore wind turbines are among the largest in the world. In the UK, 600,000 homes may be powered by the 659 megawatts of energy that the Walney Extension is capable of producing (Sadati et al. 2015).

1.8.3 Geothermal Energy

From the Greek words "GEO" and "THERM," the name "geothermal" is derived. The term "therm" in Greek signifies heat from the earth, whereas the word "geo" means the earth. The heat of the ground is used to create geothermal energy. The distance to the molten centre of the earth is around 4000 miles, and it is extremely hot. It is believed that the temperature is at least 5000 °C. The mantle or outer layers of rock are heated by the heat that radiates from the earth's core. Magma is the term for the molten state of this kind of rock. Just below the earth's surface, magma can be found. Rainwater occasionally leaks through fault lines and fissures in the earth's crust, becoming extremely hot from the hot rocks below. Some of this extremely hot water rises back to the earth's surface, where it forms geysers or hot springs. There are occasions when the hot water is trapped below as a geothermal reservoir. Drilling wells into geothermal reservoirs is one method of using geothermal energy to generate power. Steam is produced at the surface from the hot water that is rising. Steam is used to power electricity-generating turbines. Even if the water does not reach the steam-producing temperature, it may still be utilized to heat buildings and houses, conserving both gas and energy.

For thousands of years, geothermal heat and water have been utilized. Hot mineral springs were utilized for bathing, cooking, and healing by the Romans, Chinese, and Native Americans. Geothermal water is now used for a variety of purposes, such as district heating, systems that transport steam or hot water to numerous units, and the heating and cooling of single structures like offices, retail establishments, and homes utilizing geothermal heat pumps. Also, it has industrial potential for heating water in fish farms, drying crops, growing plants in greenhouses, and other industrial activities. Since roughly a century ago, geothermal energy has also been used to generate electricity. At the moment, electricity may be produced utilizing so-called Improved Geothermal Systems without the requirement for natural water resources (EGS, also known as hot dry rock). Water is pumped from the surface into boreholes to expand them and create some fractures in the hot rock in order to extract energy from hot, impermeable rock. These holes let water go through, where it heats up and is used to generate electricity as it comes out. As geothermal energy is clean, renewable, constant, and widely accessible, it is already used in a large number of thermal and electric power facilities. There were 10,715 megawatts (MW) of geothermal power in operation throughout 24 nations, with a predicted 67,246 GWh of energy output in 2010. This information comes from the International Geothermal Association (IGA). IGA projects expanded to 18,500 MW by 2015 as a result of the projects now under consideration, many of which are in areas where it was previously thought there were few exploitable resources. As per IEA report the geothermal energy generation was 2.5 GW for the year 2017–22 which is expected to rise to 4.1 GW in the year 2028.

1.8.4 Ocean Thermal Energy Conversion (OTEC)

OTEC, a marine renewable energy method, harnesses the sun's energy absorbed by the ocean to generate power. The sun's radiation heats the surface water more than the deep ocean water, creating the ocean's natural temperature gradient, or thermal energy. OTEC evaporates a working fluid, such as ammonia, which has a low boiling point, using the warm ocean surface water, which is around 25 °C (77°F). To create electricity, the vapour expands and spins a turbine attached to a generator. The vapour is then cooled by seawater that has been drawn from the deep ocean level, where the temperature is about 5 °C (41°F). The working fluid is then available for reuse after being condensed into another liquid form. This is an ongoing cycle that produces electricity. The temperature differential has a significant impact on the cycle's efficiency. The efficiency increases with the size of the temperature differential.

The approach is only useful in tropical areas where there is an annual temperature differential of at least 20 °C, or 36 degrees Fahrenheit. The oceans are the greatest solar collectors and energy storage systems in the world since they cover more than 70% of the surface of the planet and trap a significant amount of the sun's heat in the top layers. To meet the world's energy needs, just a small amount of this energy needs to be used. As long as the sun shines and there are natural ocean currents, OTEC's energy source is unrestricted, plentiful, and always being renewed. Some well-known organizations believe that 3 to 5 terawatts of base load power generation (1 terawatt = 1012 watts) could be effectively gathered without changing ocean temperatures or the global ecosystem. It is nearly two times the demand for power worldwide. So, the oceans constitute a sizable renewable resource that has the potential to provide a sustainable manner of producing power in the future energy mix (Yuksel and Hasan 2010).

The system functions best in tropical areas with year-round temperature variations of at least 20 °C (36°F). OTEC has the capacity to provide base load electricity, which means that it can do so year-round, day and night (24/7) and at all times. For example, tropical islands with limited energy networks that can't handle a lot of erratic power have a lot to gain from this. In addition to producing power, OTEC provides the chance to co-generate a variety of beneficial goods, including fresh water, nutrients for better fish farming, and seawater-cooled greenhouses that enable food production in arid regions. Last but not least, cold water may be used in building air conditioning systems. Efficiency gains of up to 90% are possible. With the help of the massive base load of OTEC resources, many tropical and subtropical (remote) areas might become more energy-independent. At present, the 65 kw OTEC-powered desalination plant is going to be operative in Kavaratti, Lakshadweep Islands of India.

1.8.5 Wave Energy

Waves are produced when the wind blows over a lake or ocean's open surface. Ocean waves have a lot of energy. The advantages of wave energy are that it is cost-free, renewable, sustainable, and waste-free. As a consequence, it could reduce our carbon footprint. It is also unusual because it is the most focused environmentally friendly energy source on the globe, with a power density that is far greater than that of wind and solar energy. Wave energy, ocean energy, or sea wave energy are all terms used to describe a sort of energy that comes from the ocean or sea waves. The strong vertical motion of surface ocean waves may be used as kinetic (motion) energy by wave energy systems to carry out important tasks. The captured wave energy is used for a range of advantageous processes, including the production of power, water desalination, and the water being pumped into reservoirs.

Wave power is produced by the up-and-down motion of floating items placed on the ocean's surface. In other words, the waves are produced by the wind, which also produces energy. When the waves pass across the ocean, modern technology exploits the swells and currents of the water to generate power. A variety of equipment may be used to measure the wave energy resource in a certain area. A tiny floating buoy that tracks the sea surface and measures its vertical displacement is often the industry standard. An appropriate telemetry system is then used to relay this record to land while simultaneously recording it outside. To provide an accurate representation of the wave conditions, these sensors typically record for 20 to 30 min. Using a network of data buoys, meteorological agencies take measurements all around the world. Longer measures are preferred because the climate does change from year to year, however, data must be taken for at least a year to represent seasonal fluctuations in sea state. Understanding inter-annual changes in the wave energy resource requires the use of long-term numerical models of prospective locations that can extend many decades. The amplitude of a wave directly affects how much energy it transports. High-energy waves are those with large amplitude (https://cwc.gov.in/).

Similar to this, a low-energy wave has a small amplitude. Let's use a slinky as an example to better grasp the energy-amplitude relationship. By giving the first coil of a particular slinky an initial amount of displacement, a transverse pulse is injected into it, stretching the slinky in a horizontal direction. The force used to move the object from its resting position is what caused the indicated displacement. The coil will demonstrate greater displacement the more force is applied to it, which will increase its amplitude. As a conclusion, we may state that the energy that a transverse pulse carries across the medium is directly correlated with the amplitude of the pulse. The energy of the transverse pulse will only modify the amplitude of the pulse; it will not alter the pulse's wavelength, frequency, or speed. At present 1 Kw wave-powered navigational buoy is operational in Chennai, Tamil Nadu, India.

1.8.6 Tidal Energy

The natural rise and fall of tides brought on by the gravitational pull of the moon, sun, and earth is known as tidal energy. When water moves more quickly through a constriction, tidal currents with enough energy to be harvested happen. Tidal energy can be transformed into useful kinds of power, such as electricity, by placing specifically designed generators in the right places. The ocean can also produce other types of energy, such as waves, enduring currents, and variations in the salinity and temperature of saltwater.

There are three primary ways to harvest tidal energy: barrages, tidal lagoons, and tidal streams. In places with strong tidal currents, underwater turbines are positioned to harness the kinetic energy of moving water to produce electricity. This technique is known as tidal streams. Large, turbine-equipped constructions known as barricades are constructed across estuaries or tidal basins to produce electricity when water rushes in and out during tidal cycles. In enclosed or partially enclosed coastal areas known as "tidal lagoons," strategically positioned turbines are used to capture the rise and fall of the tides. With each technique coming with its own set of benefits and drawbacks, the investigation into tidal energy as a dependable and sustainable energy source is expanding.

Grain mills were powered by tidal energy for the first time in Europe over a millennium ago. Grain was milled using waterwheels driven by the outgoing tidal movement, while the incoming tidewater was held in storage ponds. The nineteenth century saw the introduction of this method of producing energy using rotating

1.8 Origin of Renewable Energy Resources

turbines and falling water. An early strategy for tidal power plants used a barrage system similar to a dam. Nevertheless, the industry's primary focus has not ultimately remained on this. Between 1924 and 1977, the US Power Commission, Nova Scotia Light and Power, and the governments of the United States and Canada, respectively, carried out four preliminary feasibility studies for large-scale tidal power plants.

All of them concentrated on particular geographic areas near the Maine–Canada border. Even if there were differences in the judgments on the economic viability, not much progress was made. Built-in 1966 near La Rance, France, the massive tidal barrage generates 240 megawatts (MW) of electricity annually. Until 2011, when a 254 MW array in South Korea was installed, it was the largest in the world.

The industry has shifted its focus to in-stream tidal energy generation over the last 20 years, which involves placing one or more devices—or groups or arrays of devices—within the tidal stream. The largest facility in the world for testing and showcasing wave and tidal technology in actual maritime environments was founded in 2003 and is called the European Marine Energy Centre.

Underwater cables carry the electricity harvested by turbines positioned in tidal streams to the grid. At locations where land constrictions cause high tidal velocities, like straits or inlets, tidal stream systems can absorb energy. With a generation capacity of up to 398 MW, the MeyGen project in Scotland will be the largest tidal stream generating station globally when it is completely operational.

The barrages in South Korea and France, with their respective capacities of 254 and 240 MW for the generation of energy, are two of the largest tidal power plants in the world. With a generation capacity of only 20 MW, it is the second largest in Canada.

Around the world, tidal power arrays of various sizes are being planned or have already been installed, with a strong emphasis on producing energy from tidal streams or currents. The newest and first of its sort is a tidal stream array in Scotland's Pentland Firth, a body of water that separates the country's northern islands from the mainland. By the end of 2020, the first four turbines in the MeyGen tidal energy project had produced and supplied the grid with over 35 gigawatt-hours of electricity. The project started operating in phases in 2018. When fully deployed, the 400 MW of electricity produced by the high-speed currents in the area will come from 61 turbines submerged on the seabed.

In Wales, a developing industry hub, several initiatives are underway. A premier centre for marine engineering, authorized by the governments of Wales and the United Kingdom in 2020, will be part of this development. Among its features will be a 90 km demonstration zone to facilitate the testing of tidal energy generation technologies in the future.

1.8.7 Biomass Energy

One possible definition of biomass is "biomass energy," which is energy generated from biomass. This energy source may be produced by any material which is organic and capable of doing so in a reaction. They can include organic materials like wood, leaves, pellets, faeces, and other materials. Biomass energy was discovered when humanity was still living in caves. In the thirteenth century, Marco Polo wrote about using biomass to produce fuel after getting the idea from the Chinese, who covered sewage tanks to produce biogas. There are two types of biomass energy: renewable and non-renewable. The process of producing biomass uses the sun as its main energy source. Photosynthesis is the process by which plants convert solar energy into chemical energy for food, use it to fuel their growth, and then convert it back into fuel. The energy obtained from biomass may be used in a variety of ways, including direct processing—by burning to create heat—direct processing—by turning it into electricity, indirect processing—by turning it into biofuels.

Direct burning, pyrolysis, co-firing, gasification, and anaerobic decomposition are the many methods of heat conversion. The biomass must first be dried before burning. Torre faction is the name for the chemical process that dries out biomass. The biomass is cooked in this procedure to a temperature between 200 and 320 °C. In addition to losing all of its moisture, biomass also loses its capacity to take in moisture. Torre faction transforms biomass into a dry, dark substance that is then crushed to create briquettes. Briquettes may be stored in damp areas because of their strong hydrophobicity. Also, the briquettes have a high level of energy and are simple to burn directly or in a co-fire.

For biofuels like ethanol and biodiesel, biomass is seen as a renewable source. Such biofuels are utilized to power automobiles and other machinery in a number of nations, including Austria, Sweden, and the United States. Biomass materials high in carbs, such as corn, wheat, and sugarcane, are fermented to make ethanol. This ethanol may be converted into biodiesel by mixing it with leftover cooking fat, vegetable oil, and animal fat (Kralova and Sjöblom 2010).

About 24 EJ of bioenergy is now consumed through the traditional usage of biomass; in the NZE Scenario, this amount declines to zero by 2030 as complete access to contemporary cooking methods is attained. The amount of bioenergy used in modern times rises from about 41 EJ in 2030 to over 75 EJ in 2050. The NZE Scenario also involves a significant shift in the global energy mix due to the rapid increase of low-emission sources, which eventually replace continuous sources throughout the energy sector. Low-emission sources of supply rise by about 125 exajoules (EJ) between 2021 and 2030. When the targets for energy access are met, the conventional usage of biomass is phased out. Modern bioenergy and solar energy have the largest rise to 2030, with around 35 EJ and 28 EJ, respectively, among low-emission sources.

1.9 Impact of Renewables on Energy Sector

Renewable energy is produced from naturally replenishing resources. It consists of geothermal, biomass, solar, hydro, and wind power. Renewable energy has a huge and broad impact on the energy industry, providing both environmental advantages and economic prospects. The effect of renewables on the energy industry will be examined in this section, with a particular emphasis on the advantages of the transition to a future with a greater dependency on non-conventional sources and its difficulties (Thangamayan et al. 2018; Chilán et al. 2018).

Benefits of Renewables on the Energy Sector:

1. Environmental Benefits: The point that renewable energy is clean and emission-free is its most important advantage. For many years, fossil fuels have been the dominant source of power, but they are also the main cause of climate change. Using renewable energy sources lowers carbon emissions, which can help lessen the harmful effects of climate change.
2. Economic Benefits: The market for renewable energy has the potential to increase employment and stimulate the economy. The industry is still relatively new, but it has already created thousands of jobs worldwide. In addition, the use of renewables can also reduce energy costs for businesses and households.
3. Energy Security: By lowering reliance on imported oil and gas, the utilization of renewable energy sources can increase energy security. Renewable energy resources are abundant and widely distributed, so they can be used to power local communities and reduce reliance on imported energy.
4. Innovation and Technological Advancements: The renewable energy sector is driving innovation and technological advancements in energy storage and transmission. These advancements are critical to overcoming some of the challenges associated with integrating renewables into the grid.
5. Flexibility and Decentralization: Natural sources of energy are adaptable and suitable for a variety of uses. They can also be deployed in a decentralized manner, which allows for greater energy independence and resilience in the face of power outages and natural disasters.

Challenges of Renewables in the Energy Sector:

1. Intermittency and Storage: One of the most significant challenges associated with renewables is their intermittency. It can be challenging to balance energy supply and demand because solar and wind energy depend on the weather and time of day. Energy storage technologies are still in their infancy, which means that renewable energy cannot be stored at scale for use during periods of high demand.
2. Grid Integration: Integrating renewables into the grid can be challenging due to their variability and intermittency. Energy grids are designed to handle large, centralized power plants that operate at a constant output. However, renewables produce electricity in a more decentralized and variable manner, which can create grid stability issues.
3. Infrastructure Investment: The transition to renewable energy will require significant investment in new infrastructure, including transmission lines, energy storage facilities, and new power plants. These investments can be costly and require significant government and private sector support.
4. Policy and Regulatory Framework: Governments need to develop policies and regulatory frameworks that incentivize the transition to clean energy. The switch

to renewable energy sources can be accelerated by policies like renewable portfolio standards, feed-in tariffs, and carbon pricing.

5. Technological Maturity: Some renewable energy technologies are still relatively new and may not be fully matured. For example, the development of advanced biofuels and offshore wind turbines is still in progress. These technologies may require additional research and development to become competitive with traditional energy sources.

The effect of renewables on the power sector is significant and far-reaching. The benefits of renewable energy are clear, including environmental benefits, economic opportunities, energy security, innovation, technological advancements, flexibility and decentralization. However, the transition to a more renewable energy future also comes with challenges, including intermittency and storage, grid integration, infrastructure investment, policy and regulatory frameworks, and technological maturity. To solve these challenges, governments, corporations, and individuals must work together to accelerate the shift to an era with more renewable forms of energy (Bilgili et al. 2016; Shahzad et al. 2017).

1.10 Renewable Energy: Global Statistics and Analysis

Renewables play a critical role in clean energy transitions. The deployment of renewables for electricity generation, heat production for buildings and industry, and transport is one of the main enablers of keeping the average global temperature rise below 1.5 °C. Latest developments have been encouraging, preliminary projections indicate that 2022 will set a new record for renewable capacity additions, with an annual capacity of roughly 340 GW. The US Inflation Reduction Act and REPowerEU, two significant plans launched this year, will provide additional support for accelerating the adoption of renewable electricity in the upcoming years (https://www.eia.gov/naturalgas/crudeoilreserves/; https://www.iea.org/reports/tracking-clean-energy-progress-2023).

To reach the milestones in the Net Zero Emissions by 2050 Scenario, the renewable electricity share of generation must grow more quickly, from about 29% in 2021 to more than 60% by 2030. The average yearly generation must increase by more than 12% between 2022 and 2030, which is twice the rate of 2019–2021.

Almost 90% of the growth in renewable power generation in 2021, a record 522 TWh increase, was contributed by wind and solar PV technologies combined. The share of renewable energy in the world's electrical production rose to 28.7% in 2022 after just a 0.4% point rise in 2021. The sluggish growth of the renewables industry was influenced by the greatest level of power consumption ever seen, the rebound in economic activity after the COVID-19-induced recession and droughts in several regions. Figure 1.12 (https://www.iea.org/energy-system/renewables) shows the renewables percentage of power generation as per the NZE scenario starting from the year 2010 to the year 2030.

1.10 Renewable Energy: Global Statistics and Analysis

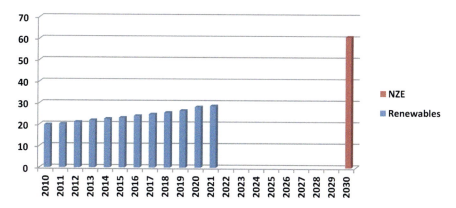

Fig. 1.12 Net zero scenario (2010–2030) renewables percentage of power generation (https://www.iea.org/energy-system/renewables)

The majority of the growth in renewable energy output in the year 2021 was attributed to China, mostly as a result of record-breaking increases in PV capacity from wind and the sun in 2020. 8% of the growth was contributed by both the European Union and the United States. Regrettably, there was a decline in renewable energy production in Turkey and Brazil as a result of ongoing dry spells that reduced hydropower production, and then in Germany and the United Kingdom as a result of protracted periods with little wind that hindered air farm production.

The production of renewable energy must continue to increase by more than 12% yearly to achieve the Net Zero Scenario throughout the years 2022–2030. Power growth fell well short of the Net Zero Scenario milestone despite huge gains in renewable capacity in 2021. All renewable technologies will need to be deployed much more quickly around the globe.

Almost 50% of the record-breaking 522 TWh growth in renewable power in 2021 output came from wind, primarily as a result of China's nearly tripling of Increases in wind capacity in 2020 and continued high expansion in 2021. Solar PV accounted for one-third of the generation before the escalation of installations in China, the European Union, and the United States in 2020 and 2021 increase. The remaining 15% of the increase came from bioenergy power generation. In contrast, Brazil, the United States, Turkey, China, India, and Canada are all experiencing droughts; hydropower production fell by around 0.4% in 2021, marking first time in twenty years that this has happened. Due to the restricted capacity increases, Geothermal, concentrated solar, and ocean technologies all continued to expand their output stagnant in 2021.

PV solar energy has recently experienced the greatest increase in performance additions; yet, indeed the historic addition of 150 GW in 2021 is only roughly three-quarters of the typical yearly increases throughout the milestone years of 2022–2030 for the Net Zero Scenario. The average annual growth rate for wind power installations must double from what it was in 2021, while the growth rates for hydropower and biofuels must be roughly twice as high as they were on average over the preceding

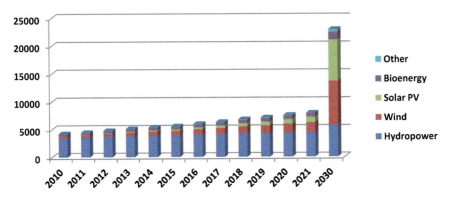

Fig. 1.13 Net zero scenario, 2010–2030, renewables electricity production in TWh (https://www.iea.org/energy-system/renewables)

five years. Although the growth rates needed for Communications Service Provider (CSP), geothermal, and ocean power to catch up to the scenario's long-term levels are still significantly below what they need to be, generally much greater efforts are needed to place these popular technologies firmly on the Net Zero Scenario trajectory. The electricity production in TWh of different renewable energy has been shown in Fig. 1.13 (https://www.iea.org/energy-system/renewables).

1.11 Renewable Energy Regulations and Policies

Energy policy refers to the approach used by a certain body, usually governmental, to tackle challenges related to energy development. This includes energy conversion, distribution, and usage, as well as reducing greenhouse gas emissions to mitigate the effects of climate change. Energy policy can take various forms, such as international treaties, taxation, investment incentives, conservation recommendations, legislation, and other public policy measures. One essential element of contemporary economies is energy. Energy is necessary for a functional economy in order to power manufacturing operations, transportation, communication, agriculture, and other activities. Labour and capital are also necessary.

A. **Global Scenario**

Most nations in the world encourage renewable energy through policy and have national goals in place. Several forms of assistance, including technologically specific measures, have been put into place. The following significant adjustments were made in 2021–2022 (Renewables—Energy System—IEA 2023; Hicks and Ison 2011).

- The ambitious goal of 33% of power generation coming by 2025, from renewable sources, rising from around 29% in 2021 was in China's fourteenth 5-Year Plan that was published in June 2022.

- The Inflation Reduction Act was submitted by the US federal government in August 2022. With tax incentives and other measures, this bill considerably increases support for renewable energy over the following ten years.
- The European Commission suggested raising the 2030 renewable energy target for the EU from 32 to 40% in July 2021. The proposed aim was altered by the REPowerEU Plan, increasing it from 40 to 45% in May 2022 (which would require 1236 GW of installed renewable capacity overall). In order to speed capacity expansion and in reaction to the energy crisis brought on by Russia's invasion of Ukraine, some European nations have already boosted their renewable support structures.
- India announced additional aims for 2030 throughout the Climate Change Conference in Glasgow (COP26), which took place in Glasgow in Nov 2021. These targets include five hundred GW of overall renewal energy-producing capacity, net zero emissions by 2070, and a 50% share of renewable power production (up from 22% in 2020).

To speed up the expansion of renewable energy globally, a number of international organizations, partnership programmes, groups, and projects are at work, including the ones listed below:

- By November 2021, 109 Parties to the Paris Agreement had filed their Nationally Determined Contributions, which included renewable energy targets.
- Corporate efforts like 100% renewable energy (RE100) and The Climate Pledge have renewable energy targets and net zero emission goals.
- The 2030 renewable energy targets imposed by the European Union.
- Several bilateral and multinational programmes, including the IEA Technological Cooperation Programmes that promote information exchange, technology advancement, and policy design help when it comes to sustainable power.

Beyond the businesses actively engaged in the development of renewable energy projects, the primary sustainable energy a private sector activity is the execution of agreements for corporate power purchases. In this manner, businesses can buy green energy straight from manufacturers, hedging the cost and guaranteeing the application of eco-friendly innovations. In 2020, nearly 25 GW of capacity was contracted through PPAs, with the majority of those contracts being signed with solar and wind power plants both in Europe and the United States.

All renewable energy technologies must have a long-term goal and consistent strategy in order to maintain investor trust and continuous expansion. The policy must constantly adapt to the changing market circumstances in order to improve cost-competitiveness and the integration of renewable energy into the system.

Administratively determined quotas, feed-in tariffs or surcharges, tradable clean card systems, renewable portfolio standards, net metering, capital grants and tax breaks are just a few of the policy tools used to boost renewable energy. In various nations, especially for wind and solar power, auctions for the centralized global purchasing of renewables have recently gained popularity and played a crucial role in determining renewable energy prices and managing policy costs. However, the effectiveness of such policies in accomplishing deployment and development goals depends on their conception and subsequent capacity to draw funding and rivalry.

Renewable energy sources are quickly changing the world's power networks due to their rising competitiveness, particularly solar and wind energy. To ensure large-scale investment in new renewable capacity as well as in the flexibility of the electric grid to include significant amounts of intermittent renewable energy dependable and budget-friendly way, adjustments to the design of the Foundations for the electricity market and policies will be required. Policies that guarantee capital in all forms of flexibility become increasingly important as the amount of variable renewable energy rises. Improving power station adaptability, enabling requirement control, assisting power reserve, and increasing grid architecture are examples of solutions.

Certain renewable energy sources still have a high cost and/or have unique technological and business difficulties, necessitating more specialized strategies. To hasten the adoption of reservoir hydropower, pumped-hydrostorage, and concentrating solar power (CSP) technologies, better compensation of the storage market value is required. To significantly reduce offshore wind costs, timely grid connections and the ongoing use of regulations that encourage competition are required.

The deployment of additional renewable energy generation is being hampered by drawn-out and difficult permitting procedures, particularly in Europe. Renewable energy project development might take up to 10 years because of complex criteria, responsibility being shared among several government bodies, and a lack of employees. Policymakers should take into account creating one-stop shops, giving developers clear instructions, and including the public in site selection in order to eliminate permission bottlenecks.

The policymakers continuously update the rules and regulations for achieving the NZE targets. The recently announced policies for different countries as per IEA have been shown in Table 1.2.

Table 1.2 Recent announced policies for different countries (https://www.iea.org/policies?topic%5B0%5D=Renewable%20Energy&type%5B0%5D=Grants&status=Announced)

Policy	Country	Year	Status	Jurisdiction
Investment support for utility-scale battery storage	Hungary	2023	Announced	National
Investment in floating offshore wind	United Kingdom	2022	Announced	National
State support for energy efficiency and renewable energy projects	Estonia	2022	Announced	National
AUD 37 million government investment in a new high-tech solar farm	Australia	2021	Announced	National
EUR 500 million investment in nuclear	Netherlands	2021	Announced	National
State Aid for electricity production from renewable sources (Recovery & Resilience Plan)	Austria	2021	Announced	National

B. **Indian Scenario**

Electricity

By March 2022, the government hopes to have 175 GW of grid-connected renewable electricity, including 60 GW of wind power, 100 GW of solar energy, 10 GW of biomass, and 5 GW of small hydropower. The Ministry of Renewable Energy (MNRE) also plans to reach a geothermal energy capacity of 1 GW by 2022. By 2027, the 275 GW of sustainable power targets in the 2018 National Electricity Plan will boost their participation by about 24% of electrical output and 44% of installed capacity, respectively (Key World Energy Statistics 2021).

Renewables Utility-scale

India uses faster depreciation for commercial and industrial renewable energy assets customers, renewable electricity certificates (RECs), renewable purchase obligations (RPOs), and most recently, competitive bidding for utility-scale renewables. The RPOs mandate that some consumers, energy producers, and DISCOMs get a portion of their electricity from renewable sources. The National Energy RPO trajectory is determined by regulatory commissions and keeps an eye on compliance. In June 2018 the RPO requirement was raised from 17 to 22%, with 10.5% from solar, up from 6.75%, and 10.5% from non-solar renewable sources by 2022, up from 10.25%. It will need to be raised for the 450 GW target in the future.

The required entities employ the RECs to fulfil their RPO obligations. To resolve the disparity between the need required for consumers and utilities to achieve under the Energy Act of 2003, their RPOs and the supply of power from renewable sources across regional markets, the Central Electricity Regulatory Commission (CERC) developed voluntary RECs in 2010 and enabled in March 2011, they began trading. The CERC enforces the REC programme. Unfortunately, due to demand, uncertainty around investments, an absence of long-term objectives, and poor compliance, the REC markets have not been able to sufficiently encourage big investments.

To be able to support the country's hydroelectric sector, the Government of India (GoI) approved various steps in March 2019. One of them was making all hydropower projects part of the renewable energy sector projects and establishing hydropurchase obligations (HPOs), which are comparable to RPOs. The economic feasibility of hydropower plants will be enhanced by HPOs.

After a two-year-long gap, the tax on accelerated depreciation incentive for builders of sustainable power plants was reinstated in 2014. Up until March 2017, it was set at 80%. Since April 1, 2017, the advantage has been reduced to 40%. Consumers of renewable energy can claim tax benefits on the value of the depreciation taken in a given year and may dramatically increase the rate of depreciation of their investment in a renewable energy plant compared to other capital assets. A variety of renewable technologies qualify for the tax break, including flat-plate solar collectors, concentrating and pipe-type solar collectors, solar power generating systems, windmills and related machinery, biogas plants, engines, electrically powered vehicles like battery- or fuel-cell-powered vehicles, and agricultural and municipal waste conversion equipment.

In order to reach the 2022 target, the government launched competitive auctions for solar PV (2010) and wind (2017) with long-term power purchase agreements containing fixed price contracts. The MNRE stated that in order to meet its goal of 100 GW of solar PV by 2022, it will tender 25 to 30 GW per year until the end of 2021. (In 2019 India's installed solar capacity was 32.5 GW).

Solar Energy Corporation of India Limited (SECI) conducts massive contracts for 47 parks with a combined capacity of more than 25 GW that have been awarded after central auctions for solar parks. The fundamental goal of solar parks is to offer "plug-and-play" interfaces for projects, allowing developers to concentrate on other areas in the development of projects and reduce project risk. With measures to lessen offtake risk, solve income deficit from curtailment, and minimize delays linked to land purchase, the MNRE recently modified its standards for competitive bidding. Nevertheless, issues with grid integration, connectivity, and land acquisition have delayed the SECI auctions.

In this regard, February 2017 was a significant milestone for assuring the off-taker's financial stability. SECI became a party to the trilateral deal between the Reserve Bank of India (RBI), state governments, and the GoI. The alternative to vital renewable energy auctions is SECI, which subsequently uses power purchase agreements to sell the electricity to DISCOMs. According to the tripartite agreement's guarantee by the RBI, SECI is a low-risk off-taker because it gets paid for any payment delays from DISCOMs. The National Thermal Power Company (NTPC), a participant in this agreement and a potential off-taker, also conducts renewable energy auctions (Key World Energy Statistics 2021).

Figure 1.14 (https://www.iea.org/energy-system/renewables) depicts the global energy prices resulting from auctions. The auction prices for solar and onshore wind have a downward trend, but the price decrease for solar PV was more pronounced, reflecting the greater maturity of onshore wind technology at the start of the period.

Renewable generators are granted must-run status under the Indian Electricity Grid Code 2010 and various state grid codes and regulations under the Electricity Act 2013. The State Load Dispatch Centres can only be restricted for grid security concerns (not for commercial purposes), and there are compensation guidelines in place.

In addition, until 2022, interstate transmission power networks that are directly connected to renewable energy projects through competitive auctions are free from paying transmission fees. India is spending money on transmission all around the nation, particularly through its Green Energy Corridors.

A policy for repowering onshore wind farms was also established by the MNRE in 2018. With the exception of tax advantages, the government does not provide any financial assistance (accelerated depreciation).

The general classification of power sector policies has been shown in Fig. 1.15.

1.11 Renewable Energy Regulations and Policies

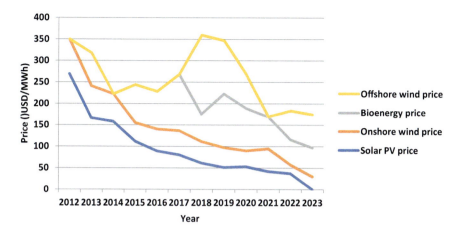

Fig. 1.14 Global average prices resulting from auctions, 2012–2022 (IEA) (https://www.iea.org/energy-system/renewables)

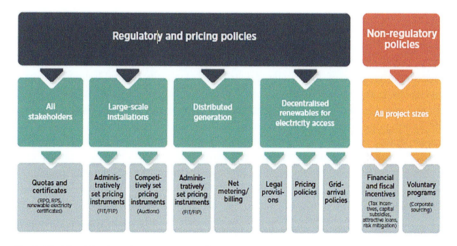

Fig. 1.15 Classification of power sector policies

Rooftop Solar PV

Within the 100 GW solar targets, the GoI has set an ambitious aim of 40 GW of rooftop solar by 2022, objectives for residential and commercial solar PV applications. The MNRE has adopted guidelines for the implementation of Phase II of its Grid-Connected Rooftop Solar Programme.

The target is supported by RPOs, rooftop auctions and programmes that facilitate the deployment of rooftop solar PV on government buildings across states. The MNRE has several policies to incentivise and facilitate rooftop installations: (a) providing central financial assistance for residential, institutional, social and government buildings; (b) advising states to implement net/gross metering regulations and tariff orders; c) providing a model memorandum of understanding, power purchase agreement (PPA) and capital expenditure (CAPEX) agreement for rooftop projects in the government sector; and d) appointing experts to support public-sector undertakings in the implementation of rooftop projects in ministries and departments.

To encourage farmers to adopt solar energy, financial approval was given by the Cabinet Committee on Economic Affairs (CCEA). With net metering programmes in 28 states and a variety of tariff structures, solar deployment is increasing in response to India's high power retail rates.

Offshore Wind

The GoI assesses offshore wind's potential in the range of 10–20 GW. Offshore wind, a good substitute for onshore wind, is still not widely used in India. Given that India lacks a significant offshore turbine sector, a 1 GW first tender is anticipated, and a white paper is being created in partnership with the European Union to address supply chain and industrial infrastructure obstacles.

Off-Grid Solar PV

Various schemes are available at both the national and state levels to support the uptake of off-grid electrification, mainly through solar technologies. In 2015 the Deen Dayal Upadhyaya Gram Jyoti Yojana (DDUGJY) scheme was launched to support the adoption of decentralized distributed electricity in rural India via off-grid installations, mainly mini-grids. In 2017 the Off-Grid and Decentralized Solar PV Programme was put in place to facilitate the uptake of various solar PV applications for lighting and water pumping in rural areas by providing financial means to the implementing agencies. The programme was extended until the end of the 2020 financial year. In December 2018 the Atal Jyoti Yojana (AJAY) Phase II programme was initiated to finance the installation of over 3 million solar street lights in selected regions. Initiated by the GoI in February 2019 (and followed by guidelines in July 2019), the KUSUM scheme will support farmers in replacing existing diesel pumps with solar PV pumps (with both on-grid and off-grid features). The scheme will allow farmers to become prosumers and sell power to the DISCOMs at a predetermined price. The scheme aims to add solar and other renewable capacity of 28 GW by 2022 (Bassam et al. 2012). It has three main components:

(a) Financing of 10 GW of renewable energy plants, each up to 2 MW capacity.
(b) Offering 1.75 million standalone solar agriculture pumps, central government to provide 30% subsidy and the state government to provide a 30% subsidy.
(c) Converting 1 million grid-connected agriculture pumps to solar-powered operations with central government and state government providing a 30% subsidy each.

India also supports off-grid EfW uptake by providing capital subsidies for the purchase and installation of biomass gasifiers in rural areas.

Table 1.3 Recent policies for India (https://www.iea.org/policies?topic%5B0%5D=Renewable%20Energy&country%5B0%5D=India)

Policy	Country	Year	Status	Jurisdiction
National green hydrogen mission	India	2023	In force	National
National bioenergy programme	India	2022	In force	National
National policy on biofuels (2022 Amendment)	India	2022	In force	National
Roadmap for ethanol blending in India 2020–25	India	2022	In force	National
Imposition of basic custom duty (BCD) on solar PV cells and modules/panels	India	2021	Planned	National
Ratle hydro power project (India)	India	2021	In force	National
Renewable energy investment	India	2021	In force	National

Bioenergy and Waste

India is on track to reach its 10 GW of bioenergy capacity goal by 2022. The use of bagasse in cogeneration facilities at sugar mills is the main contribution. The plan to encourage the use of co-generation based on biomass in sugar factories as well as other industries, which is currently in effect until 2020, provides central financial support as a capital subsidy for each extra megawatt of power supplied by making additional investments in effective technology for co-generation. As sugar mills are renovated to more effective co-generation systems, there is potential to enhance the amount of power they produce.

The policy status for India is shown in Table 1.3.

1.12 Environmental Aspects of Energy Sources

Energy is an essential component of our daily lives, and it powers everything from homes, transportation, and industries to healthcare, education, and communication. The sources of energy that we use significantly influence our surroundings, both positively and negatively. This section discusses the environmental aspects of different sources of power, including fossil fuels, nuclear, hydropower, wind, solar, and biofuels (Akella et al. 2009; Río and Burguillo 2009).

- **Fossil Fuels**:

  Coal, oil, and natural gas are examples of fossil fuels. These are the most widely used power sources worldwide. However, they are also the most substantial causes of emissions of greenhouse gases, which are accountable for causing climate change and global warming. As fossil fuels are burned, carbon dioxide is released and other pollutants into the atmosphere, causing air pollution, acid rain, and a host of other environmental problems.

In addition to greenhouse gas emissions, the extraction and transportation of fossil fuels can also have adverse environmental impacts. For instance, coal mining often results in soil erosion, land degradation, and water pollution. Oil spills from drilling and transportation can harm marine life and ecosystems. Furthermore, the extraction of natural gas through hydraulic fracturing or "fracking" can lead to water contamination, earthquakes, and other environmental problems.

- **Nuclear Energy**:

The low-carbon energy source nuclear energy does not emit greenhouse gases during power generation. Nonetheless, the creation of nuclear fuel and the management of radioactive waste present serious environmental challenges. The mining of uranium, the primary fuel for nuclear reactors, requires large amounts of energy and water, and it can cause soil erosion and water pollution. Moreover, the handling and storage of radioactive waste pose long-term environmental risks, as the waste remains hazardous for 10,000 years.

- **Hydropower**:

Hydropower is a kind of renewable energy that harnesses falling water to produce power. Hydropower does not emit greenhouse gases or other pollutants during power generation. However, the building of huge reservoirs and dams may have major negative effects on the ecosystem, such as the relocation of people and animals, changes in water flow and temperature, and the loss of biodiversity. Additionally, the development and use of hydro-power plants can lead to the release of methane, a potent greenhouse gas, from decomposing organic matter in the reservoirs.

- **Wind Energy**:

A renewable energy source called wind energy harnesses the wind's energy to make energy. Wind turbines do not emit greenhouse gases or other pollutants during power generation. However, the construction and operation of wind turbines can have environmental impacts, including habitat fragmentation, noise pollution, and bird and bat fatalities. Wind turbines can also interfere with the flight paths of migratory birds and disrupt local ecosystems.

- **Solar Energy**:

Solar energy is a renewable energy source that produces electricity by harnessing the power of the sun. Solar panels do not emit greenhouse gases or other pollutants during power generation. However, the production of solar panels requires a large value of energy, water as well and raw materials, which can lead to environmental impacts such as air pollution, water pollution, and land degradation. Additionally, the disposal of solar panels at the end of their lifespan can pose environmental risks if not properly managed.

- **Biofuels**:

Biofuels are renewable energy sources that are derived from plant-based materials such as corn, soybeans, and sugarcane. Biofuels can be compared to reduction of greenhouse gas emissions from fossil fuels, but their climate benefits depend on the type of feedstock used and the production methods employed. For example, the use of land for biofuel crops can result in deforestation and habitat loss, while the use of fertilizers and pesticides can cause water pollution and harm to wildlife. Moreover, the production of biofuels can compete with food production, leading to food security issues.

Renewable energy is important from an environmental standpoint as it helps to lessen the negative impacts lessen the impact of climate change and dependence on non-renewable sources of energy that contribute to greenhouse gas emissions. Further, the environmental aspects of renewable energy, including its benefits, drawbacks, and potential for future development have been discussed (https://www.researchgate.net/publication/273697005_Positive_and_Negative_Impacts_of_Renewable_Energy_Sources).

- **Benefits of Renewable Energy**

Renewable energy has numerous environmental benefits. One of the primary advantages is that it produces much lower greenhouse gas emissions than fossil fuel energy sources. According to the Intergovernmental Panel on Climate Change, fossil fuels such as coal, oil, and natural gas are the main causes of climate change. Sustainable energy is the opposite, producing little to no emissions, making it an important tool for mitigating climate change.

Another benefit of renewable energy is that it helps to conserve natural resources. Fossil fuel energy sources such as coal and oil are finite and will eventually run out. By using renewable energy, we can reduce our dependence on these finite resources and preserve them for future generations.

Sustainable power is also having the capacity to create jobs and boost local economies. The International Renewable Energy Agency (IRENA) estimates that 11.5 million people worldwide were engaged in the renewable energy industry in 2019, with the potential to create millions more jobs in the future. In addition, renewable energy projects can stimulate local economies by providing new business opportunities and increasing tax revenue (Frondel et al. 2010).

- **Drawbacks of Renewable Energy**

While renewable energy has many benefits, it also has some drawbacks. One of the main challenges is that it can be intermittent and unpredictable. For instance, wind turbines only generate power at night while solar panels only do so during the day when the wind is blowing. This can make it difficult to rely solely on renewable energy sources for power, especially during times of high energy demand.

Another challenge is that renewable energy infrastructure can sometimes have negative impacts on the environment. For example, large-scale wind and solar farms can have an impact on wildlife habitats and ecosystems. Additionally, the construction and maintenance of renewable energy infrastructure can require significant amounts of land, water, and other resources, which can be detrimental to the environment if not managed carefully.

Table 1.4 Social effect analysis of various renewable energy sources

Technology	Impact	Magnitude
Photovoltaic	Toxins	Minor-Major
	Visual	Minor
Wind	Bird strike	Minor
	Noise	Minor
	Visual	Minor
Hydro	Displacement	Minor-Major
	Agricultural	Minor-Major
	River damage	Minor-Major
Geothermal	Seismic activity	Minor
	Odour	Minor
	Pollution	Minor-Major
	Noise	Minor

To the advantage of the populace, several essential considerations should be made, such as the climate, the level of education and living standards, the area (whether urban or rural from an agricultural perspective), and so on. Moreover, these resources provide societal advantages including health improvement according to technical choice, consumer advancements, and employment opportunities. The primary factors in every nation's growth are social factors. Renewable power plants may lead to local job creation and better health, career prospects and client preference, among other societal advantages. According to the study's findings, the total amount of emissions reduced following the construction of renewable energy projects in distant places increases exponentially with time. Table 1.4 lists the social effects of each resource along with their intensity.

The two primary environmental issues are contamination of the air and water, which are often brought on by sewage from houses and enterprises, contaminated rain, and the disposal of waste hazardous substances and heavy metals like lead and mercury present in certain oils and liquids. Together Natural resources can be preserved, the greenhouse impact and air pollution can be reduced, and water pollution can be controlled through the right exploitation of renewable energy sources as indicated in Table 1.5.

- **Future Development of Renewable Energy**

Despite these challenges, renewable energy has enormous potential for future development. One of the main areas of focus is the development of energy storage technologies, which can help to mitigate the intermittent nature of renewable energy sources. Technologies for energy storage like batteries, pumped hydro, and thermal storage help to store excess energy generated by renewable sources during times of low demand and release it during times of high demand (Morel et al. 2015).

1.12 Environmental Aspects of Energy Sources

Table 1.5 Environmental impact summary

Impact Category	Connection to traditional sources	Remarks
	Exposure to dangerous substances	
Emission of Mercury (Hg), Cadmium (Cd), and other toxic elements	Decreased emissions	Emissions reduced a few hundred times
Emission of particles	Decreased emissions	Very little emission
	Exposure to dangerous gases	
CO_2 emission	Decreased emissions	A big advantage
Acid rain, SO, NO_x	Decreased emissions	More than 25 times reduced
Other greenhouse gases	Decreased greenhouse gases	Big advantage—Global warming
	Other	
Spouts off fossil fuels	Elimination of all or some oil spills	Spills involving heavy fuel oil and other petroleum products
Water quality	Higher standard water	Lower levels of water contamination
Soil erosion	Smaller land loss	Most of the time, there is no deep ground penetration

Another area of focus is the development of new renewable energy sources. For example, wave energy, which harnesses the power of ocean waves to generate electricity, has the potential to provide a sizeable quantity of renewable power in coastal regions. Similarly, hydrothermal power development, which utilizes the heat from the earth to produce electricity, could provide a stable and reliable source of renewable energy.

Renewable energy has significant environmental benefits and has the main role in reducing the harmful effects of climate change. While there are several difficulties associated with sustainable power, such as its intermittent nature and potential environmental impacts, ongoing research and development are addressing these issues and enhancing the effectiveness and reliability of renewable power sources. With continued investment and innovation, sustainable power has the capability to transform the power sector and help foster a more sustainable world and resilient future for all (Zeb et al. 2014).

Review Questions

1. How is the standard of living correlated with per capita energy consumption?
2. Discuss the 1973 oil crisis.
3. What kind of energy are primary and secondary sources?
4. What is conventional and non-conventional energy?
5. Provide a list of the several non-conventional energy sources. Describe their classification, relative merits, and availability

6. Go over the key characteristics of several types of renewable and non-renewable energy sources and discuss the significance of non-conventional energy sources in the context of global warming.
7. What does the term renewable energy sources mean?
8. What does the term commercial energy mean to you?
9. What are the benefits and drawbacks of traditional energy sources?
10. How much of the total energy consumed globally comes from fossil fuels?
11. How much of India's energy needs are covered by coal?
12. Describe the main characteristics of non-conventional energy sources.
13. Discuss various renewable energy sources in India.
14. What advantages and disadvantages do traditional energy sources have?
15. What does the term greenhouse effect mean to you, and what are its effects? How is it caused?
16. What exactly do you mean by green power?
17. How do greenhouse gases work?
18. What is the global primary energy usage at the moment? How fast is it expanding?
19. What are your thoughts on the global trend in the future availability of fossil fuels?
20. What is the current hydropower potential for the world, and how much of it has already been used?
21. Which nation gets the majority of its energy from nuclear sources?
22. What potential does solar energy have on a global scale? How much power, on average, is received at noon on a sunny day on the surface of the earth?
23. What are the futures of biomass and wind energy?
24. Comment about India's potential for fossil fuels
25. What are the current conditions and possibilities for non-traditional energy sources in India?
26. Comment about India's expanding energy sector.
27. Comment on the Indian government's plans for rural electrification
28. Describe several renewable energy sources, paying particular attention to the Indian context.

Objective Type Questions

1. What factor is used to measure a country's population's level of living?

    (a) Per capita energy usage
    (b) Industrial generation
    (c) Vehicles number of vehicles per house
    (d) Population density

2. How much electricity is used in India per person?

    (a) 1234 kWh
    (b) 1255 kWh
    (c) 1123 kWh

1.12 Environmental Aspects of Energy Sources

(d) 1134 kWh

3. According to reports, which year marked the beginning of extensive worldwide planning for renewable energy?

(a) 1942
(b) 1973
(c) 1991
(d) 1850

4. What proportion of the world's main energy usage is made up of fossil fuels?

(a) 84%
(b) 50%
(c) 15%
(d) 99%

5. Which energy type electrical, thermal, mechanical, or chemical is regarded as being of the highest calibre?

(a) Mechanical
(b) Chemical
(c) Thermal
(d) Electrical

6. Among energy, environment, and the money

(a) Only energy and environment are related
(b) All three are correlated
(c) Only energy and the economy are related
(d) All three are non-dependent.

7. The primary causes of global warming include

(a) Heat emission from engines
(b) CO_2 emissions due to the burning of fossil fuels
(c) Air pollution
(d) Nuclear energy utilization

8. Nuclear energy utilization is generally opposed due to

(a) High cost
(b) Depletion of uranium reserves rapidly
(c) Nuclear waste handling issue
(d) Ecological imbalance issue

9. Climate change in the future result in

(a) Acid rains
(b) Rise of agricultural production
(c) Increase in the heat engine's efficiency
(d) Climatic pattern change and its severity

10. How much of the world's fossil fuel reserves are in India?

    (a) 17%
    (b) 20%
    (c) 4%
    (d) 6.85%

11. What is India's potential for producing energy from renewable sources?

    (a) 89 GW
    (b) 175 GW
    (c) 210 GW
    (d) 257 GW

12. The growing of appropriate plants on adequate soil is the preferred method for energy farming of biodiesel

    (a) Rooftops of buildings
    (b) Good fertile land in the country
    (c) Marginal and fallow land, not suitable for normal agriculture
    (d) Sea

13. Which source has having higher potential for power generation in India?

    (a) Solar
    (b) Wind
    (c) Fossil fuels
    (d) Hydel

14. Which Ministry in India is primarily in charge of the import, distribution, and export of petroleum products and natural gas?

    (a) Ministry of Fossil Fuel
    (b) Ministry of Petroleum and Natural Gas
    (c) Ministry of Oil
    (d) Ministry of Non-Renewable Energy

15. In India, _____ is the fossil fuel that is most widely accessible.

    (a) Natural Gas
    (b) Coal
    (c) Oil
    (d) Petroleum

16. Which of the following promotes long-term expansion?

    (a) Negative economic growth, deterioration of the environment with economic expansion, and energy resources meeting energy demands
    (b) Positive economic growth, deterioration of the environment with economic expansion, and energy resources meeting energy demands

(c) Positive economic growth, deterioration of environment with economic expansion, and growth of energy resources meeting energy demands

(d) Positive economic growth, non-deterioration of the environment with economic expansion, and growth of energy resources not meeting energy demands

17. Energy resources that support sustainable development

(a) can be completely exhausted
(b) cannot be exhausted completely
(c) cannot meet the growing demand
(d) can destroy the environment

18. Which of the following factors supports economic and energy growth?

(a) Continuous capital and human resource investment
(b) Intermittent capital and human resource investment
(c) No capital and human resource investment
(d) Decrement in capital and human resource investment

19. Which statement about global energy usage is true?

(a) Total energy consumed by humanity
(b) Total energy produced and used by humanity
(c) Total energy produced by humans in the biological pyramid
(d) Total energy consumed by humans in the biological pyramid

20. Why is there such a need for renewable energy sources?

(a) Because of low or zero carbon footprint
(b) Because they emit greenhouse gases
(c) Because they are more efficient
(d) Because of the decreasing global temperatures

Answer

1. a	2. b	3. b	4. a	5. d	6. b	7. b
8. d	9. d	10. d	11. b	12. c	13. c	14. b
15. b	16. c	17. b	18. a	19. b	20. a	

References

Akella AK, Saini RP, Sharma MP (2009) Social, economical and environmental impacts of renewable energy systems. Renew Energy 34:390–396. https://doi.org/10.1016/J.RENENE.2008.05.002

Ali S, Anwar S, Nasreen S (2017) Renewable and non-renewable energy and its impact on environmental quality in South Asian Countries. Forman J Econ Stud 00:177–194. https://doi.org/10.32368/FJES.20170009

Annual Energy Outlook 2023—U.S. Energy Information Administration (EIA). https://www.eia.gov/outlooks/aeo/

Bilgili F, Koçak E, Bulut Ü (2016) The dynamic impact of renewable energy consumption on CO_2 emissions: a revisited environmental Kuznets curve approach. Renew Sustain Energy Rev 54:838–845. https://doi.org/10.1016/J.RSER.2015.10.080

Central Water Commission, Ministry of Jal Shakti, Department of Water Resources, River Development and Ganga Rejuvenation, GoII. https://cwc.gov.in/

Chilán JCH, Torres SGP, Machuca BIF, Cordova AJT, Pérez CAM, Gamez MR (2018) Social impact of renewable energy sources in the province of Loja. Int J Phys Sci Eng 2:13–25. https://doi.org/10.29332/IJPSE.V2N1.79

Demirbaş A (2006) Global renewable energy. Resources 28:779–792. https://doi.org/10.1080/00908310600718742

Editorial Board (2009) Renew Sustain Energy Rev 13:CO_2. https://doi.org/10.1016/s1364-0321(09)00050-1

"Emissions Gap Report 2022: the closing window—climate crisis calls for rapid transformation of societies [EN/AR/RU/ZH/SW]—World | ReliefWeb."[Online]. https://www.unep.org/resources/emissions-gap-report-2022

Energy Agency (2021) World energy outlook 2021. www.iea.org/weo

Energy Agency I (2021) Review 2021 assessing the effects of economic recoveries on global energy demand and CO_2 emissions in 2021 global energy

Energy Agency I (2021) Renewables 2021—Analysis and forecast to 2026

El Bassam N, Maegaard P, Schlichting ML (2012) In: Distributed renewable energies for off-grid communities: strategies and technologies toward achieving sustainability in energy generation and supply. Elsevier

Frondel M, Ritter N, Schmidt CM, Vance C (2010) Economic impacts from the promotion of renewable energy technologies: the German experience. Energy Policy 38:4048–4056. https://doi.org/10.1016/J.ENPOL.2010.03.029

Global Wind Energy Outlook 2000 gigawatts by 2030—Global Wind Energy Council. https://gwec.net/global-wind-energy-outlook-2000-gigawatts-2030/

Goudsblom J (1992) The civilizing process and the domestication of Fire. J World History 3(1):1–12. http://www.jstor.org/stable/20078510

Hafner M, Luciani G (2022) In: The Palgrave handbook of international energy economics.https://doi.org/10.1007/978-3-030-86884-0

Hicks J, Ison N (2011) Community-owned renewable energy (CRE): opportunities for rural Australia. Rural Soc 20:244–255. https://doi.org/10.5172/RSJ.20.3.244

Home | Ministry of Petroleum and Natural Gas | Government of India. https://mopng.gov.in/en

IEA (2021) Net-Zero by 2050—a roadmap for the global energy sector. https://www.iea.org/reports/net-zero-by-2050

Key World Energy Statistics 2021—Analysis—IEA. https://www.iea.org/reports/key-world-energy-statistics-2021

Khan BH (2024) In: Non-conventional energy resources. 3rd edn. TMH Pvt. Ltd., New Delhi

Kralova I, Sjöblom J (2010) Biofuels–renewable energy sources: a review. 31:409–42.https://doi.org/10.1080/01932690903119674

Ministry of Coal, Government of India. https://coal.gov.in/en/public-information/reports/annual-reports/annual-report-2021-22

Morel J, Obara S, Morizane Y (2015) Stability enhancement of a power system containing high-penetration intermittent renewable generation. J Sustain Dev Energy Water Environ Syst 3:151–162. https://doi.org/10.13044/J.SDEWES.2015.03.0012

Positive and Negative Impacts of Renewable Energy Sources. https://www.researchgate.net/publication/273697005_Positive_and_Negative_Impacts_of_Renewable_Energy_Sources

References

Proved Reserves of Crude Oil and Natural Gas in the United States, Year-End 2021. https://www.eia.gov/naturalgas/crudeoilreserves/

Renewables—Energy System—IEA. https://www.iea.org/energy-system/renewables

Del Río P, Burguillo M (2009) An empirical analysis of the impact of renewable energy deployment on local sustainability. Renew Sustain Energy Rev 13:1314–1325.https://doi.org/10.1016/J.RSER.2008.08.001

Sadati SMS, Qureshi FU, Baker D (2015) Energetic and economic performance analyses of photovoltaic, parabolic trough collector and wind energy systems for Multan Pakistan. Renew Sustain Energy Rev 47:844–855. https://doi.org/10.1016/J.RSER.2015.03.084

Shahzad SJH, Kumar RR, Zakaria M, Hurr M (2017) Carbon emission, energy consumption, trade openness and financial development in Pakistan: a revisit. Renew Sustain Energy Rev 70:185–192. https://doi.org/10.1016/J.RSER.2016.11.042

Smil V (1994) In: Energy in World history. 1st edn. Taylor & Francis

Sukhatme SP, Nayak JK (1996) Solar energy principle of thermal collection and storage. 3rd edn. TMH, New Delhi

Thangamayan S, Chithirairajan B, Sudha S (2018) Energy sector reforms and its role of economic development with special reference to Tamilnadu. Int J Eng Technol 7:424–426

Tracking Clean Energy Progress 2023—Analysis—IEA. https://www.iea.org/reports/tracking-clean-energy-progress-2023

Turney D, Fthenakis V (2011) Environmental impacts from the installation and operation of large-scale solar power plants. Renew Sustain Energy Rev 15:3261–3270. https://doi.org/10.1016/j.rser.2011.04.023

Whiten J, Widdowson P (1992) Discourse analysis. Oxford University Press

Yuksel I, Arman H (2010) Global warming and hydropower in turkey for a clean and sustainable energy future. https://doi.org/10.5772/10288

Zeb R, Salar L, Awan U, Zaman K, Shahbaz M (2014) Causal links between renewable energy, environmental degradation and economic growth in selected SAARC Countries: progress towards green economy. Renew Energy 71:123–132. https://doi.org/10.1016/J.RENENE.2014.05.012

Chapter 2
Fundamentals of Solar Energy

Abstract This chapter is concerned with the availability of solar radiation as an energy source the nature of the radiation emitted by the sun and the incident on the Earth's atmosphere. Extra-terrestrial radiation on a horizontal surface, its spectral distribution [i.e., we will be concerned primarily with radiation in a wavelength range of 0.25 to 3.0 μm, the portion of the electromagnetic radiation that includes most of the energy radiated by the sun] and the radiation at the earth's surface is discussed first. The second major topic in this chapter is solar radiation geometry that is, the direction from which beam solar radiation is received its angle of incidence on various surfaces and the quantity of radiation received over various time spaces. Various angles are also defined so that it is possible to convert the flux on one plane to an equivalent flux on another. Why solar radiation data is needed? What are the proper ways to achieve the solar radiation data? Instruments for measuring solar radiation and methods for presenting data are then described. Empirical equations for predicting/Estimating the availability of solar radiation at a location have been finally discussed.

Keywords Solar energy · Spectral distribution · Extra-terrestrial radiation · Solar radiation · Geometry · Solar radiation on tilted surface

This chapter is concerned with the availability of solar radiation as an energy source the nature of the radiation emitted by the sun and the incident on the earth's atmosphere. Extra-terrestrial radiation on a horizontal surface, its spectral distribution [i.e. we will be concerned primarily with radiation in a wavelength range from 0.25 to 3.0 μm, the portion of the electromagnetic radiation that includes most of the energy radiated by the sun] and the radiation at the earth's surface is discussed first. The second major topic in this chapter is solar radiation geometry; that is, the direction from which beam solar radiation is received its angle of incidence on various surfaces and the quantity of radiation received over various time spaces. Various angles are also defined so that it is possible to convert the flux on one plane to an equivalent flux on another. Why solar radiation data is needed? What are the proper ways to achieve the solar radiation data? Instruments for measuring solar radiation and methods for

© The Author(s), under exclusive license to Springer Nature Singapore Pte Ltd. 2024
K. Namrata et al., *Wind and Solar Energy Systems*, Energy Systems in Electrical Engineering, https://doi.org/10.1007/978-981-99-9710-7_2

54   2 Fundamentals of Solar Energy

presenting data are then described. Empirical equations for predicting/estimating the availability of solar radiation at a location have been finally discussed.

2.1 Introduction

The structure and properties of the sun govern the kind of energy it emits into space. At temperatures higher than absolute zero, all substances solids, liquids, and gases emit energy in the form of electromagnetic waves. The name of this energy is radiation. It is possible to think of heat transmission through radiation as the transportation of energy by photons, which are tiny energy bundles that are emitted from excited atoms and move along straight routes until they are absorbed or dispersed by other atoms. By its velocity of propagation, which, in a vacuum, is independent of frequency and has the value of 2.997925×10^8 m/s, and the absence of an intermediary medium needed for its transmission, radiation differs from other forms of heat transfer, such as conduction and convection.

Transverse waves include electromagnetic waves. Different electromagnetic waves have a broad variety of wavelengths. Table 2.1 displays the electromagnetic spectrum, broken down into several wavelength ranges. The radiation that is released by the sun, the planet, and the atmosphere that falls within the ultraviolet, visible, and infrared spectral areas is the radiation that is most significant to humans. The majority of solar and terrestrial radiation falls between 0.15 and 120 μm (micrometre), while the range in which the radiation that is practically relevant to solar energy consumers falls between 0.15 and 3.0 μm. The visible spectrum has wavelengths between 0.38 and 0.72 μm.

Table 2.1 Electromagnetic spectrum according to wavelength (1 cm $= 10^8$ AU $= 10^4$ μm) (Johnson 1954; Thekaekara and Drummond 1971)

Wavelength	E.M. Radiation
$< 10^{-6}$	Cosmic rays
10^{-6}–10^{-3}	X-rays and γ-rays
10^{-3}–0.2	Far ultraviolet
0.2–0.315	Middle ultraviolet
0.315–0.38	Near ultraviolet
0.38–0.72	Visible
0.72–1.5	Near infrared
1.5–5.6	Middle infrared
6.5–1000	Far infrared
> 1000	Micro and radio waves

2.2 The Sun

With a diameter of 1.39×10^9 m and an average distance of 1.5×10^{11} m from the earth, the sun is a sphere of very hot gaseous substance. The sun spins on its axis once every four weeks as viewed from the earth. It does not, however, revolve like a solid body; each revolution takes around 30 days for the northern regions and roughly 27 days for the equator.

The actual blackbody temperature of the sun is 5777 K. According to different estimates, the core interior's temperature ranges from 8×10^6 to 40×10^6 K, and its density is around 100 times that of water. The gases that make up the sun operate as the "containing vessel" of the continuous fusion reactor, which is kept in place by gravitational forces. It has been proposed that the energy emitted by the sun may be obtained from several fusion processes. The process in which hydrogen (i.e. four protons) and helium (i.e. one helium nucleus) combine is thought to be the most significant. The mass of the helium nucleus is less than that of the four protons since mass was lost in the reaction and turned into energy.

It is necessary to transmit the energy created in the solar sphere's interior, where temperatures may reach several millions of degrees, to the surface before it can be projected into space. The radiation in the sun's core is in the X-ray and gamma-ray regions of the spectrum, with the wavelengths of the radiation rising as the temperature lowers at longer radial distances. A series of radioactive and convective processes occur with sequential emission, absorption, and irradiation.

Figure 2.1 depicts the sun's schematic structure. According to estimates, 40% of the sun's mass is contained within the range of 0–0.23R (R is the sun's radius), where 90% of all energy is produced. Temperature and density have decreased to 130,000 K and 70 kg/m^3, respectively, at a distance of 0.7R from the centre. At this point, convection processes start to play a significant role and the region between 0.7 and 1.0R is referred to as the convective zone. The density is around 10^{-5} kg/m^3, and the temperature falls to almost 5000 K inside this zone.

Granules (irregular convection cells) with sizes ranging from 1000 to 3000 km and cell lifetimes of a few minutes appear to make up the sun's surface. Other characteristics of the solar surface include larger dark regions called sunspots, which can range in size, and smaller dark regions called pores, which are of the same order of magnitude as the convective cells. The outer layer of the convective zone is called the photosphere. The edge of the photosphere is sharply defined, even though it is of low density (about 10^{-4} that of air at sea level). It is essentially opaque, as the gases of which it is composed are strongly ionized and able to absorb and emit a continuous spectrum of radiation. The photosphere is the source of most solar radiation.

An atmosphere around the sun that is more or less transparent may be seen by devices that occult the sun's disc or during total solar eclipses. The reverse layer, which is several hundred kilometres thick and is composed of colder gases, is located above the photosphere. Outside of that lies a layer known as the chromosphere, with a depth of around 10,000 km. While it has somewhat greater temperatures than the

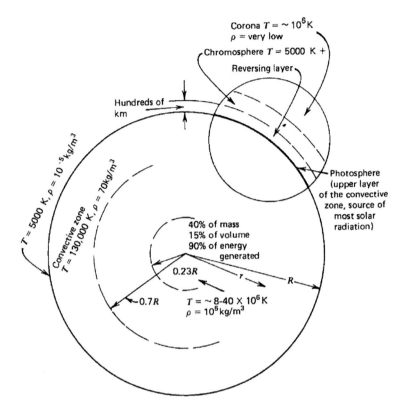

Fig. 2.1 Structure of the sun (Johnson 1954; Thekaekara and Drummond 1971)

photosphere, this layer of gas has a lower density. The corona, which is much farther out, is a zone of very low density and extremely high temperature (10^6 K).

Understanding that the sun does not operate as a blackbody radiator at a constant temperature, will help to have a simplified understanding of the sun's physical structure, temperature gradients, and density variations. Instead, the radiation from the sun is a combination of radiation from many layers that produce and absorb light of different wavelengths.

2.3 The Solar Constant

The geometry of the sun-earth relation is shown graphically in Fig. 2.2. The variation in the distance between the sun and the earth is caused by the eccentricity of the earth's orbit, which is 1.7%. The sun subtends an angle of 32′ at a distance of one astronomical unit, or 1.495×10^{11} m, the average distance between the earth and the sun. Outside of the earth's atmosphere, solar radiation has a virtually constant

2.3 The Solar Constant

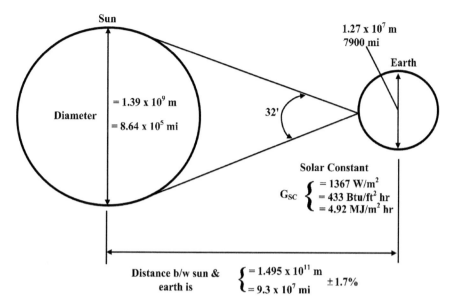

Fig. 2.2 Sun-earth relationships (Johnson 1954)

intensity due to the sun's radiation output and its geographical proximity to the planet. The solar constant I_{sc} is the amount of energy from the sun received over a unit area of surface per unit time at a mean earth-sun distance outside the atmosphere, perpendicular to the direction of radiation propagation.

Before the development of rockets and spacecraft, calculations of the solar constant had to be dependent on measurements taken from the ground of solar radiation that had already passed through the atmosphere and been partially absorbed and dispersed by atmospheric constituents. Estimates of air transmission in various sun spectrum regions were used to extrapolate from terrestrial observations taken from high mountains. C. G. Abbot and his colleagues at the Smithsonian Institute conducted ground-breaking investigations. Johnson (1954) compiled these investigations and other measurements made with rockets, and he corrected Abbot's determination of the solar constant from 1322 to 1395 W/m².

Direct measurements of solar radiation may now be made outside of most or the earth's entire atmosphere because of the availability of extremely high-altitude aircraft, balloons, and spacecraft. Nine different experimental programmes were used to conduct these measurements with a range of tools. Using an estimated error of ± 1.5% they produced a value of 1353 W/m² for the solar constant I_{sc}. The figure of 1367 W/m², with an uncertainty on the order of 1%, has been approved by the World Radiation Centre (WRC) (Thekaekara and Drummond 1971; Froelich and Brusa 1981; Froelich and Wehrli 1981).

2.4 Variation of Spectral Distribution and Extra-Terrestrial Radiation

2.4.1 Spectral Distribution

Knowing the spectral distribution of extra-terrestrial radiation, or the radiation that would be received in the absence of the atmosphere, is important in addition to understanding the total energy contained in the solar spectrum (also known as the solar constant). Based on high-altitude and space data, a standard spectral irradiance curve has been created. In Fig. 2.3, the WRC standard is displayed (Froelich and Wehrli 1981; Mani 1981). Table 2.2 gives the same details in numerical form regarding the WRC spectrum. The second column provides the average energy $I_{sc,\lambda}$ (in W/m^2) for small bandwidths centred at wavelength. The third column contains the percentage of the total energy in the spectrum that is between wavelengths zero and λ. The table is divided into two sections: the first at even fractions $f_{o-\lambda}$ and the second at regular wavelength intervals (Table 2.3).

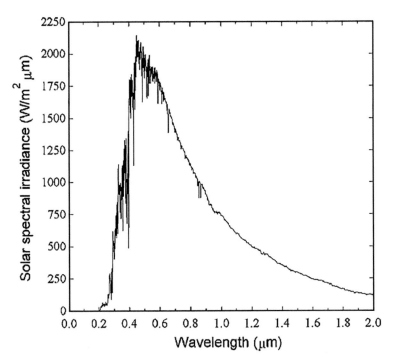

Fig. 2.3 WRC standard spectral irradiance curve at a mean earth-sun distance (Froelich and Wehrli 1981)

2.4 Variation of Spectral Distribution and Extra-Terrestrial Radiation

Table 2.2 Extra-terrestrial solar irradiance (WRC spectrum) in increments of wavelength [a] (Froelich and Wehrli 1981)

λ (μm)	$I_{sc,\lambda}$ (W/m$^2\mu$m)	$f_{0-\lambda}$ (−)	λ (μm)	$I_{sc,\lambda}$ (W/m$^2\mu$m)	$f_{0-\lambda}$ (−)	λ (μm)	$I_{sc,\lambda}$ (W/m$^2\mu$m)	$f_{0-\lambda}$ (−)
0.250	81.2	0.001	0.520	1849.7	0.243	0.880	955.0	0.622
0.275	265.0	0.004	0.530	1882.8	0.257	0.900	908.9	0.636
0.300	499.4	0.011	0.540	1877.8	0.271	0.920	847.5	0.648
0.325	760.2	0.023	0.550	1860.0	0.284	0.940	799.8	0.660
0.340	955.5	0.033	0.560	1847.5	0.298	0.960	771.1	0.672
0.350	955.6	0.040	0.570	1842.5	0.312	0.980	799.1	0.683
0.360	1053.1	0.047	0.580	1826.9	0.325	1.000	753.2	0.695
0.370	1116.2	0.056	0.590	1797.5	0.338	1.050	672.4	0.721
0.380	1051.6	0.064	0.600	1748.8	0.351	1.100	574.9	0.744
0.390	1077.5	0.071	0.620	1738.8	0.377	1.200	507.5	0.785
0.400	1422.8	0.080	0.640	1658.7	0.402	1.300	427.5	0.819
0.410	1710.0	0.092	0.660	1550.0	0.425	1.400	355.0	0.847
0.420	1687.2	0.105	0.680	1490.2	0.448	1.500	297.8	0.871
0.430	1667.5	0.116	0.700	1413.8	0.469	1.600	231.7	0.891
0.440	1825.0	0.129	0.720	1348.6	0.489	1.800	173.8	0.921
0.450	1992.8	0.143	0.740	1292.7	0.508	2.000	91.6	0.942
0.460	2022.8	0.158	0.760	1235.0	0.527	2.500	54.3	0.968
0.470	2015.0	0.173	0.780	1182.3	0.544	3.000	26.5	0.981
0.480	1975.6	0.188	0.800	1133.6	0.561	3.500	15.0	0.988
0.490	1940.6	0.202	0.820	1085.0	0.578	4.000	7.7	0.992
0.500	1932.2	0.216	0.840	1027.7	0.593	5.000	2.5	0.996
0.510	1869.1	0.230	0.860	980.0	0.608	8.000	1.0	0.999

[a] $I_{sc,\lambda}$ represents the average sun irradiation for the period between the midpoint of the wavelength interval before and after. For instance, the average value between 0.595 and 0.610 μm is 1748.8 W/m^2 μm at 0.600 μm

2.4.2 Variation of Extra-Terrestrial Radiation

Two sources of variation in extra-terrestrial radiation must be considered. The first is the variation in the radiation emitted by the sun. There are conflicting reports in the literature on periodic variations of intrinsic solar radiation. It has been suggested that there are small variations (less than $\pm$ 1.5%) with different periodicities and variations related to sunspot activities. Willson et al. (1981) report variances of up to 0.2% correlated with the development of sunspots. Others consider the measurements to be inconclusive or not indicative of regular variability. Measurements from Nimbus and Mariner satellites over periods of several months showed variations within limits of $\pm$ 0.2% over a time when sunspot activity was very low (Fröhlich 1977). Data from

Table 2.3 Extra-terrestrial solar radiation with equal energy increment (Froelich and Wehrli 1981; Mani 1981)

Energy band	Wavelength	Midpoint	Energy band	Wavelength	Midpoint
$f_i - f_{i+1}$	Range	Wavelength	$f_i - f_{i+1}$	Range	Wavelength
(−)	(μm)	(μm)	(−)	(μm)	(μm)
0.00−0.05	0.250−0.364	0.328	0.50−0.55	0.731−0.787	0.758
0.05−0.10	0.364−0.416	0.395	0.55−0.60	0.787−0.849	0.817
0.10−0.15	0.416−0.455	0.437	0.60−0.65	0.849−0.923	0.885
0.15−0.20	0.455−0.489	0.472	0.65−0.70	0.923−1.008	0.966
0.20−0.25	0.489−0.525	0.506	0.70−0.75	1.008−1.113	1.057
0.25−0.30	0.525−0.561	0.543	0.75−0.80	1.113−1.244	1.174
0.30−0.35	0.561−0.599	0.580	0.80−0.85	1.244−1.412	1.320
0.35−0.40	0.599−0.638	0.619	0.85−0.90	1.412−1.654	1.520
0.40−0.45	0.638−0.682	0.660	0.90−0.95	1.654−2.117	1.835
0.45−0.50	0.682−0.731	0.706	0.95−1.00	2.117−10.08	2.727

Hickey et al. (1982) over 2.5 years from the *Nimbus 7* satellite suggest that the solar constant is decreasing slowly, at a rate of approximately 0.02% per year. See Coulson (1975) or Thekaekara (1976) for further discussion of this topic. For engineering purposes, given the uncertainties and variability of atmospheric transmission, the energy emitted by the sun can be considered to be fixed (Iqbal 1983).

Nonetheless, changes in the earth-sun distance do cause variations in the extra-terrestrial radiation flow of up to ± 3.3%. Figure 2.4 illustrates how extra-terrestrial radiation is affected by the season. The first part of Eq. 2.1 provides a straightforward equation with precision suitable for the majority of engineering computations. Iqbal (1983), using Spencer (1971), offers a more precise equation (± 0.01%) in the form of the second part of Eq. (2.1).

Variation of the earth-sun distance, however, does lead to variation of extra-terrestrial radiation flux in the range of ± 3.3%. The dependence of extra-terrestrial radiation on time of year is shown in Fig. 2.4. A simple equation with accuracy adequate for most engineering calculations is given by Eq. 2.1a. Spencer (1971), as cited by Iqbal (1983), provides a more accurate equation (± 0.01%) in the form of Eq. 2.1b:

$$I_{on} = \begin{cases} I_{sc}\left(1 + 0.033 \cos \frac{360 n}{365}\right) \\ I_{sc}(1.000110 + 0.034221 \cos B \\ \quad + 0.001280 \sin B + 0.000719 \cos 2B + 0.000077 \sin 2B), \end{cases} \quad (2.1a, b)$$

where I_{on} is the extra-terrestrial radiation that hits the earth on the nth day of the year and B is provided by

2.5 Solar Irradiance Falling at the Earth's Surface

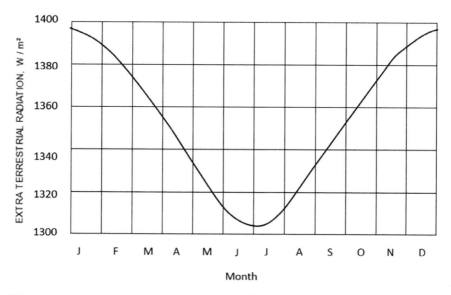

Fig. 2.4 Variation of declination over a year (Iqbal 1983)

$$B = (n-1)\frac{360}{365}. \tag{2.2}$$

In relation to the sun's orbit, the earth's rotational axis is 23.5° tilted, and throughout orbital motion, the earth maintains this orientation. Along with the planet's daily rotation and annual revolution, the tilt of the globe explains the variations in solar radiation throughout the planet's surface, the lengths of daylight and night, and the seasonal changes. Figure 2.5 provides a rough illustration of the effects of the earth's axis tilt at different times during the year.

During the winter solstice (December 21), the North Pole is inclined 23.5° away from the sun, as seen in Fig. 2.6, which also depicts the earth's position in relation to the sun's beams at the summer solstice. All points on the surface of the planet north of 66.5° north latitude are completely in the dark for 24 h, while all areas lying at 23.5° south are continuously in the light. When the summer solstice occurs (June 21), the circumstance has changed. All locations on the surface of the globe experience 12 h of darkness and 12 h of sunshine on the two equinoxes, which occur on March 21 and September 21, respectively.

2.5 Solar Irradiance Falling at the Earth's Surface

Renewable radiation reaches the following, the earth's surface passes through the atmosphere surrounding the earth (Fig. 2.7). The atmosphere is described in terms of several layers, each layer merging into the next with no distinct lines of separation.

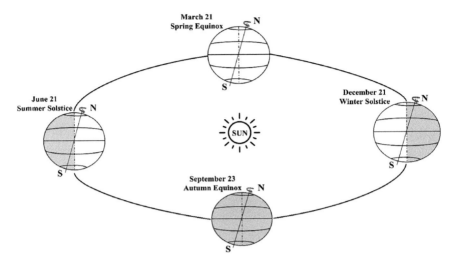

Fig. 2.5 Position of the earth with respect to the sun with solstices and equinoxes

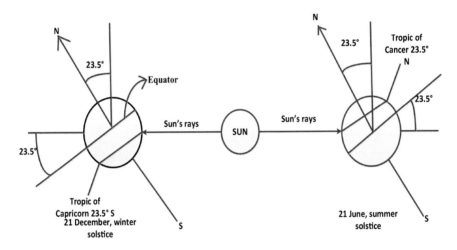

Fig. 2.6 Tropics

The first two near the surface of the earth are the troposphere and the stratosphere. The troposphere has an average depth of 15 km, around 20 km in the tropics and around 7 km in the Polar Regions. It contains 99% of the water vapour and 75% of the gases in the atmosphere. When one goes away from the earth's surface, air pressure drops and is only about 10% of the value at sea level at the top of the troposphere. Similarly, the temperature also decreases. The next layer, the stratosphere extends to an altitude of about 50 km above the earth's surface with the upper part comprising a thin layer of ozone. Above the troposphere and stratosphere are the mesosphere and the ionosphere, respectively.

2.5 Solar Irradiance Falling at the Earth's Surface

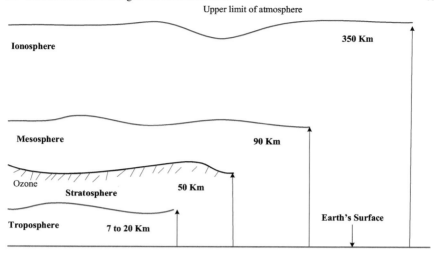

Fig. 2.7 Atmosphere of the earth (Mani and Chako 1973)

Solar radiation is received at the earth's surface in an attenuated form because it is subjected to the mechanism of absorption and scattering as it passes through the earth's atmosphere as shown in Fig. 2.8. The troposphere and stratosphere are where the scattering and absorption mostly take place. The stratospheric ozone layer is crucial in absorbing the dangerous ultraviolet portion of solar energy. Water vapour partly absorbs other wavelengths, along with gases (such as CO_2, NO_2, CO, O_2, and CH_4) and particulate particles, to a lesser amount (aerosols). The atmosphere's internal energy rises as a consequence of absorption. The interaction of the radiation with atmospheric particulate matter causes scattering. The scattered radiation is redistributed in all directions, some going back into space and some reaching the earth's surface (Mani and Chako 1973).

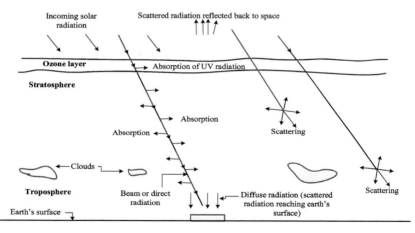

Fig. 2.8 Schematic representation: (i) Stratospheric and tropospheric absorption and scattering processes (ii) Radiation, both diffuse and beam, that strikes earth's surface (Mani and Chako 1973)

Clear sky and clouded sky are two different kinds of skies that may be found everywhere on the surface of the world. A clean sky is devoid of clouds and has very little turbidity, while a cloudy sky has partial clouds or is fully clouded. The term "low turbidity" is used to describe an atmosphere in which there is very little absorption and scattering of solar radiation. The main reason for distinguishing between a clear sky and a cloudy sky is that it is easier to analyse and develop correlations for "clear sky" situations. Marty and Philipona have defined a term called Index to the Clear Sky distinguishes between a sky that is clear and an overcast sky on a quantitative basis. Both kinds of sky have a similar absorption and scattering process. Nonetheless, less attenuation occurs when the sky is clear. As a result, a clear sky is necessary for the surface of the earth to receive the maximum radiation.

Solar radiation received at the earth's surface without change of direction, i.e. in line with the sun, is called beam or direct radiation. The radiation that reaches the surface of the earth from all parts of the sky hemispheres after being dispersed by the environment is called diffuse radiation. The Sum of the beam and diffuse radiation is referred to as total or global radiation.

Diffuse radiation's intensity does not always remain constant as it travels across the sky in different directions. As a result, diffuse radiation is described as having an anisotropic character. However, in many situations (e.g. part or heavy cloud cover), the intensity from all directions tends to be reasonably uniform. It is then modelled as being perfectly uniformed and is set to be isotropic.

A term called the airmass (AM) is often used as a measure of the distance travelled by beam radiation through the atmosphere before it reaches a location on the earth's surface. It is defined as the ratio of the mass of the atmosphere through which the beam radiation passes to the mass it would pass through if the sun is directly overhead (i.e. at its zenith). The zenith angle θ_z is the angle made by the sun's rays with the normal to a horizontal surface. It can be shown approximately that for locations at sea level and the zenith angle from zero to 70°, the airmass is equal to the secant of the zenith angle. Thus, airmass zero (AM0) corresponds to extra-terrestrial radiation, airmass one (AM1) corresponds to the case of the sun at its zenith, and airmass two (AM2) corresponds to the case of a zenith angle of 60°.

The design of devices like concentrating collectors for solar thermal application or the determination of the area of a solar PV array at a specified location demands knowledge of the variation with time of direct radiation and/or global radiation at the location. In the past, designers of solar equipment resorted to one of the following options:

1. Monitoring the area where the solar equipment will be deployed for radiation over an extended length of time.
2. Making use of data from a different site whose climate is known to be somewhat comparable to the one under examination.
3. Making use of empirical prediction equations that relate solar radiation levels to other weather prediction factors whose morals are accepted in the area in issue. These characteristics include things like the amount of sunlight received each day and the amount of cloud cover, among others.

2.6 Solar Radiation Geometry

However, in the one to two decades before now, significant research has been done on the mechanisms of atmospheric scattering and absorption, and the ability to calculate the attenuation coefficients for different constituents more accurately has increased. This has been possible because of observations made by sensors on satellites and also because of the relative ease with which computations involving large amounts of data can now be carried out. Consequently, many models have been created, and these models include being increasingly used for calculating the availability of global/beam radiation at a specified location for a particular application.

2.6 Solar Radiation Geometry

The geometric relationships between a plane of any particular orientation relative to the earth at any time (whether that plane is fixed or moving relative to the earth) and the incoming beam solar radiation, that is, the position of the sun relative to that plane, can be described in terms of several angles (Cooper 1969; Braun and Mitchell 1983). Some of the angles are indicated in Fig. 2.10. The angles and a set of consistent sign conventions are as follows:

If θ is the angle between an incident beam of flux I_{bn} and the normal to a plane surface, then the equivalent flux falling normal to the surface is given by $I_{bn} \cos \theta$. The angle θ can be related by a general equation to ϕ the latitude, β the slope, γ the surface azimuth angle, δ the declination, and, ω the hour angle. First, each of them shall be defined.

The **latitude** ϕ of a location is the angle made by the radial line joining the location to the centre of the earth with the projection of the line on the equatorial plane. Conventionally, the northern hemisphere's latitude is expressed as positive. It may vary from $-90°$ to $+90°$.

The **slope** β is the angle made by the plane surface with the horizontal. It varies from 0 to 180°.

The **surface azimuth angle** γ is the angle made in the horizontal plane between the horizontal line due south and the projection of the normal to the surface on the horizontal plane. It can vary from $-180°$ to $+180°$. We adopt the convention that the angle is positive if the normal is east of south and negative if west of south.

The angle is called the **declination** δ formed by the projection of the line connecting the centres of the earth and the sun on the plane of the equator and is measured in degrees. It results from the fact that the earth spins on an axis that forms a tilt concerning the direction of its spin around the sun roughly 66.5 degrees. From a higher value of $+23.45°$ on June 21 to a lower value of $-23.45°$ on December 21, the declination angle fluctuates. During the March 21 and September 22 equinox days, it is zero. Cooper (1969) has provided the straightforward relation below for determining declination.

$$\delta(\text{in degrees}) = 23.45 \sin\left[\frac{360}{365}(284 + n)\right], \tag{2.3}$$

where n is the day of the year. Equation (2.3) is plotted in Fig. 2.9. Its accuracy of prediction is adequate for engineering purposes

The **hour angle** ω is an angular measure of time and is equivalent to $15°$ per hour. It also varies from $-180°$ to $+180°$. We adopt the convention of measuring it from

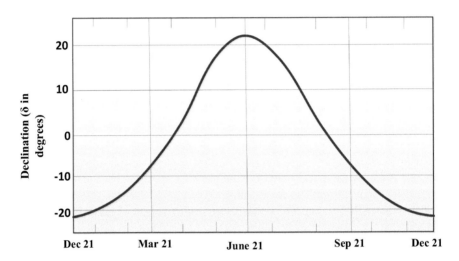

Fig. 2.9 Declination variation over a year Eq. (2.3) (Braun and Mitchell 1983)

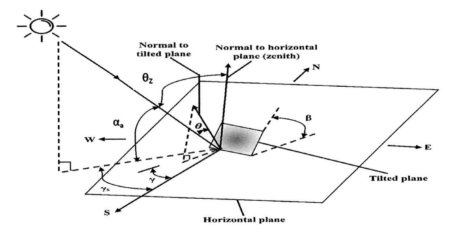

Fig. 2.10 Diagram illustrating the zenith angle θ_z, the angle of incidence θ, the slope β, the solar altitude angle α_a, the solar azimuth angle γ_s, and the surface azimuth angle γ (Braun and Mitchell 1983)

2.6 Solar Radiation Geometry

noon based on local apparent time (LAT), being positive in the morning and negative in the afternoon.

Based on the preceding definitions. It may be represented as:

$$\begin{aligned}\cos\theta = {} & \sin\phi(\sin\delta\cos\beta + \cos\delta\cos\gamma\cos\omega\sin\beta)\\ & + \cos\phi(\cos\delta\cos\omega\cos\beta - \sin\delta\cos\gamma\sin\beta)\\ & + \cos\delta\sin\gamma\sin\omega\sin\beta.\end{aligned} \tag{2.4}$$

In most circumstances, special cases of Eq. (2.4) are necessary. Here are a few of them:

Vertical surface $\beta = 90°$,

$$\cos\theta = \sin\phi\cos\delta\cos\gamma\cos\omega - \cos\phi\sin\delta\cos\gamma + \cos\delta\sin\gamma\sin\omega \tag{2.5}$$

Horizontal surface $\beta = 0°$,

$$\cos\theta = \sin\phi\sin\delta + \cos\phi\cos\delta\cos\omega. \tag{2.6}$$

The angle θ in this case is the zenith angle θ_z. Calculations also often employ the zenith angle complement. Its name is the solar altitude angle, and the symbol for it is α_a. Figure 2.10 illustrates a few of the aforementioned angles.

The inclined surface facing due south $\gamma = 0°$,

$$\begin{aligned}\cos\theta = {} & \sin\phi(\sin\delta\cos\beta + \cos\delta\cos\omega\sin\beta)\\ & + \cos\phi(\cos\delta\cos\omega\cos\beta - \sin\delta\sin\beta)\\ = {} & \sin\delta\sin(\phi - \beta) + \cos\delta\cos\omega\cos(\phi - \beta).\end{aligned} \tag{2.7}$$

Vertical surface facing due south $\beta = 90°$, $\gamma = 0°$,

$$\cos\theta = \sin\phi\cos\delta\cos\omega - \cos\phi\sin\delta. \tag{2.8}$$

The inclined surface facing due north $\gamma = 180°$,

$$\cos\theta = \sin\delta\sin(\phi + \beta) + \cos\delta\cos\omega\cos(\phi + \beta). \tag{2.9}$$

The β slope, γ the surface azimuth angle, γ_s the solar azimuth angle, and θ_z the zenith angle may all be used to represent θ the angle of incidence. Michael and Barun (1983) have shown that

$$\cos\theta = \cos\theta_z\cos\beta + \sin\theta_z\sin\beta\cos(\gamma_s - \gamma). \tag{2.10}$$

The solar azimuth angle (γ_s) is the angle made in the horizontal plane between the horizontal line due south and the projection of the line of sight of the sun on the horizontal plane. As a result, it indicates the direction of the shadow that a vertical

68   2 Fundamentals of Solar Energy

rod casts in a horizontal plane. Conventionally, if the line of sight is projected east of south, the solar azimuth angle is assumed to be positive, and it is negative if it is projected west of south. In order to apply Eq. (2.10), it is essential to first determine θ_z and γ_s. θ_z is obtained from Eq. (2.6), while γ_s is obtained from the expression (Iqbal 1983).

$$\cos \gamma_s = (\cos \theta_z \sin \phi - \sin \delta)/\sin \theta_z \cos \phi. \tag{2.11}$$

Example 2.1

(a) Determine the angle made by beam radiation with the normal to a flat-plate collector on July 21 at 09:00 h (LAT). The collector is located at Jamshedpur (22.8046° N, 86.2029° E). It is tilted at an angle of 46° with the horizontal and is pointing due south.
(b) For 1200 h, repeat the calculation (local apparent time).

Solution

(a) For the given case ($\gamma = 0°$), Eq. (2.7) is applicable. On September 23, $n = 201$. So, from Eq. (2.3) we have

$$\delta = 23.45 \sin\left[\frac{360}{360}(284 + 265)\right] = 20.44°.$$

At 0900 h, $\omega_s = (12–09)*15° = 45°$. On substituting the ω_s in Eq. (2.7) we have

$$\cos \theta = \sin 20.64° * \sin(22.80° - 46°)$$
$$+ \cos 20.64° * \cos 45° * \cos(22.80° - 46°) = 0.47$$
$$\theta = 61.87°.$$

(b) At 1200 h, $\omega_s = (12–12) * 15° = 0°$. Following the steps of the part (a) we have

$$\cos \theta = \sin 20.64° * \sin(22.80° - 46°)$$
$$+ \cos 20.64° * \cos 0° * \cos(22.80° - 46°) = 0.7237$$
$$\theta = 43.64°.$$

Alternatively, consider Fig. 2.11 which shows a cross section through the local meridian. Since $\omega = 0°$ and $\gamma = 0°$, all the other angles involved, viz. θ, ϕ, δ, and β lie in this plane. The answer for part (b) is obvious from Fig. 2.11. Considering the parallel sun rays OS1 and PS2 and the intersecting line OPZ, we have

$$\angle POS_1 = \angle ZPS_2,$$
$$\text{i.e. } \phi - \delta = \beta - \theta$$
$$\text{or, } \theta = \beta - \emptyset + \delta = 43.64°$$

2.7 Sun's Apparent Motion

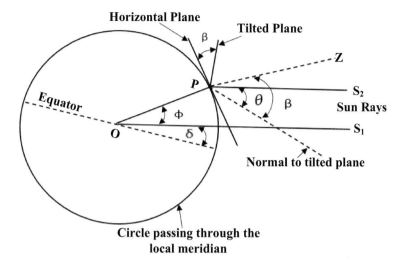

Fig. 2.11 Example 2.1

2.7 Sun's Apparent Motion

Having a description of the sun's apparent motion as viewed from the earth would be helpful at this point. To an observer on the earth, on any given day, the sun rises in the east, moves in a plane tilted at an angle of 90° − Ø with the horizontal and finally sets in the west. As a result, a line running east–west connects the sun's perceived position in space and travels to the plane's surface. Nevertheless, The E-W line does not connect with this line here that passes by way of viewer O due to the declination angle as shown in Fig. 2.12. For a location in the northern hemisphere, this line is to the south in the winter (declination negative), to the north in summer (declination positive) and coincides with the E-W line passing through the observer on the two equinox days of March 21 and September 22 (declination zero).

2.7.1 Day Length, Sunrise, and Sunset

- **Horizontal Surface**—If the zenith angle is substituted with a value of 90°, Eq. (2.6) may be used to calculate the angle of the hour that corresponds to sunrise or sunset (ω_s) on a horizontal surface. We discover

$$\cos \omega_s = -\tan \phi \tan \delta$$
$$\omega_s = \cos^{-1}(-\tan \phi \tan \delta). \tag{2.12}$$

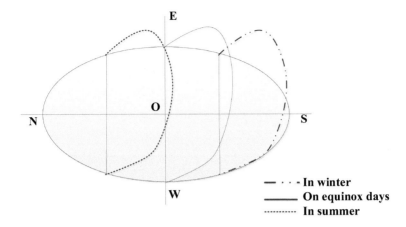

Fig. 2.12 An area in the northern hemisphere's apparent sun motion plane

Equation (2.12) returns two values, with the positive value for ω_s corresponding to sunrise, and a negative value corresponding to sunset. Because 1 h is equal to 15° of the hour angle, the equivalent day length is (in hours)

$$s_{\max} = \frac{2}{15}\omega_s = \frac{2}{15}\cos^{-1}(\tan\phi\tan\delta), \tag{2.13}$$

where ω_s is in degrees.

Inclined Surface Facing Due South—Between September 22 and March 21 in the northern hemisphere, Eq. (2.12) accurately determines the solar angle at the beginning or end of the day for an inclined surface facing due south ($\gamma = 0°$). During this period, the Declination is unfavourable, and the sun's apparent plane of motion crosses an E-W line south of the observer's inclined plane. For sunrise or sunset, Eq. (2.12) yields a lower hour angle (ω_{st}) compared to days between March 21 and September 22. During the latter period, when the Declination is favourable, the hour angle (ω_{st}) can be obtained by substituting $\theta = 90°$ into Eq. (2.7), maintaining accuracy in calculations. This yields

$$\omega_{st} = \cos^{-1}[-\tan(\phi - \beta)\tan\delta]. \tag{2.14}$$

Thus, the magnitude of ω_{st} for a south-facing inclined surface ($\gamma = 0°$) is

$$|\omega_{st}| = \min\left[|\cos^{-1}(-\tan\phi\tan\delta)|, |\cos^{-1}\{-\tan(\phi-\beta)\tan\delta\}|\right]. \tag{2.15}$$

- **Inclined Surface Facing Due North**—Proceeding in the same manner as for a sloping slope that faces south, it can be shown by using Eqn. that for a sloped northward-facing surface

2.7 Sun's Apparent Motion

$$|\omega_{st}| = \min\left[\left|\cos^{-1}(-\tan\phi\tan\delta)\right|, \left|\cos^{-1}\{-\tan(\phi-\beta)\tan\delta\}\right|\right]. \quad (2.16)$$

In general, for a plane surface not symmetrically oriented, hour angles at sunrise and sunset would be unequal in magnitude apart from having opposite signs. The general procedure would be to calculate $\omega_s t$ by substituting $\theta = 90°$ in Eq. (2.4) and by using Eq. (2.12). Depending upon the day of the year and the orientation of the surface, proper judgement would need to be exercised in selecting the correct values from the solutions thus obtained.

Example 2.2 Determine the hour angle for a surface inclined at an inclination of $10°$ and facing straight south ($\gamma = 0°$) on January 21 and June 21. Ranchi is the surface's location ($23° \ 20'$ N, $85° \ 17'$ E).

Solution

On January 21, $\delta = 23.45 \ \sin\left[\frac{360}{365}(284 + 121)\right] = -20.138°$

$$|w_{st}| = \min[\left|\cos^{-1}\{-\tan 23.34° * \tan(-20.138°)\}\right|,$$
$$\left|\cos^{-1}\{-\tan(23.34° - 10°) * \tan(-20.138°)\}\right|$$
$$= \min[80.89°, 85.01°] = 80.89°$$
$$w_{st} = \pm 80.89°.$$

On June 21, $\delta = 23.45 \ \sin\left[\frac{360}{365}(284 + 172)\right] = 23.45°$

$$|w_{st}| = \min[\left|\cos^{-1}\{-\tan 23.34° * \tan(23.45°)\}\right|,$$
$$\left|\cos^{-1}\{-\tan(23.34° - 10°) * \tan(23.45°)\}\right|$$
$$= \min[100.79°, 95.91°] = 95.91°$$
$$w_{st} = \pm 95.91°.$$

2.7.2 Local Apparent Time

The time used for calculating the hour angle (ω) is the local apparent time. This can be obtained by making two modifications to the standard time on a clock. The first correction arises because of the difference between the longitude of a location and the meridian on which the standard time is based.

The correction has a magnitude of 4 min for every degree difference in longitude.

The second adjustment, sometimes known as the equation of time correction, results from slight differences in the earth's rotational speed and orbit. This adjustment, which is presented in Fig. 2.13, is based on experimental findings.

Thus,

Local apparent time = Standard time

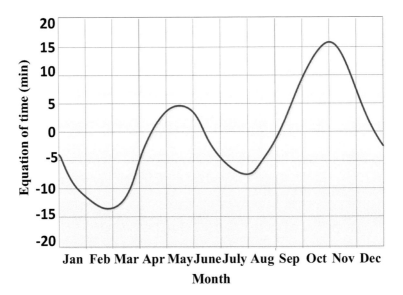

Fig. 2.13 Time correction equation

$$\pm 4(\text{Standard time longitude} - \text{longitude of location})$$
$$+ (\text{Equation of time correction}).$$

The negative sign of the first correction refers to the eastern hemisphere, whereas the positive sign refers to the western hemisphere.

The following empirical correlation may also be used to obtain the time correction equation (in minutes) (Iqbal 1983).

$$E = 229.18(0.000075 + 0.001868 \cos B - 0.032077 \sin B \\ - 0.014615 \cos 2B - 0.04089 \sin 2B), \tag{2.17}$$

where n is the day of the year, and $B = (n-1)360/365$.

Example 2.3 Calculate the local apparent time (LAT) in Mumbai $(19°07'N, 72°51'E)$ that corresponds to 14:30 IST on July 1. In India, standard time is based on $82.50°E$.

Solution

From Eq. (2.17), the July 1 time equation adjustment is (-3.5) minutes.
Therefore,

Local apparent time = 1430 h−4(82.50−72.85)minutes + (−3.5 minutes)
= 1430 h−38.6 minutes−3.5 minutes = 1348 h.

2.7.3 Why Solar Radiation Data is Needed?

Data on solar radiation shows how much of the sun's energy falls on a certain spot on the surface of the globe throughout time.

These data play a key role in effective research into solar energy utilization or we can say that without having the proper information regarding the solar radiation data, a research program in the field of renewable energy-based systems can't be started.

Data on solar radiation are utilized in many ways and for many different things. The following details of radiation statistics are crucial for understanding and using them:

The measurement equipment, the direction of the receiving surface (usually plane surface, occasionally angled at a fixed angle or perpendicular to the radiation from the beam), the measurement time or period, if the observations are of a beam, a diffuse field, or total radiation; if the measurements are immediate (irradiance); or whether the readings are averaged over a certain amount of time; (irradiation), typically an hour or a day; and, if averaged, the length of time they're averaged across (for instance, monthly averages of daily radiation).

The amount on a light beam and diffuse radiation horizontal plane each hour provides the most precise information. On sloping planes and smaller periods, there are a few metrics available. Daily statistics are often accessible, and they may be used to predict hourly radiation).

Figure 2.14 displays an average everyday report of the radiation intensity overall and diffuse that was observed on a bright day. On a clear day, it is evident that a pretty smooth fluctuation is produced, with the maximum happening about mid-day. In contrast, a foggy day may provide an uneven fluctuation with several peaks and troughs. The current values of the diffuse and global flux depicted in Fig. 2.14 will be denoted by the symbols I_g and I_d, respectively, and expressed in W/m^2. Although solar radiation fluxes often do not fluctuate much over time, hourly quantities are also represented by the same symbols, I_g and I_d. These amounts will be given in kWh/m^2 or kJ/m^2. The global and diffuse flux incident for a day is shown by the shaded regions below the graphs. The symbols H_g and H_d are used to represent these quantities and are expressed in kWh/m^2 day or kJ/m^2 day.

Sometimes the amount of solar radiation is expressed every day or hour in Langley (1 Langley = 1 cal/cm^2 = 1.163×10^{-2} kWh/m^2). In memory of Samuel Langley, who made the first assessment of the spectral composition of the sun, the unit "Langley" has been chosen.

The mean values for a place are what a solar designer is most interested in. The average is often calculated over each day of the month, and a bar is placed above the sign to show this. Hence, the hourly global average for the monthly and diffuse radiation is shown by the symbols $\overline{I_g}$ and $\overline{I_d}$, respectively, while the daily global average for the monthly and diffuse radiation is indicated by the symbols, $\overline{H_g}$ and $\overline{H_d}$, respectively.

India's solar radiation statistics may be found from a variety of sources. The Indian Meteorological Department (IMD) collects data at around 45 sites, and it is possible

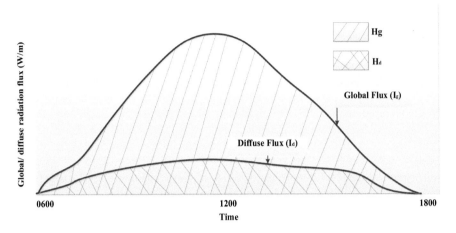

Fig. 2.14 Record of diffuse and global radiation flux measured on a sunny day

to get data from the agency. A guidebook containing information for a selected few places is also available, and it may be purchased originating with the Ministry of New and Renewable Energy (MNRE) website. Based on Global Positioning System (GPS) information collected solar maps with monthly and yearly values of DNI and GHI have been created. Presently, the National Institute of Wind Energy at Chennai is carrying out an extensive solar resource mapping programme by acquiring data at 111 stations across the country. The data for some stations in India is also available from the World Radiation Data Centre.

The increasing amount of aerosols in the atmosphere in recent years in India has generated considerable worry owing to pollution from vehicle emissions, industrialization, biomass burning, and dust storms. According to reports, this rise is what's causing the observed values of the worldwide horizontal irradiance to decrease.

What are the ways to achieve solar radiation data?

- By measurement
- With the help of the empirical model (where measurement can't be done/or measured data is not available)

2.8 Measurement of Solar Radiation

The instruments used for measurement will now be described. For measuring the flux of sun radiation, people frequently utilize pyrometers or pyrheliometers.

Detecting radiation that is falling on a horizontal surface through a hemispherical perspective, either direct or diffuse, is the function of pyrometers. Figure 2.15 is a schematic of one kind of pyranometer that has been set up to monitor ambient radiation. Basically, the "black" surface of the pyranometer warms up when it is

2.8 Measurement of Solar Radiation

exposed to sunlight. When the rate of heat loss through convection, conduction, and reradiation is equal to the rate of heat absorption from solar radiation, the temperature rises. The cold junctions of a thermopile are protected from direct radiation exposure by a guard plate, while the hot junctions are attached to the black surface. As a result, an emf is generated (often between 0 and 10 mV), which may then be read, documented, or incorporated over time to give an accurate value of the total radiation.

It has its junctions arranged in the form of a horizontal circular disc of diameter 25 mm and coated with a special black lacquer having a very high absorptivity in the solar wavelength region. The disc is placed on a large diameter guard plate which may be horizontal (Fig. 2.15). Two concentric hemispheres, 30 and 50 mm in diameter, respectively, made of optical glass having excellent transmission characteristics, are used to protect the disc surface from the weather using the device, an accuracy of roughly ± 2% may be attained.

The pyrometer is also utilized to measure diffuse radiation. Place it in the centre of a semicircular shadow ring, with its plane parallel to the plane of the sun's daily rotation as it passes across the sky per day. Also, it always protects the thermopile element and the two glass domes of the pyrometer from the sun. As a result, the pyrometer simply records the diffuse radiation that the sky provides.

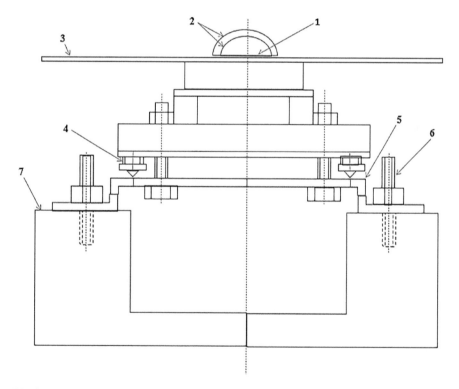

Fig. 2.15 Pyranometer for detecting global radiation (1. black surface, 2. glass domes, 3. guard plate, 4. levelling screws, 5. mounting plates, 6. grouted bolts, 7. platform)

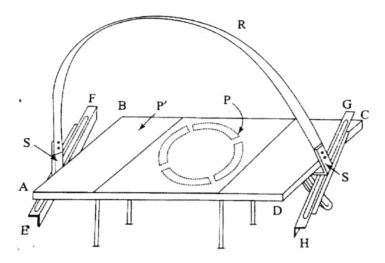

Fig. 2.16 Shading arrangement for the measurement of diffuse radiation

Figure 2.16 illustrates how one kind of shading ring is built. ABCD is a horizontal rectangular frame that measures 35 by 80 cm and has long sides that go east to west. The frame's sides AB and CD are pivotally joined to two angle-iron arms, EF and GH, each 70 cm long and having slots spanning the length of them to accommodate sliders, SS. On these sliders is fixed the semicircular shade ring R.

The arms may be adjusted to pivot along a horizontal axis that passes through the rectangular frame's centre and is orthogonal to the horizontal at an angle equal to the station's latitude. The shade ring is 450 mm in diameter, 50 mm broad, and composed of aluminium. The inner portion of the ring is coated with dull black paint, while the remaining shading ring design is coated with dull matt white paint. The bottom of the frame ABCD is attached to a thick metal plate P with a circular groove so that, when the frame is positioned on a masonry platform with nuts and bolts, it may be turned into the proper position along a vertical axis. On a substantial metal plate P' that is fastened to the top of the frame, the pyranometer is set up.

The solar beam radiation flux that reaches a surface at a right angle to the sun's rays is measured using a pyrheliometer. This is referred to as the Direct Normal Irradiance (DNI). In contrast to a pyranometer, a thermopile's hot junctions are connected to the black absorber plate, which is located at the bottom of a collimating tube (Fig. 2.17). The tube is aligned with the direction of the sun's rays using a two-axis tracking system and an alignment indicator. As a result, only a small amount of diffuse radiation within the instrument's "acceptance angle" is also absorbed by the black plate.

2.8 Measurement of Solar Radiation

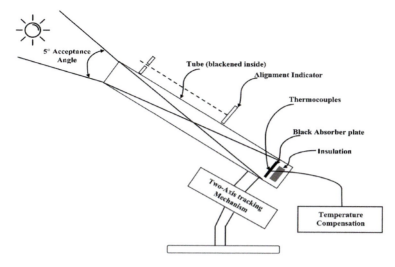

Fig. 2.17 Pyrheliometer for measuring beam radiation

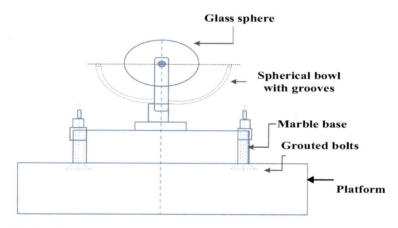

Fig. 2.18 Sunshine recorder

A sunlight recorder like the one in Fig. 2.18 is used to measure how long each day's bright sunshine lasts. A glass sphere placed concentrically with the sphere directs sunlight onto a spot on a card strip that is kept in a groove in the bowl. The image created whenever there is strong sunlight is potent enough to leave a burnt area on the card strip. As a result, the strip of paper acquires a burnt trace whose length is proportionate to the duration of the sunlight.

2.9 Calculating the Sun Radiation Availability

The method for determining average radiation statistics for a location is to monitor solar radiation there over an extended period. In the location where it is not feasible, information from surrounding areas with a comparable topography and climate may be utilized. If none of these techniques is feasible, one may utilize empirical correlations to relate radiation (global or diffuse) levels to weather variables like sunshine hours or cloud cover.

Even when radiation measurements are available, the data may not always be in the proper format. For instance, it may be possible to quantify the daily global radiation while needing hourly data for another reason. Once again, the beam and diffuse components may be specifically required outside of the daily global radiation. Several researchers have tried to create empirical relations for such circumstances. The equations developed are given in Sects. 2.9.1–2.9.4 and their use is illustrated with the help of numerical examples. The relationships given are generally valid for cloudy skies.

2.9.1 Monthly Average Daily Global Radiation

Angstrom, who claimed that it may be connected to the quantity of sunshine by a straightforward linear equation of the form, is responsible for the first attempt to estimate solar radiation

$$\frac{\overline{H}_g}{\overline{H}_c} = a + b\left(\frac{\overline{S}}{S_{\max}}\right), \tag{2.18}$$

where

$\overline{H}_g$ = monthly average of the daily global radiation on a horizontal surface at a location (kJ/m^2 day),

$\overline{H}_c$ = monthly average of the daily global radiation on a horizontal surface at the same location on a clear day (kJ/m^2 day),

$\overline{S}$ = Monthly average of the sunshine hours per day at the location (h),

$\overline{S}_{max}$ = monthly average of the maximum possible sunshine hours per day at the location, i.e. the day length on a horizontal surface (h), and

a, b = Constants obtained by fitting data.

It is recommended that $(\overline{H}_c)$ in Eq. (2.18) be substituted by $(\overline{H}_o)$, the average monthly amount of extra-terrestrial radiation that would fall on a horizontal surface at the prospective location, since it may be difficult to determine what constitutes a clear day (Page 1961).

2.9 Calculating the Sun Radiation Availability

$$\frac{\overline{H}_g}{\overline{H}_o} = a + b\left(\frac{\overline{S}}{S_{max}}\right) \tag{2.19}$$

Values of a and b have been derived using regression analysis of observed values of global solar radiation and sunshine duration of various cities in the globe (Lof et al. 1966) whose values are given in Table 2.4. A similar set of values has also been obtained for 17 Indian cities (Modi and Sukhatme 1979) are given in Table 2.5.

In the above computations, the quantity $\overline{H}_o$ is the mean of the value (H_o) for each day of the month. H_o is obtained by integrating over the day length as follows:

$$H_o = I_{sc}\left(1 + 0.033 \ \cos \ \frac{360 \, n}{365}\right)\int (\sin \phi \ \sin \delta + \cos \phi \ \cos \delta \ \cos \omega)\mathrm{d}t. \tag{2.20}$$

Now, $t = \frac{180\omega}{15\pi}$,
where t is in hours and ω is in radians.
Hence, $\mathrm{d}t = \frac{180}{15\pi}\mathrm{d}\omega$.
Substituting in Eq. (2.20)

Table 2.4 Constants a and b in Eq. (2.19) for many cities of the world (Lof et al. 1966)

Location	Range of $\overline{S}/S_{max}$	a	b
Albuquerque, New Mexico, USA	0.68–0.85	0.41	0.37
Atlanta, Georgia, USA	0.45–0.71	0.38	0.26
Blue Hill, Mass., USA	0.42–0.60	0.22	0.50
Brownsville, Texas, USA	0.47–0.80	0.35	0.31
Buenos Aires, Argentina	0.47–0.68	0.26	0.50
Charleston, S.C., USA	0.60–0.75	0.48	0.09
Dairen, Manchuria	0.55–0.81	0.36	0.23
El Paso, Texas, USA	0.78–0.88	0.54	0.20
Ely, Nevada, USA	0.61–0.89	0.54	0.18
Hamburg, Germany	0.11–0.49	0.22	0.57
Honolulu, Hawaii, USA	0.57–0.77	0.14	0.73
Madison, Wisconsin, USA	0.40–0.72	0.30	0.34
Malange, Angola	0.41–0.84	0.34	0.34
Miami, Florida, USA	0.56–0.71	0.42	0.22
Nice, France	0.49–0.76	0.17	0.63
Pune, India	$\begin{cases} 0.25 - 0.49 \\ 0.65 - 0.89 \end{cases}$	0.30	0.51
		0.41	0.34
Stanleyville, Congo	0.34–0.56	0.28	0.39
Tamanrasset, Algeria	0.76–0.88	0.30	0.43

$$H_o = \frac{12}{\pi} I_{sc} \left(1 + 0.033 \cos \frac{360n}{365}\right) \int_{-\omega_s}^{+\omega_s} (\sin \phi \sin \delta + \cos \phi \cos \delta \cos \omega) d\omega$$

$$H_o = \frac{24}{\pi} I_{sc} \left(1 + 0.033 \cos \frac{360n}{365}\right) (\omega_s \sin \phi \sin \delta + \cos \phi \cos \delta \sin \omega_s).$$

$$(2.21)$$

The calculation of $\overline{H}_o$ has been simplified by Klein (1977), who has recognized the specific day each month when extra-terrestrial radiation is almost equal to the average amount for that month. January 17, February 16, March 16, April 15, May 15, June 11, July 17, August 16, September 15, October 15, November 14, and December 10 are the dates when H_o equals $\overline{H}_o$. These dates are about in the middle of the month, as to be anticipated.

When statistics on sunshine hours are available, Eq. (2.19) may be used to determine the average daily global radiation for a location. Values of a and b are known for a neighbouring area with a comparable topography and climate can be considered.

Example 2.4 Calculate the monthly mean of the everyday global radiation on a flat surface at Kolkata (22° 34′ N, 88° 21′ E) during the month of February, if the mean sunshine hour per day is 8.5 h.

Solution

For the Kolkata region, the values of a and b are assumed as given in Table 2.5.

Table 2.5 Constants a and b in Eq. (2.19) for Indian cities (Modi and Sukhatme 1979)

Location	a	b	Mean error (%)
Ahmedabad	0.28	0.48	3.0
Bengaluru	0.18	0.64	3.9
Bhavnagar	0.28	0.47	2.8
Kolkata	0.28	0.42	1.3
Goa	0.30	0.48	2.1
Jodhpur	0.33	0.46	2.0
Kodaikanal	0.32	0.55	2.9
Chennai	0.30	0.44	3.5
Mangalore	0.27	0.43	4.2
Minicoy	0.26	0.39	1.4
Nagpur	0.27	0.50	1.6
New Delhi	0.25	0.57	3.0
Pune	0.31	0.43	1.9
Shillong	0.22	0.57	3.0
Srinagar	0.35	0.40	4.7
Thiruvananthapuram	0.37	0.39	2.5
Vishakhapatnam	0.28	0.47	1.2

2.9 Calculating the Sun Radiation Availability

$a = 0.28$ and $b = 0.48$.

For Kolkata, $\Phi = 22. (34/60)° = 22.57°$

Based on Kelvin's recommendation, H_o is equal to the value of $\overline{H}_o$ on February 17, $n = 48$.

Also given the average sunshine hours, $\overline{S} = 8.5$ h.

Using Eq. (2.3), $\delta = 23.45 \sin\left[\frac{360}{365}(284 + 48)\right] = -12.62°$.

The sunshine hour may be calculated from Eq. (2.10)

$$w_s = \cos^{-1}\{-\tan(22.57°)^* \tan(-12.62°)\} = 84.66° \text{ or } 1.478 \text{ radians.}$$

Day length, $S_{max} = (2/15) *84.66° = 11.29$ h.

From Eq. (2.21),

$$H_o = \frac{24}{\pi} \times 1.367 \times 3600 \times \left(1 + 0.033 \times \cos \frac{360 \times 48}{365}\right) \times \left(\sin 22.57° \times \sin(-12.62°)\right)$$

$$+ \cos 22.57° \times \cos(-12.62°) \times \sin 84.66°) = 29721 \text{ kJ/m}^2 \text{ day.}$$

Therefore, from Eq. (2.19),

$$\overline{H}_g = \left(0.28 + 0.48 * \left(\frac{8.5}{11.29}\right)\right) * 29721 = 19062 \text{ kJ/m}^2 \text{ day.}$$

Example 2.5 Estimate the monthly average daily global radiation on a horizontal surface at Vadodara (22°00′ N, 73° 10′ E) during the month of March, if the average sunshine hours per day is 9.5.

Solution

For the Vadodara region, the values of a and b are assumed as given in Table 2.5.

$a = 0.28$ and $b = 0.48$.

For Kolkata, $\Phi = 22. (34/60)° = 22.57°$

Based on Kelvin's recommendation, H_o is equal to the value of $\overline{H}_o$ on February 17, $n = 48$.

Also given the average sunshine hours, $\overline{S} = 8.5$ h.

Using Eq. (2.3), $\delta = 23.45 \sin\left[\frac{360}{365}(284 + 48)\right] = -2.42°$.

The sunshine hour may be calculated from Eq. (2.10)

$$w_s = \cos^{-1}\{-\tan(22.57°) * \tan(-2.42°)\} = 89.02° = 1.554 \text{ radians.}$$

Day length, $S_{max} = (2/15) *89.02° = 11.87$ h.

From Eq. (2.21),

$$H_o = \frac{24}{\pi} \times 1.367 \times 3600 \times \left(1 + 0.033 \times \cos \frac{360 \times 75}{365}\right)$$

$$\times (1.554 \sin 22° \times \sin(-2.45°) + \cos 22° \times \cos(-2.42°) \times \sin 89.02°)$$

82   2 Fundamentals of Solar Energy

$$= 34,206 \text{ kJ/m}^2 \text{ day}.$$

Therefore, from Eq. (2.19),

$$\overline{H}_g = \left(0.28 + 0.48 * \left(\frac{9.5}{11.87}\right)\right) * 34206 = 22718 \text{ kJ/m}^2 \text{ day}.$$

Other meteorological parameters have also been used for predicting solar radiation. These include cloud cover (the amount of sky dome covered by clouds) and precipitation (the number of days in the month with precipitation greater than 0.3 mm). However, in general, the sunshine ratio parameter $\left(\frac{\overline{S}}{\overline{S}_{max}}\right)$ is the most reliable predictor.

Correlations similar to Eq. (2.18) have been suggested by many investigators based on data for specific locations and countries. Some of these include additional parameters and help to generalize the applicability of the correlations. Gopinathan (1988) has suggested the correlation

$$\frac{\overline{H}_g}{\overline{H}_o} = a_1 + b_1 \left(\frac{\overline{S}}{\overline{S}_{max}}\right). \tag{2.22}$$

Based on data from 40 places throughout the globe, Eq. (2.22) was created. The constants a_1 and b_1 are related to three parameters, the latitude, the elevation, and sunshine hours as follows:

$$a_1 = -0.309 + 0.539 \cos \phi - 0.0693 E_L + 0.290 \left(\frac{\overline{S}}{\overline{S}_{max}}\right)$$

$$b_1 = 1.527 - 1.027 \cos \phi + 0.0926 E_L - 0.359 \left(\frac{\overline{S}}{\overline{S}_{max}}\right), \tag{2.23}$$

where $\phi = $ latitude (in degrees) and
$E_L = $ elevation of the location above mean sea level (in kilometres).

For forecasting the daily global radiation at places throughout the globe, including locations in India, Eq. (2.23) is recommended.

Example 2.6 Use Gopinathan's correlation, to calculate the value of $\overline{H}_g$ for Vadodara for the month of March (assume $E_L = 34$ m).

Solution

From the above Example 2.4:
    $\Phi = 22.00°$, $\overline{S} = 9.5$ h, $\overline{S}_{max} = 11.87$ h and $\overline{H}_o = 34,206$ kJ/m^2 day.
    On substituting the above data in Eq. (2.23),

$$a_1 = -0.309 + 0.539 \times \cos(22.00) - 0.0693 \times 0.034 + 0.290$$

2.9 Calculating the Sun Radiation Availability

$$\times \, (9.5/11.87) = 0.421$$

$$b_1 = 1.527 - (1.027 \times \cos(22.00)) + (0.0926 \times 0.034) - (0.359 \times (9.5/11.87))$$
$$= 0.291$$

Therefore,

$$\overline{H}_g = \left(0.421 + 0.291 \times \left(\frac{9.5}{11.87}\right)\right) \times 34206 = 22367 \text{ kJ/m}^2 \text{ day}$$

2.9.2 Monthly Average Daily Diffuse Radiation

Liu and Jordan (1960) demonstrated that it was possible to connect the daily diffuse-to-global radiation ratio versus the daily global-to-extra-terrestrial radiation factor using data based on studies for a few countries. The following cubic equation is used as an expression for the correlation.

$$\frac{\overline{H}_d}{\overline{H}_g} = 1.390 - 4.027 \left[\frac{\overline{H}_g}{\overline{H}_o}\right] + 5.531 \left[\frac{\overline{H}_g}{\overline{H}_o}\right]^2 - 3.108 \left[\frac{\overline{H}_g}{\overline{H}_o}\right]^3, \qquad (2.24)$$

where $\overline{H}_d$ = monthly average of the daily diffuse radiation on a horizontal surface (kJ/m^2 day).

The significance of the additional symbols is the same as that previously. The ratio $(\overline{H}_g/\overline{H}_o)$, also known as the monthly average clearness index, is often represented by the letter $\overline{K}_T$. Kreith and Kreider (1978) have pointed out that Eq. (2.24) has been obtained with a value of 1394 W/m^2 for the solar constant.

Several researchers have created empirical formulae for calculating the diffuse-to-global radiation ratio for different regions of the world. Gopinathan and Soler (1995) have examined radiation data for 40 widely spread places throughout the globe in the 36° S to 36° N latitude range. They have suggested the following formula using the sunshine ratio and the clearness index:

$$\frac{\overline{H}_d}{\overline{H}_g} = 0.87813 - 0.33280\overline{K}_T - 0.53039 \left[\frac{S}{S_{\max}}\right]. \qquad (2.25)$$

It is advised to utilize Eq. (2.25) to forecast the daily diffuse radiation at places all over the globe since it is based on more current data than that which Liu and Jordan had access to. Modi and Sukhatme (1979) found the following linear equation after analysing the available Indian data:

$$\frac{\overline{H}_d}{\overline{H}_g} = 1.411 - 1.696\left[\frac{\overline{H}_g}{\overline{H}_o}\right]. \tag{2.26}$$

Garg and Garg (1985) have examined radiation data for 11 Indian cities and proposed the equation

$$\frac{\overline{H}_d}{\overline{H}_g} = 0.8677 - 0.7365\left[\frac{\overline{S}}{\overline{S}_{max}}\right]. \tag{2.27}$$

Equations (2.26) and (2.27) agree well with each other. For Indian regions, any equation may be utilized. However, when compared to Gopinathan's and Soler's correlation, it is observed that there are significant differences. Equation (2.25) predicts values for the diffuse radiation which are lower than the predictions of Eqs. (2.26) and (2.27). This is because India has a considerably greater diffuse component.

Example 2.7 Using the data from Example 2.5, determine the monthly average of the daily diffused and beam radiations on a horizontal surface for Vadodara.

Solution

As per the calculation of Example 2.4,
$\overline{H}_o = 34{,}206$ kJ/m^2 day, and $\overline{H}_g = 22{,}718$ kJ/m^2 day.

Therefore, the average clear index $\overline{K}_T = \frac{\overline{H}_g}{\overline{H}_o} = \frac{22718}{34206} = 0.6642$.

From Eq. (2.26),

$$\frac{\overline{H}_d}{\overline{H}_g} = 1.411 - \left(1.696 \times \overline{K}_T\right)$$

$$\overline{H}_d = 22718(1.411 - 1.696 \times 0.664) = 6465 \text{ kJ/m}^2 \text{ day}.$$

From Eq. (2.27),

$$\frac{\overline{H}_d}{\overline{H}_g} = 0.8677 - \left(0.7365 \times \left(\frac{\overline{S}}{\overline{S}_{max}}\right)\right)$$

$$\overline{H}_d = 22718\left\{0.8677 - 0.7365 \times \left(\frac{8.5}{11.87}\right)\right\} = 6321 \text{ kJ/m}^2 \text{ day}.$$

Thus, we see that the predictions made by Eqs. (2.26) and (2.17) differ by only 3.79%.
From Eq. (2.25),

$$\frac{\overline{H}_d}{\overline{H}_g} = \left\{0.87813 - 0.33280 \times \overline{K}_T - 0.53039 \times \left(\frac{\overline{S}}{\overline{S}_{max}}\right)\right\}$$

2.9 Calculating the Sun Radiation Availability

$$\overline{H}_d = 22718\left\{0.87813 - 0.3328065 \times 0.64 - 0.53039\left(\frac{9.5}{11.87}\right)\right\}$$

$$= 5284 \text{ kJ/m}^2 \text{ day}.$$

The prediction made by Eq. (2.6) is 18.35% less than the prediction made by Eq. (2.26). The prediction made by Eq. (2.17) is 15.13% less than the prediction made by Eq. (2.27). Concludes, as per earlier statements Eq. (2.25) significantly does not predict the diffuse component of Indian regions. Therefore, it is desirable to use either Eq. (2.26) or (2.27) for Indian regions.

The beam component can be calculated as,

$$\overline{H}_b = \overline{H}_g - \overline{H}_d = 22718 - 5284 = 17434 \text{ kJ/m}^2 \text{ day}.$$

2.9.3 Monthly Average Hourly Global Radiation

Many investigations have also been made to establish relationships that may be used to forecast the daily shift in the average hourly monthly global radiation at a given site. The relationship described below was established by Collares-Pereira and Rabl (1979):

$$\frac{\overline{I}_g}{\overline{H}_g} = \frac{\overline{I}_o}{\overline{H}_o}(a + b \cos \omega), \tag{2.28}$$

where

$a = 0.409 + 0.5016 \sin(\omega_s - 60°)$,

$b = 0.6609 + 0.4767 \sin(\omega_s - 60°)$,

$\overline{I}_g$ = monthly average of the hourly global radiation on a horizontal surface (kJ/m²-h).

$\overline{I}_o$ = monthly average of the hourly extra-terrestrial radiation on a horizontal surface (kJ/m²-h).

The other symbols in Eq. (2.28) have been defined earlier, and this equation has been developed based on the following facts:

1. Measured data generally show a similarity between the diurnal variation of $\overline{I}_g$ and $\overline{I}_o$.
2. There is a close correlation between the values of the ratios $(\overline{I}_g/\overline{H}_g)$ and $(\overline{I}_o/\overline{H}_o)$.

It will be noted from Eq. (2.28) that the symbol I has been used for denoting an hourly value (kJ/m²-h), while earlier it was used for denoting an instantaneous value, i.e. a flux (kW/m²). As stated earlier in Sect 2.7.3., for most situations involving solar radiation, transient processes occur at a slow pace. It is, therefore, not necessary to make a distinction between the two quantities. Whenever required, an instantaneous value (kW/m^2) at the midpoint of the hour may be used to derive an hourly value $(kJ/m^2\text{-}h)$ by multiplying it by 3600.

Gueymard (1986) modified Eq. (2.28) as suggested by Collares-Pereira and Rabl (1979) by incorporating a normalizing factor f_c. Thus Eq. (2.28) becomes

$$\frac{\overline{I}_g}{\overline{H}_g} = \frac{\overline{I}_o}{\overline{H}_o}(a + b \cos \omega)f_c, \tag{2.29}$$

where $f_c = a + 0.5b\left[\frac{\frac{\pi \omega_s}{180} - \sin \omega_s \cos \omega_s}{\sin \omega_s - \frac{\pi \omega_s}{180} \cos \omega_s}\right]$.

Equation (2.29) is the simplest and most satisfying correlation for forecasting monthly average hourly global radiation for sites all over the globe within latitudes 65° N to 65° S, according to Gueymard (2000), who assessed a variety of prediction models using a huge data set for 135 locations.

Satyamurty and Lahiri (1992) have done experiments for the predictions of Eq. (2.29) against the measured data of 14 locations in India and proved that there is a good correlation between the two sets of values. The RMS difference between the predicted and measured values of $(\overline{I}_g/\overline{H}_g)$ lie between 2.6 and 5.5% for 13 of the 14 locations. There was just one region where a significant variation of 10.4% was found.

2.9.4 Monthly Average Hourly Diffuse Radiation

Liu and Jordan (1960) have proposed the following relation for calculating the monthly average hourly diffuse radiation in a way similar to that used for developing Eq. (2.28).

$$\frac{\overline{I}_d}{\overline{H}_d} = \frac{\overline{I}_o}{\overline{H}_o}. \tag{2.30}$$

The predictions of Eq. (2.30) have also been validated against measured data from 14 sites in India by Satyamurty and Lahiri (1992). The agreement in this instance is quite low, with the RMS difference between the predicted and actual values of $(\overline{I}_d/\overline{H}_d)$ varying from 5.7 to 13.4%. Thus, Satyamurty and Lahiri have proposed the following better relation

$$\frac{\overline{I}_d}{\overline{H}_d} = (a' + b' \cos \omega)\frac{\overline{I}_o}{\overline{H}_o}, \tag{2.31}$$

where $a' = 0.4922 + \{0.27/(\overline{H}_d/\overline{H}_g)\}$ for $0.1 \le (\overline{H}_d/\overline{H}_g) \le 0.7$

or $a' = 0.76 + \{0.113/(\overline{H}_d/\overline{H}_g)\}$ for $0.7 \le (\overline{H}_d/\overline{H}_g) \le 0.9$

and $b' = 2(1 - a')(\sin \omega_s - \omega_s \cos \omega_s)/(\omega_s - 0.5 \sin 2\omega_s)$.

2.9 Calculating the Sun Radiation Availability

Example 2.8 Use the prediction equations to calculate the monthly average hourly global and hourly diffuse radiation during the month of April on a horizontal surface at New Delhi (28° 35′ N, 77° 12′ E, elevation 216 m). Time 9:00 to 10:00 h (LAT). The average number of sunshine hours per day is 8.6.

Solution

Daily Radiation

The representative day in April is the 15th On April 15, $n = 105$.
Using Eq. (2.3), $\delta = 23.45 \sin\left[\frac{360}{365}(284 + 105)\right] = 9.42°$.
The sunshine hour may be calculated from Eq. (2.10)

$$w_s = \cos^{-1}\{-\tan(28.58°) * \tan(9.42°)\} = 95.18° \text{ or } 1.661 \text{ radians.}$$

Day length, $S_{max} = (2/15) *95.18° = 12.69$ h.
From Eq. (2.21),

$$H_o = \frac{24}{\pi} \times 1.367 \times 3600 \times \left(1 + 0.033 \times \cos \frac{360 \times 105}{365}\right)$$
$$\times (1.661 \sin 28.58° \times \sin(9.42°) + \cos 28.58° \times \cos(9.42°)$$
$$\times \sin 95.18°) = 37034 \text{ kJ/m}^2 \text{ day.}$$

From Table 2.5, $a = 0.25$ and $b = 0.57$.
Putting $\overline{H}_o$ equal to H_o on April 15,
Therefore, from Eq. (2.19),

$$\overline{H}_g = \left(0.25 + 0.57 * \left(\frac{8.6}{12.69}\right)\right) * 37034 = 23564 \text{ kJ/m}^2 \text{ day}$$

Alternatively using Eq. (2.33) suggested by Gopinathan, we get $a_1 = 0.3459$, $b_1 = 0.4019$ and $\overline{H}_g = 22{,}895 \text{ kJ/m}^2$-day which is in close agreement with the value of 23,564 kJ/m²-day. We proceed with the value of $\overline{H}_g = 23564 \text{ kJ/m}^2$-day.
From Eqs. (2.26) and (2.27)

$$\overline{H}_d = (1.411 - 1.696 \times (23564/37034)) \times 23564 = 7820 \text{ kJ/m}^2 \text{ - day}$$
$$\overline{H}_d = \left\{0.8677 - 0.7365 \times \left(\frac{8.6}{12.69}\right)\right\} \times 23564 = 8685 \text{ kJ/m}^2 \text{ - day.}$$

Hourly Radiation

The hourly extra-terrestrial radiation on a horizontal surface on April 15 between 9:00 to 10:00 h is obtained by calculating the instantaneous value at 09:30 h and multiplying by 3600 s.
From Eq. (2.5),

$$I_o = 1.367\left(1 + 0.033 \times \cos \frac{360 \times 105}{365}\right) \times (\sin 28.58° \times \sin(9.42°)$$

$$+ \cos 28.58° \times \cos(9.42°) \times \cos 37.5°)$$
$$= 1.0384 \text{ kW/m}^2 = 3738 \text{ kJ/m}^2 \text{ - h.}$$

Substituting in Eq. (2.28)

$$a = 0.409 + 0.5016 \sin(95.18° - 60°) = 0.69800$$
$$b = 0.6609 - 0.4767 \sin(95.18° - 60°) = 0.38625$$
$$f_c = 0.9931.$$

Once again putting $\overline{I}_o$ equal to I_o on April 15, we get

$$\frac{I_g}{23564} = \frac{3738}{37034}(0.69800 + 0.38625 \cos 37.5°)$$
$$\overline{I}_g = 2406 \text{ kJ/m}^2 \text{ h}$$

Finally, from Eq. (2.30), with $\overline{H}_d = 7820 \text{ kJ/m}^2$-day

$$\overline{I}_d = 7820 \times \left(\frac{3738}{37034}\right) = 789 \text{ kJ/m}^2 \text{ - h.}$$

Instead of Eq. (2.30), if we use Eq. (2.31), a slightly different value is obtained. We get

$$\overline{I}_d = 780 \text{ kJ/m}^2 \text{ - h.}$$

Alternatively, putting $\overline{H}_d = 8685 \text{ kJ/m}^2$-day, we have

$$\overline{I}_d = 866 \text{ kJ/m}^2 \text{ - h.}$$

2.9.5 Hourly Global Beam and Diffuse Radiation Under Clear Skies

Now let's focus on the radiation forecast for days with a clear sky. ASHRAE (1972) created a technique for calculating the hourly values of the global horizontal irradiance and direct normal irradiance based on an investigation of US data. The calculations depend on an exponentially decaying model, in which the beam irradiation falls as the distance through the atmosphere is increased. The global radiation I_g reaching a horizontal surface on the earth is determined by

$$I_g = I_b + I_d,$$

2.9 Calculating the Sun Radiation Availability

where

I_g = hourly global horizontal irradiance
I_b = hourly beam component
I_d = hourly diffuse component.

Now,

$$I_b = I_{bn} \cos \theta_z, \tag{2.32}$$

where I_{bn} = direct normal irradiance
and θ_z = zenith angle

Thus,

$$I_g = I_{bn} \cos \theta_z + I_d \tag{2.33}$$

In the ASHRAE model, it is postulated that

$$I_{bn} = A \exp(-B/\cos \theta_z) \tag{2.34}$$

and

$$I_d = C I_{bn}. \tag{2.35}$$

Here, on a monthly basis, the values of the constants A, B, and C were established. These constants change throughout the year as a result of seasonal fluctuations in the atmosphere's dust and water vapour content as well as the shifting earth-sun distance. Threlkeld and Jordan (1958) originally provided the values for A, B, and C and Iqbal (1983) later changed them which have been given in Table 2.6.

Table 2.6 Value of the constants A, B, and C for predicting hourly solar radiation on clear days in the ASHRAE model (1972)

	A (W/m^2)	B	C
January 21	1202	0.141	0.103
February 21	1187	0.142	0.104
March 21	1164	0.149	0.109
April 21	1130	0.164	0.120
May 21	1106	0.177	0.130
June 21	1092	0.185	0.137
July 21	1093	0.186	0.138
August 21	1107	0.182	0.134
September 21	1136	0.165	0.121
October 21	1136	0.152	0.111
November 21	1190	0.144	0.106
December 21	1204	0.141	0.103

2.10 Solar Radiation on Tilted Surfaces

It is very clear from the previous sections that measuring instruments typically provide results for solar radiation that is incident on a horizontal surface. To absorb radiation, most solar devices (such as flat-plate collectors or PV modules) are tilted at an angle from the horizontal. So, it is important to evaluate the flux that incident on a tilted surface. This flux is caused by radiation that falls directly on the surface as well as radiation that is diffused and reflected from the surroundings onto the surface (Ineichen 2006; Independent Databanks 2004; Liu and Jordan 1961; Sukhatme and Nayak 2012).

Beam Radiation

The tilt ratio for beam radiation is the ratio of the flux of beam radiation falling on a tilted surface to that falling on a flat surface which is normally represented by the symbol r_b. In the scenario of a south-facing (i.e. $\gamma = 0°$), tilted surface

$$\cos\theta = \sin\delta\,\sin(\phi - \beta) + \cos\delta\,\cos\omega\,\cos(\phi - \beta)$$

while for a horizontal surface

$$\cos\theta_z = \sin\phi\,\sin\delta + \cos\phi\,\cos\delta\,\cos\omega.$$

Hence,

$$r_b = \frac{\cos\theta}{\cos\theta_z} = \frac{\sin\delta\,\sin(\phi - \beta) + \cos\delta\,\cos\omega\,\cos(\phi - \beta)}{\sin\phi\,\sin\delta + \cos\phi\,\cos\delta\,\cos\omega}. \qquad (2.36)$$

Expressions for r_b may also be generated for additional scenarios where the tilted surface is oriented differently with respect to $\gamma \neq 0°$.

Diffuse Radiation

The tilt factor (r_d) for diffuse radiation is the ratio of the flux of diffuse radiation falling on a tilted surface to that falling on a horizontal surface. The value of this tilt factor depends upon the distribution of diffuse radiation over the sky and on the portion of the sky dome seen by the tilted surface. Assuming that the sky is an isotropic source of diffuse radiation, we have a tilted surface with a slope β.

$$r_d = (1 + \cos\beta)/2 \qquad (2.37)$$

since $(1 + \cos\beta)/2$ represents the radiation shape factor for a surface that is tilted.

The sky has an anisotropic distribution of diffuse radiation. Experimental evidence suggests that forward scattering of a portion of the beam radiation causes a small area of the sky close to the solar disc to have a higher intensity of diffuse radiation. Similarly, the intensity of diffuse radiation near the horizon is higher because of

2.10 Solar Radiation on Tilted Surfaces

multiple Rayleigh scattering by a larger air mass keeping these factors in mind; some sky models taking account of the anisotropic distribution have been developed. Yet, it is common for engineering calculations to assume that the diffuse radiation's distribution is isotropic.

Reflected Radiation

It follows that $(1 + \cos \beta)/2$ is the radiation shape factor for the surface associated with the surrounding ground because $(1 + \cos \beta)/2$ indicates the radiation form factor for a tilted surface with respect to the sky. The tilt factor for reflected radiation is determined by assuming that the reflectivity of the beam and diffuse radiations incident to the ground is ρ, and that the reflectivity of the diffuse and beam irradiance striking on the ground is diffuse and isotropic, and it is given by

$$r_r = \rho(1 - \cos \beta)/2. \tag{2.38}$$

Flux on Tilted Surface

The flux I_T falling on a tilted surface at any instantaneous time can be expressed as:

$$I_T = I_b r_b + I_d r_d + (I_b + I_d) r_r, \tag{2.39}$$

where the values of r_b, r_d, and r_r are as given in Eqs. (2.36), (2.37), and (2.38). It should be noted that Eq. (2.36) is valid only for a tilted surface with $\gamma = 0°$, whereas Eqs. (2.37) and (2.38) for any tilted surface with a slope β are valid. When we divide Eq. (2.39) by I_g, both sides, we obtain the ratio between the incident solar flux on a tilted surface at any moment or point of time to that on a horizontal surface.

$$\frac{I_T}{I_g} = \left(1 - \frac{I_d}{I_g}\right) r_b + \frac{I_d}{I_g} r_d + r_r. \tag{2.40}$$

The value of the diffuse reflectivity ρ is typically unknown, which creates a problem when using Eq. (2.39). A value around 0.2, generally expected with surfaces of concrete or grass, can be used. Fortunately, the reflected radiation term does not very often contribute much to the total.

Equation (2.40) may be utilized for estimation of the hourly radiation incident on a tilted surface if the ω value is taken at the exact midpoint of the hour. If estimations are made for the representative day of the month as indicated in Sect. 2.9.1, it may also be applied to compute the monthly average hourly value $(\overline{I}_T)$. Then, the modified form of Eq. (2.40) is applied.

$$\frac{\overline{I}_T}{\overline{I}_g} = \left(1 - \frac{\overline{I}_d}{\overline{I}_g}\right) \overline{r}_b + \frac{\overline{I}_d}{\overline{I}_g} \overline{r}_d + \overline{r}_r, \tag{2.41}$$

where on the representative day $\bar{r}_b = r_b$, $\bar{r}_d = r_d = (1 + \cos \beta)/2$ and $\bar{r}_r = r_r = (1 - \cos \beta)/2$.

In many applications, the radiation that strikes a tilted surface daily is also a crucial factor. As per Liu and Jordan (Liu and Jordan 1961; Sukhatme and Nayak 2012) the ratio of the daily radiation incident on a tilted surface (H_T) to the daily global radiation falling on a horizontal surface (H_g) can be expressed by an equation which is approximately similar to Eq. (2.40). Hence,

$$\frac{H_T}{H_g} = \left(1 - \frac{H_d}{H_g}\right)R_b + \frac{H_d}{H_g}R_d + R_r. \tag{2.42}$$

For a surface facing south direction ($\gamma = 0°$), Liu and Jordan show that

$$R_b = \frac{\omega_{st} \sin \delta \, \sin(\phi - \beta) + \cos \delta \, \sin \omega_{st} \cos(\phi - \beta)}{\omega_s \sin \phi \, \sin \delta + \cos \phi \, \cos \delta \, \sin \omega_s} \tag{2.43}$$

$$R_d = r_d = (1 + \cos \beta)/2 \tag{2.44}$$

$$R_r = r_r = \rho(1 - \cos \beta)/2. \tag{2.45}$$

The sunset or sunrise hour angles (expressed in radians) for the tilted surface and a horizontal surface, respectively, are given in Eq. (2.43) as ω_{st} and ω_s.

If the necessary values are established for the typical day of the month, Eq. (2.42) may also be used to compute the monthly average daily radiation falling on a tilted surface. Equation (2.42) is then used in the modified form

$$\frac{\overline{H}_T}{\overline{H}_g} = \left(1 - \frac{\overline{H}_d}{\overline{H}_g}\right)\overline{R}_b + \frac{\overline{H}_d}{\overline{H}_g}\overline{R}_d + \overline{R}_r, \tag{2.46}$$

where

$\overline{R}_b = R_b$ on the representative day,
$\overline{R}_d = R_d = \frac{1+\cos \beta}{2}$.
$\overline{R}_r = R_r = \rho(1 - \cos \beta)/2$.

Example 2.8 Calculate the monthly average, total daily radiation falling on a flat-plate collector facing south ($\gamma = 0°$) and titled by 30° from the ground, at New Delhi (28°35' N, 77° 12' E) for the month of November. Assume ground reflectivity as 0.2, $\overline{H}_g = 16282.8$ kJ/m2-day, $\overline{H}_d = 4107.6$ kJ/m^2-day.

Solution

Given data,

$$\Phi = 28°.58'$$
$$\beta = 30°.$$

2.10 Solar Radiation on Tilted Surfaces

The representation day for the month of November is 14°.

Therefore, for the day of the year on November 14, $n = 318$.

Monthly average of the daily global radiation for the month of November in New Delhi $\overline{H}_g = 16{,}282.8$ kJ/m2-day.

Monthly average of the daily diffused radiation for the month of November in New Delhi, $\overline{H}_d = 4107.6$ kJ/m2-day.

Using Eq. (2.3) $\delta = -18.91°$.

Using Eq. (2.10), $w_s = 79.245°$, or 1.383 radians.

As the day under consideration lies between September 22 and March 21, the hour angle at sunrise will be the same as the one obtained for a horizontal surface. Thus, $w_{st} = w_s = 79.245°$ or 1.383 radians.

Using Eq. (2.43)

$$R_b = \frac{1.383 * \sin(28.58 - 30) + \cos(-18.91) \sin 79.245 \cos(28.58 - 30)}{1.383 * \sin 28.58 \sin(-18.91) + \cos(-18.91) \sin 79.24 \cos 28.58}$$

$$= 1.56.$$

Using Eq. (2.44),

$$R_d = \frac{1 + \cos 30°}{2} = 0.933.$$

Using Eq. (2.45),

$$R_r = 0.2 \times \frac{1 - \cos 30°}{2} = 0.0134.$$

Now, monthly average total daily radiation, $\overline{H}_T$ on a tilted surface may be calculated using Eq. 2.46,

$$\frac{\overline{H}_T}{\overline{H}_g} = 1 - \frac{\overline{H}_d}{\overline{H}_g} R_b + \frac{\overline{H}_d}{\overline{H}_g} R_d + R_r$$

$$\overline{H}_T = 16282.8 \times \left[1 - \frac{4107.6}{16282.8} \right] \times 1.56 + \frac{4107.6}{16282.8} \times 0.933 + 0.0134$$

$$= 23043.89 \text{ kJ/m}^2 \text{ - day.}$$

Review Questions

(1) What are the merits and demerits of solar energy?

(2) How does the sun create energy continuously?

(3) Define and explain the earth's albedo.

(4) Describe the solar constant, the solar irradiance, and the extra-terrestrial and terrestrial radiations. what is the standard value of the solar constant?

(5) Describe the distribution of different extra-terrestrial radiation constituents in terms of percentages.
(6) Describe how solar energy gets reduced as it travels through the atmosphere to the earth's surface.
(7) What do you mean by solar time, and explain how it is distinct from a standard clock time?
(8) What are the basic features required in an ideal pyranometer?
(9) How does a flat-plate solar collector benefit from sun tracking in terms of energy production?
(10) How does the collection of solar energy get affected by tilting a flat-plate collector with respect to the ground?
(11) Describe the components and basic principles of a sunshine recorder.
(12) Explain global, diffuse, and beam radiation. Calculate the total radiation on a tilted surface by deriving an equation. Demonstrate that no radiation from the earth is reflected by a horizontal surface.
(13) Explain and formulate an equation for the solar day length.
(14) Give definitions for the terms "incidence angle," "angle of declination," "hour angle," "zenith angle," and "solar azimuth angle."
(15) Determine the rate at which the sun releases energy into the atmosphere. In what proportion does this energy become absorbed by the earth? What is the amount that was intercepted?

Problems

(1) Determine the angle of incidence of beam radiation on a flat-plate collector for the following situation:

Location	Nagpur (21°06′ N, 79°03′E)	
Surface azimuth angle	15°	
Slope of collector	31°	
Date	December 1, 1979	
Time	0900 h (IST)	

Ans: 38.4 degree

(2) Determine and plot the variation of the angle of incidence of beam radiation on May 1st on a flat surface located in Mumbai (19°07′ N, 72°51′ E). The surface is tilted at an angle of 10° with the horizontal and is pointing due south. Would you prefer to plot the variation against IST or LAT?

Ans:

Local apparent time (LAT)	0550	0600	0800	1000	1200
θ (deg)	90	87.7	58.8	29.9	5.8

2.10 Solar Radiation on Tilted Surfaces

(3) Calculation of the angle of incidence of beam radiation on photo voltaic array located in Mumbai (19°07′ N, 72°51′ E). Given the following data:

Date	August 18, 2005
Time	1030 h (LAT)
Slope of PV array	28°
Surface Azimuth angle	20°

Ans: 24.01 degree

(4) Determine the days of the year when the sun is directly overhead in Pune (18°32′ N) at 1200 h (LAT).

Ans: May 14th and 30th July

(5) An array of flat-plate collectors is facing due south and is inclined at an angle β to the horizontal. Reflecting mirrors of the same length L as the collectors are used to boost the output (Fig. 2.19). Using the criterion that the tops of the mirrors are being hit by the sun's rays (line AA) at 1200 h (LAT) should be reflected to the top of the collectors (line BB), derive an expression for the correct inclination ψ of the booster mirrors in terms of β, δ and φ.

Ans: $\psi = (\pi - \beta - 2\phi + 2\delta)/3$.

(6) Plot the variation of the day length on a horizontal surface through the year for the following locations:

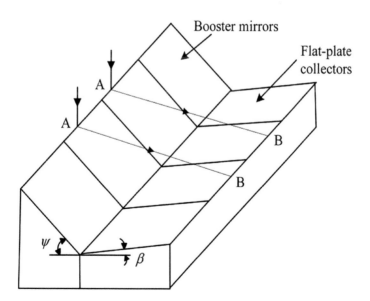

Fig. 2.19 Array of flat-plate collector facing due south

Srinagar	(34°06′ N, 74°51′ E)
Kolkata	(22°39′ N, 88°27′ E)
Vishakhapatnam	(17°43′ N, 83°14′ E)
Thiruvananthapuram	(8°29′ N, 76°57′ E)

Ans:

	Srinagar	Kolkata	Vishakhapatnam	Thiruvananthapuram
21st June	14.28	13.39	13.06	12.49 h
21st Dec	9.72	10.61	10.94	11.51 h

(7) Determine the day length on May 1 and December 1 for a surface facing south that is inclined at a 40-degree angle and situated in New Delhi (28°35′ N, 77°12′ E).

Ans: 11.59, 10.29 h

(8) Determine the hour angle at sunset and sunrise on a surface that is inclined at a 40-degree. Given $\varphi = 28°$ N, $\delta = -21°$ and $\gamma = 48°$.

Ans: Sunrise at 78.2 degree and Sunset at -62 degree

(9) A long vertical wall (4.2 m high) is built along the southern boundary of an open ground. A solar collector system is installed in the ground north of the wall. The collectors are tilted at an angle of 25° and face due south. Calculate the minimum distance that should be left on the ground between the wall and the collectors, so that the wall does not shade the collectors for two hours before as well as two hours after solar noon during the month of April. The latitude and longitude of the location are 27°50' N and 85°32' E.

Ans: 1.85 metre

(10) Derive an expression for the daily extra-terrestrial radiation which would fall on a surface having a slope β and facing due south (i.e. $\gamma = 0°$).

Ans:
$$H_o = I_{sc}[1 + 0.033 \cos(360n/365)]$$
$$[\omega_s \sin \delta \sin(\phi - \beta) + \cos \delta \cos(\phi - \beta) \sin \omega_s]$$

(11) Use the prediction equations to calculate the monthly average daily global and diffuse radiation falling on a horizontal surface at Bhavnagar (21°45' N, 72°11' E) during the month of January. The monthly average sunshine hour is 9.8. Compare the predicted values with the following measured values given in the Handbook of Solar Radiation Data for India, $\overline{H}_g = 18,511$ kJ/m^2-day, $\overline{H}_d = 4198$ kJ/m^2-day.

Ans: 18323 kJ/m^2-day, 3937 kJ/m^2-day

(12) Calculate the variation of the tilt factor r_b with slope β varying from 0° to 90° for a surface facing south ($\gamma = 0°$), given the following data,

2.10 Solar Radiation on Tilted Surfaces

Location	Nagpur (21°06' N, 79°03' E)
Date	March 1
Time	1200 h (LAT) and 1700 h (LAT)

Ans: At 1200 h (LAT):

β (Deg)	0	21.1	29.4	60	90
r_b	1	1.136	1.148 (max)	0.988	0.563

At 1700 h (LAT):

β (Deg)	0	21.1	50.5	60	90
r_b	1	1.370	1.572 (max)	1.550	1.213

Objective Questions

1. A vertical surface gets

   a. No reflected radiation component
   b. Hundred per cent of the diffused radiation component
   c. Fifty per cent of the reflected radiation component
   d. Fifty per cent of the beam radiation component

2. A horizontal surface gets

   a. Fifty per cent of the diffused radiation component
   b. Fifty per cent of the reflected radiation component
   c. Fifty per cent of the beam radiation component
   d. No reflected radiation component

3. On the representative day of each month, the extra-terrestrial daily radiation may be taken as equal to

   a. Diffused radiation at the site
   b. Beam radiation at the site
   c. Global radiation at the site
   d. Monthly average, daily extra-terrestrial radiation at the site

4. At solar noon, the hour angle will be

   a. $-90\ °C$
   b. $+90\ °C$
   c. $+180\ °C$
   d. Zero

5. On 21 September, the declination angle

    a. $+ 23.45\ °C$
    b. $+ 180\ °C$
    c. $-23.45\ °C$
    d. Zero

6. For one degree change in longitude, the change in solar time is

    a. Four seconds
    b. Four minutes
    c. One hour
    d. One minute

7. At the 30 °C inclination angle, what is the zenith angle?

    a. $150°$
    b. $30°$
    c. $120°$
    d. $60°$

8. The air-to-mass ratio is minimum

    a. Sun is at the zenith
    b. At Sunrise
    c. At 06:00 GMT
    d. At Sunset

9. The amount of incoming radiation that the earth reflects into space is

    a. 20%
    b. 10%
    c. 40%
    d. 30%

10. When the earth's atmosphere blocks some of the sun's incoming energy

    a. Different molecules selectively absorb the radiation of different wavelength
    b. Different types of molecules evenly absorb radiation of all wavelengths.
    c. Complete radiation absorption
    d. No absorption of radiation by the atmosphere

11. Which mechanism is responsible for the sun's energy production?

    a. Exothermal chemical reaction
    b. Nuclear fission reaction
    c. Nuclear fusion reaction
    d. All above mentioned

12. Which of the following assertions about solar energy is untrue??

    a. Its availability is diurnal

b. The energy is diluted.
c. It's harnessing at large scale is easy
d. There is no guarantee of availability at any given moment

13. Diffused radiation generally has

   a. A unique direction
   b. No unique direction
   c. Large magnitude as compared to beam radiation
   d. Short wavelength as compared to beam radiation

14. What is the infrared component's approximate % content in extra-terrestrial radiation?

   a. 55.5%
   b. 45.5%
   c. 80%
   d. 20%

15. Terrestrial radiation has a wavelength in the range of?

   a. $0.2\ \mu m$ to $0.5\ \mu m$
   b. $0.2\ \mu m$ to $4\ \mu m$
   c. $0.29\ \mu m$ to $2.3\ \mu m$
   d. $0.380\ \mu m\ 0.760\ \mu m$

16. What is the standard value of the solar constant?

   a. $1.367\ kWm^2$
   b. $1\ kWm^2$
   c. $5\ kWm^2$
   d. $1.5\ kWm^2$

Answers

1. c	2. d	3. c	4. d	5. c	6. b	7. d	8. a
9. d	10. a	11. c	12. c	13. d	14. b	15. c	16. d

References

ASHRAE (1972) Handbook of fundamentals, American society of heating, refrigerating and air-conditioning engineers, pp 385–443

Braun JE, Mitchell JC (1983) Solar geometry for fixed and tracking surfaces. Sol Energy 31:439

Collares-Pereira M, Rabl A (1979) The average distribution of solar radiation correlations between diffuse and hemispherical and between daily and hourly insolation values. Sol Energy 22:155

Cooper PI (1969) The absorption of solar radiation in solar stills. Sol Energy 12:3

Coulson LK (1975) Solar and terrestrial radiation: Methods and measurements. Nyap

Fröhlich C (1977) Contemporary measures of the solar constant. Soiv 93

Froelich C, Wehrli C (1981) Spectral distribution of solar irradiance from 25000 to 250 nm. World Radiation Centre, Davos, Switzerland

Froelich C, Brusa RW (1981) Solar radiation and its variation in time. Sol Phys 74:209

Garg HP, Garg SN (1985) Correlation of monthly-average daily global, diffuse and beam radiation with bright sunshine hours. Energy Convers Manage 25:409

Gopinathan KK (1988) A general formula for computing the coefficients of the correlation connecting global solar radiation to sunshine duration. Sol Energy 41:499

Gopinathan KK, Soler A (1995) Diffuse radiation models and monthly–average, daily diffuse data for a wide latitude range. Energy 20:657

Gueymard C (1986) Mean daily averages of beam radiation received by tilted surfaces as affected by atmosphere. Sol Energy 37:261

Gueymard C (2000) Prediction and performance assessment of mean hourly global radiation. Sol Energy 68:285

Hickey JR, Alton BM, Griffin FJ, Jacobowitz H, Pellegrino P, Maschhoff RH, Smith EA, Vonder Haar TH (1982) Extraterrestrial solar irradiance variability: two and one-half years of measurements from nimbus 7. Solar Energy 28(5):443–445

Independent databanks K, Ineichen P, Schroedter-Homscheidt M, Cro S, Dumortier D, Kuhlemann R, Olseth JA, Piernavieja G, Reise C, Wald L, Heinemann D (2004) Rethinking satellite-based solar irradiance modelling the SOLIS clear-sky model. Remote Sensing of Environ 91:160

Ineichen P (2006) Comparison of eight clear sky broadband models against. Sol Energy 80:468

Iqbal M (1983) An introduction to solar radiation. Academic Press, Canada

Johnson FS (1954) The solar constant. J Meteor 11:431–439

Klein SA (1977) Calculation of monthly average insolation on tilted surfaces. Sol Energy 19:325

Kreith F, Kreider JF (1978) Principles of solar engineering. Chapter 2, McGraw-Hill Book Co., New York

Liu BYH, Jordan RC (1960) The interrelationship and characteristic distribution of direct, diffuse and total solar radiation. Sol Energy 4:1

Liu BYH, Jordan RC (1961) Daily insolation on surfaces tilted towards the equator. ASHRAE Trans 67:526

Lof GOG, Duffie JA, Smith CO (1966) World distribution of solar radiation. Sol Energy 10:27

Mani A (1981) Handbook of solar radiation data for India. Allied Publishers, New Delhi

Mani A, Chako O (1973) Solar radiation climate of India. Sol Energy 14:139

Modi V, Sukhatme SP (1979) Estimation of daily total and diffuse insolation in India from weather data. Sol Energy 22:407

Page JK (1961) The estimation of monthly mean values of daily total short-wave radiation on vertical and inclined surfaces from sunshine records for latitudes 40°N–40°S. In: Proceedings UN conference new sources of energy, vol 4. pp 378

Satyamurty VV, Lahiri PK (1992) Estimation of symmetric and asymmetric hourly global and diffuse radiation from daily values. Sol Energy 48:7

Spencer JW (1971) Fourier series representation of the position of the sun. Search 2:162–172

Sukhatme SP, Nayak JK (2012) Solar energy, principles of thermal collection and storage. McGraw-Hill Book Co., New Delhi

Thekaekara MP (1976) Solar radiation measurement: Techniques and instrumentation. Solar Energy 18(4):309–325

Thekaekara MP, Drummond AJ (1971) Standard values for the solar constant and its spectral components. Nat Phys Sci 229:6

Threlkeld JL, Jordan RC (1958) Direct solar radiation available on clear days. ASHRAE Trans 64:45

Willson RC, Gulkis S, Janssen M, Hudson HS, Chapman GA (1981) Observations of solar irradiance variability. Science 211(4483):700–702

Chapter 3
Introduction to Photovoltaic Solar Energy

Abstract The chapter provides a thorough overview of photovoltaic (PV) solar energy, covering its fundamentals, various PV cell types, analytical models, electrical parameters, and features. Beginning with the fundamentals, it discusses photon energy, P-N junctions, the photovoltaic effect, and the semiconductor nature of photovoltaics in addition to exploring various materials for solar cells.

Subsequently, the various types of solar cells—monocrystalline, polycrystalline, and amorphous are examined, and their efficiencies are described. Analytical models of solar cells study the single and two-diode models as well as electrical properties including fill factor, maximum power, open-circuit voltage, and short-circuit current—all of which are crucial for understanding solar cell efficiency.

V-I and P-V characteristics, among other electrical parameters of PV cells, are described. Next, the effects of atmospheric variables and parameters on PV cell characteristics are discussed, along with maximum power point tracking (MPPT).

It also includes a review of power converter topologies, such as DC/DC and DC/AC converters, and their control strategies, as well as applications for both standalone and grid-connected solar systems.

Keywords Photovoltaics · Amorphous PV cells · Maximum power point tracking (MPPT) · Power converter topologies

3.1 Fundamentals of Photovoltaic

The use of renewable energy sources is crucial in electrical power production. There are many ways to create electrical energy using sustainable sources of energy such as solar, wind, and hydroenergy. The sun's energy is getting considerable interest due to its numerous advantages. Photovoltaic cells or so-called solar cell is the heart of solar energy conversion to electrical energy (Kabir et al. 2018). Without any involvement in the thermal process, the photovoltaic cell can transform solar energy directly into electrical energy. Compared to conventional methods, PV modules are advantageous in terms of reliability, modularity, durability, maintenance, etc.

© The Author(s), under exclusive license to Springer Nature Singapore Pte Ltd. 2024
K. Namrata et al., *Wind and Solar Energy Systems*, Energy Systems in Electrical
Engineering, https://doi.org/10.1007/978-981-99-9710-7_3

Fig. 3.1 Selenium PV cell

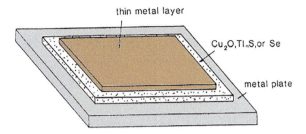

In the nineteenth century, during 1839 Edmond Becquerel discovered the photovoltaic effect and it came to be known as the Becquerel effect. He is known as the Father of Photovoltaics. Nearly 35 years after the discovery of the photovoltaic effect, Adams and Day made a selenium photovoltaic cell and published it in the year 1877. In the year 1883, Fritts C E described the first thin film selenium solar cell which is shown in Fig. 3.1. Although experiments on photovoltaic cells were reproducible and repeatable the classic physics was not able to explain the main theory and operation of the solar cell. Later in 1900, Max Planck introduced Quantum mechanics and in 1905, Albert Einstein published an article in "Annalender Physik" where he explained the concept of photon packets and through this concept, he explained the photovoltaic principle. It led to the foundation of semiconductor materials (Kannan and Vakeesan 2016). In 1933 Grondahl L O published an article on "The Copper-cuprous-oxide Rectifier and Photoelectric Cell." In 1941 Russell Ohl made the first silicon solar cell US patent with efficiency much less than 1% but it was the landmark point for solar cells as far as concerned.

The schematic diagram of the photovoltaic system in in present scenario has been shown in Fig. 3.2. Since there are no moving parts involved in the energy conversion process, there is no mechanical loss. Solar photovoltaic cells are reliable, durable, maintenance free, and modular. The average life span of solar PV cells is around 20 years or even more. Solar energy can be used as distributed generation with less or no distribution network because it can installed where it is to be used. However, the solar PV cell has some sorts of disadvantages the installation cost is expensive (Duffie and Beckman 2006). At present situation effectiveness of solar cells is less compared with alternative sources of energy. Solar energy is not available for 24 h, so there is a requirement for energy storage which makes the overall setup expensive.

Despite these disadvantages, solar energy has found some special applications where it is the best option to use it. The applications of solar cells are for power in space vehicles and satellites, remote radio communication booster stations, rooftop PV, and solar-powered vehicles. In the coming years, most of the conventional energy sources are to be replaced by solar energy sources.

Fig. 3.2 Photovoltaic system

3.1.1 Semiconductor Materials

All the materials available on the earth are divided into three categories based on the ability to carry current through it. Those types are conductors, semiconductors and insulators. Conductors are the materials which can carry the current through them because of the existence of free electrons in the outer shell of its atoms. A metal's conductivity is in the range of 10^6 mho/cm. The majority of conductors used in electrical and electronics applications are metals, such as copper, aluminium, and steel. These materials have an extremely low resistance and adhere to Ohm's law. As a result, they may transport electric current from one location to another without causing many other currents to dissolve. Insulators are materials which cannot carry current through them due to the absence of free electrons and high resistivity. An insulator's conductivity is in the range of 10^{-10} mho/cm. The insulators are used as protection in electrical circuits and household items etc. Some commonly used insulators are glass, plastic, wood, air, etc. The semiconductor materials are the materials which exhibit both the properties of conductors and insulators. At absolute zero temperature, the semiconductors behave like insulators and at room temperature, the semiconductors behave like conductors but with less conductivity than conductors. Their conductivity may be raised by including little amounts of contaminants in the pure semiconductor material and this process is known as doping. Pure semiconductors are silicon; germanium and compounds can be gallium arsenide, and cadmium selenide (Garg and Prakash 2012).

The semiconductor is an extrinsic semiconductor and an intrinsic semiconductor of two kinds. Intrinsic semiconductor is the clean form of the semiconductor where no doping process is done. An extrinsic semiconductor is a semiconductor in which intentionally some impurities are added or doping is done to alter the electrical properties of the semiconductor. Doping is done to boost the amount of unoccupied holes or electrons to make the semiconductor conductive. Based on the kind of additional impurity, the extrinsic semiconductor is characterized as an *n*-type semiconductor and a *p*-type semiconductor as shown in Figs. 3.3 and 3.4, respectively. When an

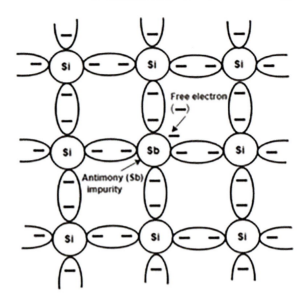

Fig. 3.3 *n*-type semiconductor

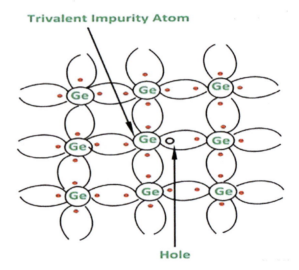

Fig. 3.4 *p*-type semiconductor

impurity of pentavalent, having five free electrons in its valence shell such as arsenic, antimony and phosphorous is included in a clean semiconductor then it's referred to as an n-type semiconductor (Garg and Prakash 2012). The conduction in *n*-type semiconductors is mainly due to the large number of free electrons in semiconductor crystal donated by the impurities added in it and the impurities that result in n-type semiconductors considered to be donor impurities. When a trivalent impurity, having three free electrons in its valence shell such as aluminium, boron, gallium and indium is included in a clean semiconductor then it's referred to as a p-type semiconductor.

The conduction in *p*-type semiconductors is mainly due to the massive amount of holes in the semiconductor crystal formed by the impurities introduced in it and those impurities are known as acceptor impurities.

The amount of doping defines the resistance offered by a semiconductor. If a semiconductor is lightly doped then the resistance offered by it is more and if a semiconductor is heavily doped then the resistance offered by it is less. To understand the nature of semiconductors one should have a brief knowledge about quantum theory. In quantum physics, the electron's potential energy is judged by the size of the orbit. An atom has the fourth quantum number namely spin, azimuthal, magnetic, and principal quantum number. The energy levels are given by the number of orbits present in the structure of an atom. The electrons present in the outside shell are called valence electrons. An energy level can contain two electrons at most with reverse spin due to the Pauli Exclusion Rule. Fermi level is the name given to this energy level and is denoted by E_F. If the electrons are excited by some external source like heat and light, then the electrons gain energy. If the energy is more than the Fermi level then the electron jumps to the conduction band and is ready to conduct. If the electron is in the valence band, then the power needed to excite it is more than the power needed to excite it when it is near to Fermi energy level. By doping impurities in the semiconductor the electrons are moved from the valence band to the band nearer to the Fermi level and hence its electrical properties changes.

3.1.2 Photon Energy

The semiconductor materials are used to form a *p-n* layer so that the manufacture of diodes, transistors, and thyristors is possible. A *p-n* layer contains both holes and electrons and the conventional direction of current flowing through the *p-n* device is either in the same direction as electrons or the other way of holes. The bulk of carriers in holes make up a p-type semiconductor, hence the holes are the major cause of conduction and because electrons make up the bulk of carriers in n-type semiconductors, conduction is mainly due to electrons. That's why the conduction in a semiconductor is due to either holes or electrons (Khan 2010).

In a solar PV cell, a silicon p–n junction with a huge surface area is present. In the solar cell, the conduction is due to electrons that jump the conduction band to the valence band when the sunlight strikes the electron present in the lattice of the cell. The sunlight is composed of photons and these photons have energy carried by themselves. The energy held by a single photon is termed photon energy. When the photon hits the electron in the lattice of the solar cell, then the energy is moved from the photon particle to the electron and this energy excites from the conduction band to the valence band by an electron. The amount of energy carried depends on electromagnetic frequency to which it is directly proportional and on wavelength to which it is negatively correlated. The larger the increase in photon frequency is its power and similarly, when a photon's wavelength is longer, its energy is lower. The unit of a kind of photon energy may be the same as that of energy. The most widely used units to indicate photon energy is the electron volt (eV) and the joule or microjoule. Like 1 J is equal to 6.24×10^{18} eV, Compared to fewer energy photons, such as radio frequency photons portion of the electromagnetic spectrum, the bigger units may be more effective for indicating higher frequency photon's energy, as well as greater energy, like gamma rays. The energy of the radiation emitted by the photon, E is given by the relation given below

$$E = hc/\lambda, \tag{3.1}$$

where h is the constant of Planck. λ is the wavelength of the radiation, and c represents the speed of light.

The photon energy must be higher than the energy band gap present in the semiconductor so that electron–hole pair generation is possible due to the absorption of photon energy. If the energy of the photons is less than the energy band gap then no energy absorption is possible, no electron–hole pair will be generated and the material will seem to be transparent for the respective photons. The photon energy should not be much greater because there will be more heat loss along with electron–hole pair generation. The semiconductor to be selected must absorb the maximum percentage of the solar spectrum efficiently.

Example 3.1 Calculate the energy of a photon with a frequency of 5×10^{14} Hz.

Solution

Given frequency, $f = 5 \times 10^{14}$ Hz. (where $f = c/\lambda$).
 Using the photon energy formula, $E = h*f$.
 Plug in the values: $E = (6.626 \times 10^{-34}$ J·s$) * (5 \times 10^{14}$ Hz$) = 3.313 \times 10^{-19}$ J.

Example 3.2 Convert the energy calculated in Example 3.1 to electron volts (eV).

Solution

Given energy in joules, $E = 3.313 \times 10^{-19}$ J.
 Use the conversion factor: 1 eV $= 1.602 \times 10^{-19}$ J.
 Divide the energy in joules by the conversion factor: E (eV) $= (3.313 \times 10^{-19}$ J$)/(1.602 \times 10^{-19}$ J/eV$) \approx 2.07$ eV.

Example 3.3 Calculate the wavelength of a photon with energy 3 eV.

Solution

Given photon energy, E(eV) $= 3$ eV.
 Convert the energy to joules: $E = 3$ eV $* 1.602 \times 10^{-19}$ J/eV $\approx 4.806 \times 10^{-19}$ J.
 Use the wavelength formula, $\lambda = c/f$.
 $\lambda = (3 \times 10^8$ m/s$)/(E/h)$, where h is Planck's constant.
 Plug in values: $\lambda \approx (3 \times 10^8$ m/s$)/(4.806 \times 10^{-19}$ J$/6.626 \times 10^{-34}$ J·s$) \approx 4.497 \times 10^{-7}$ m or 449.7 nm.

Example 3.4 Calculate the kinetic energy of a photon with an energy of 4 eV striking a material with a work function of 2 eV.

Solution

Given photon energy, $E(\text{eV}) = 4$ eV.
Given work function, $\varphi(\text{eV}) = 2$ eV.
Use the kinetic energy formula, $K = E - \varphi$.
$K = (4 \text{ eV}) - (2 \text{ eV}) = 2$ eV.

3.1.3 A P–N Junction

Two types of semiconductors: *n*-type and *p*-type combined and manufactured, and then a *p–n* junction is formed due to diffusion of both the carriers (electrons and holes) from the higher concentration region to the lower concentration region as shown in Fig. 3.5. The holes which are majority carrier in *p* region are diffused from *p* to *n* and the electrons which are majority carrier in n region are diffused from *n* top. During the diffusion process, the hole and electrons recombine themselves at the junction and form a depletion layer. This depletion layer acts as a barrier for other mobile charge carriers which exist side by side with the depletion layer or depletion area. Now whenever an external electric field is put on the *p–n* junction with proper bias and whose strength is greater than the barrier potential, then the mobile charge carriers on each side will flow to conduct and finish the loop (Khan 2010).

If the battery's positive terminal is linked to the battery's negative terminal on the *p* side, to the *n* side of the *p-n* semiconductor, then the *p-n* diode is said to be in forward bias. If the battery's positive terminal is linked to the *n* side and its negative terminal to the *p* side of the *p-n* diode, and after, the diode is said to be in reverse bias. In forward bias the *p-n* diode allows high current and is in conduction and in reverse

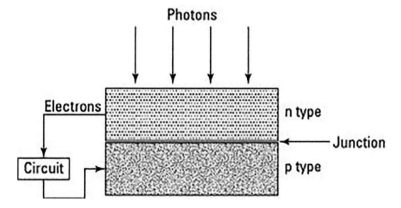

Fig. 3.5 A *p–n* junction solar diode

bias the diode blocks high reverse voltage and is in the off state. The external electric field is required to excite the electrons and holes to overcome the barrier potential and diffuse from one region to another region. Here the external electric field used is a battery and for a solar diode, the external field is the photoelectric field. The solar diode is a solar photovoltaic cell. If conduction is due to sunlight or photons then the conduction is called photoconduction (Augusto et al. 2017).

In a photodiode, the diode electric current flows from the diode to the load when there is a difference between the current produced by light and the normal diode current. This is given by the relation given as

$$I = I_L - I_D \tag{3.2}$$

where I is the electric current, I_L is the light-generated current and I_D is the diode current and given as

$$I_D = I_o\left[\exp(qV/KT) - 1\right] \tag{3.3}$$

where q is the electronic charge and I_0 is the saturation current which is calculated when the diode is reverse-biased and given by Eq. (3.4)

$$I_0 = (qD_h p - n_o/L_h) + \left(qD_e n_{po}/L_e\right), \tag{3.4}$$

where

$$D_e \text{ is the electron diffusion constant} = (KT/q) * \mu_e \tag{3.5}$$

$$D_h \text{ is the hole diffusion constant} = (KT/q) * \mu_h \tag{3.6}$$

$$L_e \text{ is the length of electron diffusion on } p \text{ side} = (D_e \tau_e)^{1/2} \tag{3.7}$$

$$L_h \text{ is the length of hole diffusion on } n \text{ side} = (D_h \tau_h)^{1/2} \tag{3.8}$$

$$p - n_o \text{ is the thermal equilibrium density of holes on the } n \text{ side} = n_i^2/N_A \tag{3.9}$$

$$n_{po} \text{ is the thermal equilibrium density of electrons on the } p \text{ side} = n_i^2/N_D, \tag{3.10}$$

where μ_e and μ_h are the mobility carrier constants for electrons and holes, respectively; $\tau_{e \text{ and }} \tau_h$ are the lifetimes of minority carrier holes on the n side and electrons on the p side, respectively.

3.1.4 Photovoltaic Effect

Earlier we have seen that the sunlight is comprised of tiny particles known as photons. These photons have energy and if this energy is more than the forbidden semiconductor material's gap energy then the electrons will excite and jump to the conduction band. Once the light falls on the semiconductor material or solar cell, then the photons strike with the mobile charge carriers and excite them to higher energy states within the material. These electrons or holes move from the semiconductor material to the load making the circuit a closed one. The excited electrons are driven by the generated potential difference or electromotive force from the solar light. This phenomenon was first analysed by French physicist Alexander E Becquerel in the year 1839 and is termed as photovoltaic effect (Augusto et al. 2017). The photovoltaic effect can be defined as the potential difference generated or the electric current generated in a material when it is exposed to sunlight.

3.1.5 Photovoltaic Cell Materials

In the year 1939 Russell Ohl built the first photovoltaic device by using a Si *p–n* junction diode. The photovoltaic cell material must need to work for a spectral range specifying the solar spectrum. The solar spectrum ranges from the infrared region to the ultraviolet region and it has non-uniform intensity. For maximum exposure to the sunlight the solar cells are wide-area devices. Conventional photovoltaic cells or solar cells are built with Si single crystal which has an efficiency of around 21 to 24% and also made of polycrystalline Si cells which have a productivity of 17 to 19%. The different types of photovoltaic cell materials are shown in Fig. 3.6. The effective solar cells are related to the band gap of the semiconductor material.

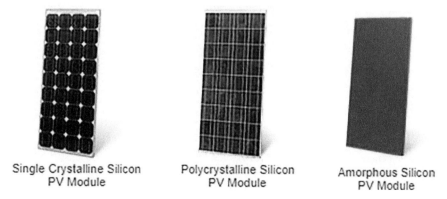

Fig. 3.6 Types of photovoltaic cell materials

Polycrystalline solar cells (Zweibel and Bernett 1993) have lower efficiency because of the introduction of defects in the cell material due to the microstructure but the manufacturing cost of the cell is cheaper compared to other cells. The amorphous silicon solar cells can be grown on glass substrates directly by techniques like glow discharge, and sputtering which makes the overall cost lower but the efficiency is also lowered. The efficiency can be improved by improving designs in the solar cells. Some of the improved design structures is an inverted pyramid structure made on the surface that enhances the absorption in passivated emitter rear locally diffused (PERL) cells, also known as passivated emitter rear locally dispersed cells.

3.2 Types of Photovoltaic Cells and Efficiency

3.2.1 Amorphous PV Cells

Among the most advanced solar cell technology is found in amorphous silicon (Carlson and Wagner 1993) and thin film solar cells. The diode structure formed in amorphous is a PIN type where p-layer and n-layer are separated by an internal electrical layer, i.e. i-layer which comprises amorphous silicon. I-layer thicknesses typically vary from 0.2 to 0.5 um due to their high absorption capacity. The amorphous silicon solar cell or a-Si solar cell is manufactured by glow discharge, evaporation, or sputtering. Small amounts of material are required in making a-Si cells because there must be tiny layers which are placed by glow discharge on glass or stainless steel surfaces. These cells have an efficiency of around 15% theoretically but 6–7% practically and thus their applications are in the range of low electric power and indoor applications. Some of those applications are pocket calculators, electronic watches, etc.

The crystal structure of a-Si is not predictable as shown in Fig. 3.7. Most of the covalent bonds in the structure are incomplete due to the randomness of atoms. The incomplete bond present in the crystal gives rise to a large number of equivalent impurity charge carriers which further bond with the mobile carriers. As a result, the number of mobile charge carriers is less compared to other silicon cells and thus it is considered a poor semiconductor material. To improve its performance the a-Si cell can be alloyed with other elements like hydrogen to produce a-Si: H material which has better properties compared to a-Si cell.

3.2.2 Monocrystalline PV Cells

The greatest efficiency solar cells of conversion are monocrystalline solar cells or single crystalline solar cells. The disadvantage of monocrystalline solar cells is the costly manufacturing due to the production of single crystalline silicon wafers. The

3.2 Types of Photovoltaic Cells and Efficiency

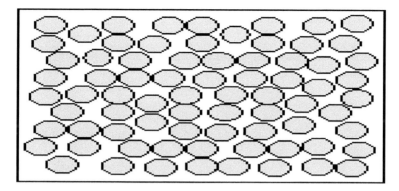

Fig. 3.7 Crystal structure of a-Si cell

manufacturing process begins with the extraction of silicon from the sand. The silicon extracted from the sand contains a large number of unwanted impurities that are referred to as metallurgical-grade silicon. Liquid compound trichlorosilane, $SiHCl_3$ is manufactured in the process of refining so that the purification of silicon is easier in a liquid state compared to a solid state. After the purified $SiHCl_3$ is obtained, it is mixed with H_2 and heated which gives the solid form of polysilicon and HCl. By the process of Czochralski, monocrystal is produced from the polysilicon. The monocrystalline silicon solar cells obtained as a circular bar are converted into wafers by the cutting process. A large amount of material is wasted during the cutting process. That's why the manufacturing process of monocrystalline PV cells is costly. The crystal design of a monocrystalline solar cell is shown in Fig. 3.8.

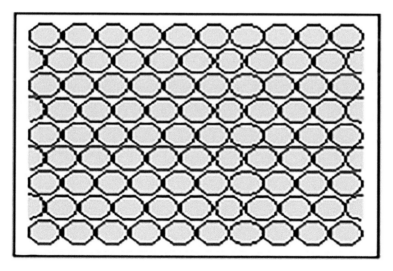

Fig. 3.8 Crystal structure of monocrystalline cell

3.2.3 Polycrystalline PV Cells

Solar cells with cost-efficient and less efficient than monocrystalline PV cells are polycrystalline solar PV cells. Polycrystalline silicon is generally used to prepare three categories of solar cell architecture, namely *p–n* junction cells, MIS (semiconductor with metal insulator) cells and semiconductor cells with conducting oxide insulation. Employing different techniques and depending on different purposes the different categories are manufactured. One of the common crystal structures for polycrystalline solar cells is shown in Fig. 3.9.

If the polycrystalline silicon film is deposited on a substrate like sapphire, graphite, metal, ceramic, glass, or metallurgical-grade silicon, then the cell prepared is called a *p–n* junction solar cell. The deposition can be done by vacuum evaporation or dipping. However the cell effectiveness of the *p–n* junction in a solar cell is low, so the concept of metal insulator semiconductor has come to light where an insulating layer is put between both the semiconductor and the metal. If the insulating layer is made thinner, then the voltage on the open circuit can be increased and be almost equal to that of the *p–n* solar cell. In the process of development, the insulators were being replaced by oxide layers but the result was overall a poor performance. A performance improvement strategy controlled oxide layer is used and a thin film oxide window is used as a window substrate on the active semiconductor substrate. This cell was known as a conducting oxide insulator semiconductor cell.

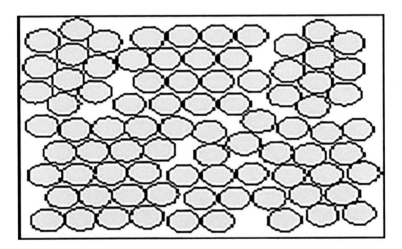

Fig. 3.9 Crystal structure of polycrystalline cell

3.3 Analytical Model of the Solar Cell

After developing a solar cell, one has to connect it to a circuit and check its real-time performance. Then the solar cell is given its rating accordingly. While manufacturing solar cells some predictions are made on the solar cell activity. The prediction of solar cells for various atmospheric and environmental conditions is possible by modelling solar cells. The model of a solar cell must be able to explain the behaviour of the solar cell, and its efficiency furthermore to get the P–V and I–V characteristic curves. For the simulation of solar cells, one should have prior knowledge of the analytical model of solar cells. The modelling of the solar cell acts as a significant element that influences the precision of the design of the solar cell. The representation of The PV system's nonlinear feature is possible by the designing of solar cells.

The common model approach for a solar PV cell is to connect a parallel current source that produces light with a p-n diode junction and then the load. Several models have been suggested for the model of a solar cell at various solar irradiance, and solar intensities as single, double, and triple diode designs, etc. The popular designs are single-diode models and the two-diode design. These two models deliver a superior understanding of the mathematical relations and maximum power point.

3.3.1 Analysis of the Single-Diode Model

The simplest is the single-diode model form of a solar photovoltaic cell where a source of current produced by light is linked in parallel with a single p–n junction diode (Garg and Prakash 2012).

The model shown in Fig. 3.10 is an ideal form of a solar cell with infinity shunt resistance and zero series resistance. Figure 3.11 shows the single-diode form with practical shunt resistance as well as series resistance. A current source serves as a representation of the light-generating diode and the light coming out from the source is represented by I_{ph} in the circuit and the diode current is shown by I_D. The output current flowing through the load is given by Eq. 3.2.

Equation 3.2 does not include the shunt resistance and series resistance, so the equation is rewritten as

$$I = I_{ph} - I_D - I_{sh}, \tag{3.11}$$

where

$$I_{sh} = (V + I R_s)/R_{sh}. \tag{3.12}$$

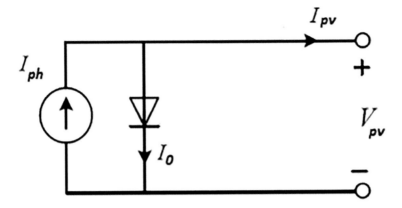

Fig. 3.10 Single-diode model with ideal state

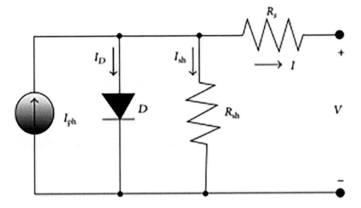

Fig. 3.11 Single-diode model with practical considerations

3.3.2 Analysis of the Two-Diode Model

The dual-diode design form is an updated version of the single-diode system where the current source is linked parallel to two diodes, where the second diode is taken into account of the recombination effect. The number of equations involved here will be increased making overall calculations more complex. Figure 3.12 shows the two-diode models where D_1 has the same role as the single diode present in the model with a single diode and diode D_2 is taken for recombination effects (IshaqueKashif and Hamed 2011).

Now the electric current flowing through the load from the PV cell is given as

$$I = I_{ph} - I_{d1} - I_{d2} - I_{sh}, \qquad (3.13)$$

3.4 Electrical Parameters of Solar Cell

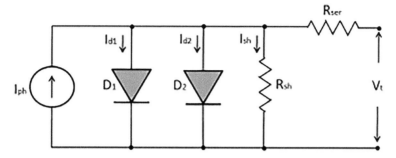

Fig. 3.12 Model of solar cell with two diodes

where I is the electric current, I_{ph} is the light-produced current; I_{d1} and I_{d2} are the diode currents; I_{sh} is the shunt current flowing through R_{sh} and is given as

$$I_{d1} = I_{o1}\left[\exp(qV/KT) - 1\right] \quad (3.14)$$

$$I_{d2} = I_{o2}\left[\exp(qV/KT) - 1\right], \quad (3.15)$$

where q represents the electrical charge, I_{01} and I_{02} are the saturation currents which are calculated when the diode is reverse-biased and given by the general Eq. (3.4) and

$$I_{sh} = (V_1 + I R_{ser})/R_{sh}. \quad (3.16)$$

3.4 Electrical Parameters of Solar Cell

Till now we have studied the solar cell's characteristics and also analysed the model of solar cell. To understand the cell's electrical properties, first, we have to understand some topologies related to the I-V properties of solar cells. Those topologies can be termed solar cell's electrical features. Those electrical specifications are open circuit voltage fill factor, short circuit current, and maximum power. These parameters are calculated and then electrical characteristics are drawn accordingly so that we can choose the efficient solar PV cells. These parameters help us to choose a suitable solar cell depending on our requirements. These parameters are discussed briefly in this section.

3.4.1 Current in a Short Circuit

Using a silicon p–n junction or a solar cell, the current in a short circuit is defined as the maximum possible current that flows via the solar cell when the output terminals of the solar cell are either cut short or there is no voltage across the cell. It is denoted by I_{sc} and in solar PV cell v-i features, it is the intercept on the y-axis when the voltage is zero, or when a circuit is shorted. The equation for short circuit current is given in Eq. 3.17 where V_{oc} is calculated from Eq. 3.18

$$I_{sc} = I_0\left[\exp\left((qV_{oc}/KT) - 1\right)\right]. \tag{3.17}$$

3.4.2 Voltage on an Open Circuit

In the sun cells the p-n junctions are made of silicon, the voltage in an open circuit is defined as the highest possible output of the solar cell's voltage when the solar cell's output terminals are open-circuited or the current through the cell becomes zero. It is denoted by V_{oc} and v-i features of a solar cell, it is the intercept on the x-axis when the current is zeroing, i.e. open-circuited. The voltage of an open circuit equation is given in Eq. 3.18 where I_{sc} is being calculated from Eq. 3.17. These two equations are interdependent.

$$V_{oc} = A_0(KT/q)\ln[(I_{sc}/I_0) + 1]. \tag{3.18}$$

3.4.3 Fill Factor

To define the solar cell's quality, we use the factor called fill factor which is indicated by the greatest power-to-volume ratio calculated (multiplication of the maximum current and voltage) to the computed theoretical peak power (multiplication of short circuit current and open circuit voltage) and is given mathematically by Eq. 3.19. It is denoted by FF. An ideal solar PV cell has an FF $= 1$ and a commercial solar PV cell has an FF in the range from 0.5 to 0.83.

$$FF = (V_m I_m)/(V_{oc} I_{sc}). \tag{3.19}$$

3.4.4 Maximum Power

In the solar cell *i-v* characteristics, for the short circuit condition the output power is zero due to zero voltage and the output power is also zero for the open circuit condition due to zero current. At the knee point of solar PV cell characteristics, the peak power can be obtained (Etienne et al. 2011) using the corresponding maximum voltage and maximum current, respectively. Mathematically the maximum power is given as the product of the maximum current and voltage of the cell, as given by Eq. 3.20

$$P_{\max} = V_m I_m. \tag{3.20}$$

3.4.5 Solar Cell Efficiency

The total solar power is not utilized in the transition procedure of energy conversion from solar to electrical. The amount of energy from the sun transformed into electricity with respect to total solar energy is given by the solar cell efficiency. Efficiency is characterized as the proportion of the solar cell's output energy to its input energy falling on the solar cell from the sun. To compare solar cells from one another the most widely used parameter is solar cell efficiency.

$$\eta = \frac{V_m I_m}{P_{\text{in}}}, \tag{3.21}$$

where

$$P_{\text{in}} = \int_0^\infty \left(\frac{hc}{\lambda}\right) f(\lambda) d\lambda. \tag{3.22}$$

3.5 Electrical Characteristics of PV Cells

From the knowledge of equivalent circuits and different parameters involved in the effectiveness of solar cells, we can now obtain the electrical features of solar cells. Certain electrical properties are not constant and they keep on changing with respect to temperature and solar radiation. The electrical characteristics of a PV cell are the I–V characteristics as well as P–V traits which are obtained from the single-diode design of the solar cell.

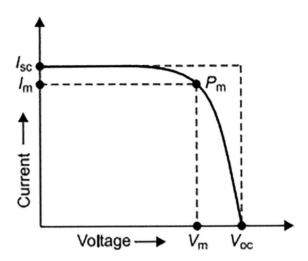

Fig. 3.13 PV cell I-V characteristics

3.5.1 PV Cell I-V Characteristics

The I-V characteristics for a PV cell are shown in Fig. 3.13 where I_{sc} and I_m represent the current in a short circuit and peak current of the solar cell, correspondingly. The notations V_{oc} and V_m represent open circuit voltage and peak of the solar cell's voltage, respectively. It can be seen from the curve that the power output short circuit current is zero at open circuit voltage. It tells that there is a condition for obtaining maximum output power in between these two instants viz. both an open and a short circuit condition. The corresponding maximum current and voltage output power are known as maximum current and maximum voltage, respectively.

3.5.2 P–V Characteristics of PV Cell

The solar cell's P–V properties are used for finding the maximum power point. The peak power point keeps varying with respect to the varying conditions. To monitor the highest power point, a number of algorithms are built. The base for all the algorithms is the solar P–V curve. The solar P–V curve shown in Fig. 3.14 has two slopes. The slope on the left-hand side is positive which shows a linear relation between voltage and the solar cell's power. If the output power lies within this slope, then the voltage is to be increased to obtain the maximum power point. The slope on the right side is a negative slope which shows the inverse proportionality relation between peak power and voltage. If the power output lies within this slope, then the corresponding voltage is to be decreased to acquire the maximum power point.

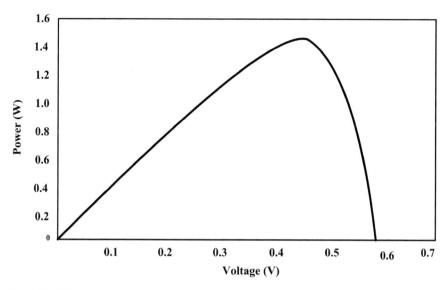

Fig. 3.14 P–V characteristics of solar PV cell

3.6 Maximum Power Point Tracking (MPPT)

The I-V characteristics may be shown to be different for different solar insulations and temperatures. If the I-V characteristics change from time to time then the respective P–V characteristic curves also change and so as the peak power point. A solar cell's peak power point is shown in Fig. 3.15. A solar cell's efficiency is stated to be best if the output power from the solar cell is equivalent to the maximum power point (Etienne et al. 2011). If the highest power is to be removed from the solar cell, then the load must adjust itself accordingly, either mechanically changing the position of the panel with respect to the sun or electrically tracking the operating point by changing the load. The process of changing load to maximize the output inside a solar cell changing situations of temperature, irradiance and insolation is called peak tracking for power point. An electronic device designed to monitor the highest power point with respect to changing load or environmental conditions is called a peak power point tracker or peak power-for-point monitoring system (MPPT). It is always interconnected with the PV system and load and acts as a feedback system making the whole PV system a closed-loop control system.

The tracker is able to monitor the highest power point using pre-described algorithm within it. Those algorithms are known as MPPT algorithms and they can be the P&O (perturb and observe) method, IC (the methods of (incremental conductance), temperature, current sweep, etc. Among the entire algorithms, P&O method algorithm is popular due to the easier calculations and simple design compared to other algorithms. The percentage of error is more in the P&O method.

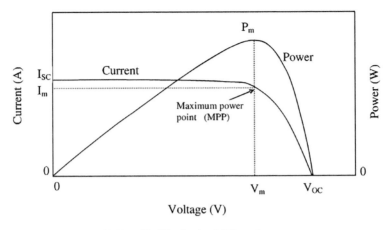

Fig. 3.15 Characteristics of I-V and P–V indicating MPP on the curve

3.7 Effect of Parameters and Atmospheric Conditions on PV Cell Characteristics

The solar intensity is not the same for all the days, weeks, months and years because of the elliptical the earth's orbit around the sun. the radiation from the sun that the planet receives is assumed to be constant and is given by I_{sc}, solar constant = 1367 W/m^2. It is taken to be constant for the calculation of global radiation, diffuse radiation, beam radiation, hourly radiation, etc. In a practical application view, the solar radiation falling on PV cells or the earth is not the same. The solar irradiance also depends on the geographical topology of the place or where the PV system is located or solar farm is to be set up (Bayrak et al. 2017). The parameters that influence the performance are solar irradiance in solar cells and temperature. The result of solar radiance on the solar PV features is shown in Fig. 3.16.

Shading is also a phenomenon that affects the solar cell characteristics. Sometimes the birds or any other creature will be sitting on the solar PV panel resulting in partial shading. This partial shading degrades the performance of the solar PV panel like that in Fig. 3.17

3.8 Photovoltaic Modules and Array

Solar cell or photovoltaic cell is the structure block of the photovoltaic system. Several solar cells are wired together in parallel or sequence to form modules whereas some sections are combined to form a PV panel and a number of panels are related to one another in sequence and parallel to form an array (Fig. 3.18). Solar cells individually provide very low electric power but when combined to form a module the output

3.8 Photovoltaic Modules and Array

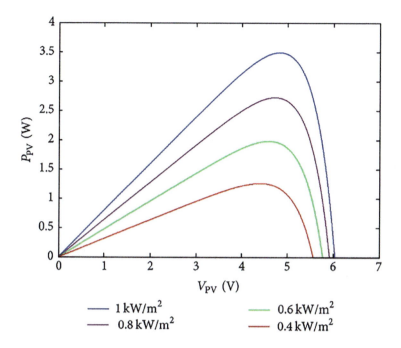

Fig. 3.16 Effect of irradiance on PV cell characteristics

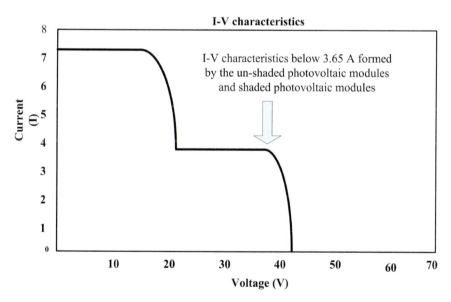

Fig. 3.17 Effect of partial shading on PV cell characteristics

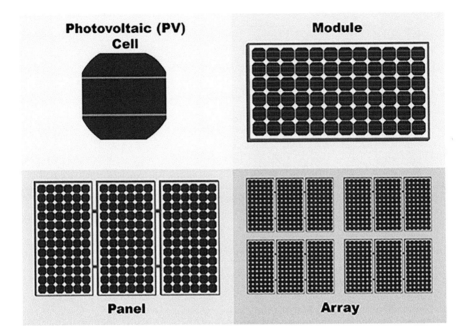

Fig. 3.18 Photovoltaic cell, module, panel, and array

power increases from a few millivolts to a few volts. For higher voltages, the modules are connected to form panels and arrays.

3.8.1 Theory and Construction

The working of solar cells is possible due to the photovoltaic effect in solar diodes as described in previous sections. The solar cell may be an a-Si cell, a monocrystalline cell, or a polycrystalline cell. When the solar light falls on the solar cell then due to the photon energy the diode current passage from a diode to the load. The output voltage for a single cell is in the range of μV to mV. The solar cells are either linked in series or parallel to improve the output voltage. For example, if 12 V of solar module has 24 solar cells in a series, then 24 V of the solar module will have 48 solar cells in a series (Cucchiella et al. 2017). Similarly, for higher voltages, solar arrays are constructed by connecting a number of solar modules in sequence or parallel. In series, solar cells are linked then the current through them will be the same and if they are parallel linked, the voltage across the solar cells will be the same.

The shape of a solar cell can be circular, semisquare or square depending on the type of cell. The monocrystalline solar cells are usually round or semisquare and the polycrystalline silicon wafers are usually square-shaped. The solar cells are interconnected and are encapsulated into a single stable unit. Encapsulating is primarily done to protect the electrically connected solar cells from the harsh environment in which they are placed and also from the water vapours or vapour present in the atmosphere. If solar cells are not encapsulated, then they are prone to mechanical damage because the solar cells are relatively thin.

The construction of solar PV modules or solar PV arrays includes bypass diodes. The bypass diodes are used to prevent hot-damaging spots consequences of heating. The hotspot heating occurs if a malfunctioning solar cell or a bad cell is present among the proper solar cells in a module. During forward bias, the current flows through the short circuit current in a solar cell. The short circuit current for the improper solar cell is low compared to the proper solar cell. If the operating current of the series-connected solar cells becomes equivalent to the short circuit current of the defective solar cell, then the total current is restricted by the defective solar cell. The additional current generated by the proper solar cells the forward-biased unshaded proper solar cells and the forward bias across all of these cells will be zero if the series string is shorted. Reverse skews the coloured cells. When several series-connected cells result in a significant backward bias across the shaded cell, a hotspot heating occurs. To prevent the reverse bias a bypass diode is linked to anti-parallel with the solar diode and therefore hotspot heating is prevented.

3.8.2 Packing Factor of PV Module

The individual solar cells are connected electrically with one another for the construction of solar PV modules. To protect the solar PV module from environmental conditions and to avoid electrical shocks from the module, the module is packaged. Once the solar cells are packed to form a solar module, the packing factor describes the output power of the module and the operating temperature of the module. The packing factor sometimes referred to as packing density is defined as the ratio of the area of the module covered with the solar cells to the total area of the module. It is a comparison of the area of the module filled with solar cells with the area of the module left blank. If the solar cells are round and packed, then the packing factor will be less or the space left in the module is more when compared to the solar cells with a square shape where the packing factor is more or the space left in the module is less.

3.8.3 Efficiency of PV Module

The PV modules or PV arrays have so many effects. The important effects are the losses due to the joining of incompatible solar cells, the temperature of solar cells, and the failure modes of PV modules. The efficiency of the PV module is different from the calculated solar cell efficiency. The solar cell efficiency describes the volume of solar light collapsing upon the converted cell into utilizable electricity. The module efficiency defines the percentage of sunlight falling on the overall panel that is converted into utilizable electricity. The higher the efficiency rating, the lesser the number of solar cells required to make the method to achieve the goal output. The module efficiency ranges from 15 to 18%.

3.9 Overview of Photovoltaic System Applications

The photovoltaic system will have vast applications in future generations in terms of electricity generation, electric vehicles, etc. The photovoltaic system is used as power-based space satellites where the ultimate energy source is sun. Photovoltaic power systems have important applications as grid-connected and standalone PV systems. Photovoltaic thermal hybrid solar collectors, telecommunication and signalling, and rural electrification are major applications of photovoltaic systems.

3.10 Overview of Photovoltaic-Based Power System

The photovoltaic-based power system can be connected to the electric grid and provided to the large number of customers or it can be connected to individuals as a standalone system as a backup plan in case of a power outage. The photovoltaic-based power system has a special interest in solar power satellites. Standalone systems are not linked to the power grid and are virtually self-sufficient, have one backup system and require no maintenance or regular fuel. In grid-connected systems, the solar PV array is a DG and supplies power to the load when there is sufficient sunlight and the grid supplies the power to the load when the sunlight is not enough.

3.10.1 Standalone Photovoltaic System

The standalone PV system is a photovoltaic power system which can be easily installed by a customer in his locality or by a group of consumers. It is similar to conventional DG which acts as a backup plan during power outages, i.e. a standalone diesel generator for a building. With the emerging trend, the customer is in the interest of low-cost electricity utilization. Although the installation cost of a standalone solar PV system may be expensive the maintenance cost is very low and durability is more. During the day time the load can be directly connected to the solar PV panel through an inverter and during the night time the stored energy can be utilized and is connected as shown in Fig. 3.19.

3.10 Overview of Photovoltaic-Based Power System

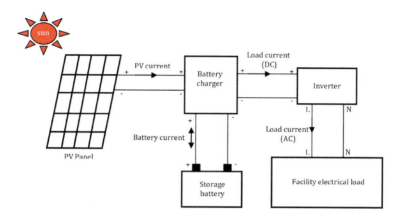

Fig. 3.19 Standalone photovoltaic system

3.10.2 Grid-Connected Photovoltaic System

Sometimes a consumer or a group of consumers are also interested in saving money by supplying power to the grid. Grid-connected photovoltaic system does the same job by supplying power to the grid and the customer benefits from the utility grid services. It can be a consumer or other electric companies which can support the government's electric generation and distribution units by providing solar power to the grid at reasonable prices with effective efficiency. To check the strength of power supplied to the grid and the strength of power utilized by the customer, a utility metre is linked between the inverter and grid as shown in Fig. 3.20.

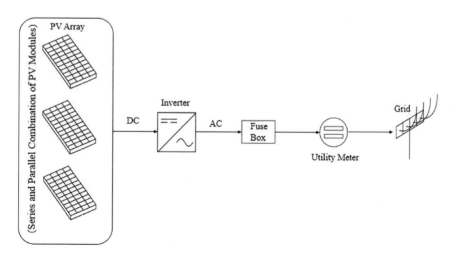

Fig. 3.20 Grid-connected photovoltaic system

3.11 Power Converter Topologies for PV-Based Power System

The output of the conventional energy sources is an AC. The electricity is transverse from the grid to the consumer location by means of transmission lines. The power levels at different substations and different consumers are changed by utilizing transformers. The output of PV-based energy sources is a DC output and this output is to be integrated with the existing grid. The integration involves the conversion of unregulated DC to regulated DC and DC to AC. After the transition of DC into AC, the converted AC is integrated into the distribution grid and the solar-based energy source is used as DG. The conversion of unregulated DC to regulated DC is known as switch mode regulator and DC to AC conversion is called inverter. In a switch mode regulator, the control is achieved with the use of pulse width modulation at a set frequency and by power electronics switching devices like BJT, MOSFET or IGBT. An inverter is used in standby power supplies and uninterruptible power supplies. To optimize the output of arrays and safeguard different electric components from harm, solar PV systems need a variety of controls. Electricity is managed and regulated using power conditioners. A blocking diode, a voltage regulator, and an inverter or converter often make up a power conditioner. When there is no sunshine, the blocking diode stops current from spilling back into the array when the system is not producing power. The voltage controller safeguards the batteries and keeps the voltage consistent across the load.

3.11.1 DC/DC Converters

A Converter DC/DC changes the DC voltage of one stage to the required DC output voltage stage. The output of a solar cell is an unregulated DC and if it is required to store it in batteries then we require a DC-DC converter. The converters are mostly used in solar batteries. The batteries need a regulated input voltage to store energy, so the output from the solar converter connects the PV system to the battery. The modern converters contain the power semiconductor switches like MOSFET, transistors and IGBTs and high frequency-based choppers are used. The maximum power point tracker, the array always runs at its peak power point regardless of changes in solar insolation, ambient temperature, or load thanks to a unique converter. To maintain the array voltage at or close to the maximum power point, an MPPT incorporates internal control logic. They are controlled by microprocessors, which sense and record the voltage and current of the array at regular intervals to calculate and modify the power output.

3.11.2 DC/AC Converters

A DC/AC converter usually known as an inverter converts the DC input voltage into the required AC voltage output. The voltage of the output AC can be single phase or three phases depending on the requirement. The solar PV array's inverter transforms the DC to electricity or from the solar battery to single-phase or three-phase AC supply appropriate for AC loads. In terms of voltage, frequency, and harmonics clarity of the pulse for the grid-interactive systems, the output must satisfy the essential standards of the electrical authority. These are done by the additional transformers and special filtering, respectively. In standalone systems, the inverter for PV systems should automatically turn off if the array output voltage is too high or too low. Whenever the output voltage of the array falls within the specified range. The inverter must automatically restart. The overall circuit must be provided with protection against short circuits and overloading.

3.12 Control of Photovoltaic-Based Power Systems

The output power from a photovoltaic-based power system should be maintained constant, and it is possible with certain controllers such as the MPPT controller, control of the DC/DC converter, and the inverter. The control of MPPT is done with a pre-defined algorithm where the instantaneous current and instantaneous voltage are measured and compared with the reference current and voltages, respectively, and the corresponding error signal is sent to the MPPT as shown in Fig. 3.21. The DC/DC converter has a controller which changes the duty cycle ratio accordingly with respect to the work of the converter in the PV system.

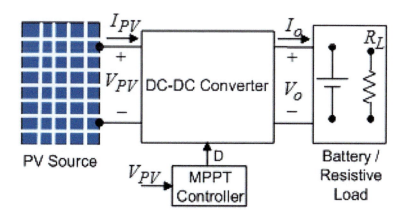

Fig. 3.21 A PV-based power system with MPPT controller

3.12.1 Maximum Power Point Tracking (MPPT) Control

MPPT is a digital electronic tracking with a charge controller in it. The voltage regulator views the output from the panel voltage and relates it with the batteries' voltage. It then estimates the highest output the panel is capable of producing to charge the battery. It takes the estimated value and changes it to the appropriate voltage to allow the battery to receive the most current. Modern MPPT controllers are having efficiency of around 93–96% in this conversion (Etienne et al. 2011). The power gain obtained from the panel varies with respect to the seasons. During winter seasons the power gain is typically 20–45% and during summer season the average power gain is typically 10–15%. The actual gain is affected by the weather, temperature, battery state of charge, and other factors.

3.12.2 DC/DC Converter Control

The control of the DC/DC converter is a crucial component as well to be studied. The DC/DC converter's output must be maintained constant for energy storage in the battery. For this purpose, the converter is provided with a feedback system. The DC/DC converter provides a feed link to the photovoltaic array. The PV array has its own I-V characteristic that depends on the illumination, temperature, and other factors. The DC-DC converter has some I-V characteristics, and the I-V characteristic of the system depends on the feed link provided by the DC/DC converter. The intersection of the I-V characteristics of the array and the DC/DC converter defines the system's operating point. For the desirable operating point or often maximum power point, the DC/DC converter I-V characteristics can be adjusted accordingly by including a control loop for the converter (Network and for the 21st century (REN21) 2010; Palz 2013). The connection diagram of the DC/DC converter has been shown in Fig. 3.22.

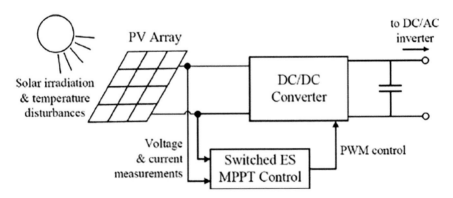

Fig. 3.22 Position of DC/DC converter in a PV system

3.12 Control of Photovoltaic-Based Power Systems

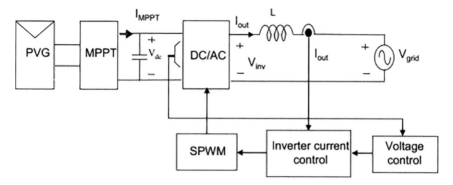

Fig. 3.23 A DC/AC inverter control in PV system

3.12.3 DC/AC Inverter Control

The output DC power cannot be utilized directly for the home appliances and the DC is converted into a single-phase AC supply for the home appliances and other loads. This purpose can be fulfilled with the help of the device used to convert DC to AC. The output AC should be controlled by feedback. In the DC/DC converter controller, the intended value may be compared to the feed-backed DC bus voltage or reference value but for the DC/AC converter a proportional integrator (PI) the reference current signal may be generated by adding a regulator to control the error between the target voltage and the actual DC bus voltage. To improve the performance of the PI controller used in the system shown in Fig. 3.23 the ability of the boost control loop, inverter, and DC bus capacitor to respond quickly enough to cancel the voltage ripples caused by changes in the instantaneous power flow through the solar system will rely on changes in the irradiance and temperature of the surrounding atmosphere. On the other hand, at 50 Hz, the output voltage (also known as the mains voltage) constitutes a significant external disturbance for the system.

Review Questions

(1) What are the types of materials according to electrical conductivity?
(2) Explain intrinsic and extrinsic semiconductors with suitable diagrams.
(3) Explain how the solar p–n junction diode works with a suitable diagram.
(4) How photon energy is used in explaining the working of solar diodes?
(5) How photoelectric effect is related to the solar PV cell?
(6) Why silicon is used mostly in the manufacturing of photovoltaic cell materials?
(7) What are the types of photovoltaic cell materials? Explain.
(8) Why analysis of two-diode model is required in spite of single-diode model? Explain.
(9) What are the different electrical parameters involved in solar cell characteristics?
(10) Explain the characteristics of solar cells in terms of I-V and P–V.

130   3 Introduction to Photovoltaic Solar Energy

(11) Why MPPT is an essential factor for the performance of solar cells?

(12) How standalone PV system is better than a conventional standalone DG?

(13) Is grid-connected photovoltaic system necessary? If yes then why?

(14) Explain the converter topologies used in solar-based power systems.

(15) Explain the DC/AC converter control in a photovoltaic-based power system.

(16) Explain the applications of photovoltaic systems in different sectors.

17) Why DC/DC inverter control is also an important aspect of a photovoltaic-based power system?

18) Explain the concept of photon energy in solar cells.

19) Explain the theory and construction of photovoltaic modules and arrays.

Problems

(1)

 (a) In a PV system, 120 solar modules, each of 200 W_p and an area of 1.67 m^2 are connected. The efficiency of the system is 0.70, as well as the typical yearly solar radiation, is 1487 kWh/m^2. Determine the projected yearly energy production.

   Ans: 33,457 kWh

   Ans: 36,802 kWh

 (b) What would the power output be if the correction factor for the sun's inclination and orientation was 1.1?

(2) When a solar cell's saturation current is 1.7 × 10–8 A/m^2, the temperature of the cell is 27 °C, and the short circuit current density is 250 A/m^2, determine the open circuit voltage, V_{oc}; voltage at maximum power, V_{max}; current density at maximum power, I_{max}; maximum power, P_{max}; and maximum efficiency, η_{max}. In a situation where there is 820 W/m^2 of solar energy available, what cell area is needed to produce 20 W?

  Ans: 0.605V, 0.47, 237 A/m^2, 111.4 W/m^2, 13.58 %, 0.18 m^2

(3) 250 W at 24 V must be produced by a PV system. Create the PV panel with the solar cells from question 2 operating at their peak power, assuming that each cell has an area of 9 cm^2.

  Ans: No. of cells = 2500, No. of cells in series = 46, No. of cells in parallel = 55

(4) A PV cell has an open circuit voltage of 0.6 V and a short circuit current of 250 A/m^2 when the temperature of the cell is 40 °C. Determine the voltage and current density which maximize the cell power and also find the maximum output power per unit cell area.

  Ans: 0.519 V, 237.6 A/m^2, 123.3 W/m^2

(5) A silicon cell at the temperature of 40 degrees centigrade has a dark current density of 3.6 × 10^{-8} A/m^2 and a short circuit current density is 220 A/m^2. Determine the voltage and current density which increase the cell power to

3.12 Control of Photovoltaic-Based Power Systems 131

its peak. Also, determine the maximum output power of a unit cell when the value of global solar irradiance is 850 W/m^2 and the cell area needed if output requirement is 36 W.

Ans: 0.5265 V, 209 A/m^2, 110.2 W, 0.327 m^2

Objective Type Questions

(1) The term photovoltaic is derived from _____

    (a) Greek
    (b) English
    (c) Latin
    (d) Spanish

(2) Solar cells cannot be used in calculators.

    (a) False
    (b) True

(3) The solar cell's current output may be provided by_____

    (a) $I_L + I_D + I_{Sh}$
    (b) $I_L - I_D + I_{Sh}$
    (c) $I_L - I_D * I_{Sh}$
    (d) $I_L - I_D - I_{Sh}$

(4) The area across the p–n junction diode where the holes and electrons dispersed is known as_____

    (a) Depletion Junction
    (b) Depletion point
    (c) Barrier potential
    (d) Depletion region

(5) A modest increase in increases the quantity of photogenerated current _____
____

    (a) Diode current
    (b) Shunt current
    (c) Temperature
    (d) Photons

(6) _____ is one of the most crucial components, also known as silicon of solar quality.

    (a) Powdered silicon
    (b) Silicon
    (c) Crushed silicon
    (d) Crystalline silicon

132   3 Introduction to Photovoltaic Solar Energy

(7) Materials in large quantities are sliced into wafers with a thickness of _____
_____.

(a) 120–180 μm
(b) 180–200 μm
(c) 180–220 μm
(d) 120–240 μm

(8) In a solar panel, a module is a

(a) Series and parallel arrangement of solar cells.
(b) Parallel arrangement of solar cells.
(c) Series arrangement of solar cells.
(d) None of the above.

(9) The solar cell's efficiency is about

(a) 40%
(b) 60%
(c) 25%
(d) 15%

(10) The energy source for satellites is

(a) Edison cells
(b) Cryogenic storage
(c) Solar cell
(d) Fuel cells

(11) The solar cell's output is in the range of

(a) 5.0 W
(b) 10.25 W
(c) 0.5 W
(d) 1.0 W

(12) What may a solar array produce at its highest level?

(a) 250 W/m^2
(b) 500 W/m^2
(c) 300 W/m^2
(d) 100 W/m^2

(13) A photovoltaic cell's current density varies from

(a) 20–40 mA/cm^2
(b) 60–100 mA/cm^2
(c) 10–20 mA/cm^2
(d) 40–50 mA/cm^2

(14) A solar cell's typical output is

3.12 Control of Photovoltaic-Based Power Systems

133

    (a) 0.1 V

    (b) 0.26 V

    (c) 1.1 V

    (d) 2 V

(15) What kind of substance is utilized in solar cells?

    (a) Barium

    (b) Silicon

    (c) Silver

    (d) Selenium

(16) The primary use of solar cells is to produce energy from_____

    (a) Wind

    (b) Biomass

    (c) Water

    (d) Sunlight

(17) Electrical connections to solar photovoltaic cells are often made in _____

    (a) Parallel

    (b) Series

    (c) Randomly

    (d) Neither series nor parallel

(18) Solar power is a good _____ renewable source.

    (a) Economical

    (b) Commercial

    (c) Commercial and economical

    (d) Neither commercial nor economical

19) When sunlight is not the source of the light then the photovoltaic cell is used as _____

    (a) Phototransmitter

    (b) Solar cell

    (c) Photodetector

    (d) Photodiode

Answers

1. a	2. a	3. b	4. d	5. c	6. d	7. d
8. a	9. d	10. c	11. d	12. a	13. d	14. b
15. b	16. d	17. b	18. c	19. c		

References

Augusto A, Herasimenka SY, King RR, Bowden SG, Honsberg C (2017) Analysis of the recombination mechanisms of a silicon solar cell with low bandgao voltage offset. J Appl Phys 121(20):205704

Bayrak F, Ertürk G, Oztop HF (2017) Effects of partial shading on energy and exergy efficiencies for photovoltaic panels. J Clean Prod 164:58–69

Carlson DE, Wagner S (1993) Amorphous silicon photovoltaic systems. In: Burnham L (ed) Chapter 9 of renewable energy, Island press, Washington, D.C.

Cucchiella F, D'Adamo I, Gastaldi M (2017) Economic analysis of a photovoltaic system: a resource for residential households. Energies 10:814

Duffie JA, Beckman WA (2006) In: Solar engineering of thermal processes. 3rd edn Wiley & Sons, INC.

Etienne S, Alberto T, Mikhaïl S (2011) Explicit model of photovoltaic panels to determine voltages and currents at the maximum power point. Sol Energy 85(5):713–22

Garg HP, Prakash J (2012) Solar energy fundamentals and applications, Tata Mcgraw- Hill education private limited New Delhi, First revised Edition

IshaqueKashif SZ, Hamed T (2011) Simple, fast and accurate two diode model for photovoltaic modules. Sol Energy Mater Sol Cells 95(2):586–594

Kabir E, Kumar P, Kumar S, Adelodun AA, Kim K (2018) Solar energy: potential and future prospects. Renew Sustain Energy Rev 82:894–900

Kannan N, Vakeesan D (2016) Solar energy for future world: a review. Renew Sustain Energy Rev 62:1092–1105

Khan BH (2010) Non-conventional energy resources, Tata Mcgraw Hill education private limited New Delhi, 2nd edn

Renewable Energy Policy Network for the 21st century (REN21) (2010) Renewables 2010 Global Status Report, Paris, pp 1–80

Palz W (2013) In: Solar power for the World: what you wanted to know about photovoltaics. CRC Press. pp 131. ISBN 978-981-4411-87-5

Zweibel K, Bernett AM (1993) Polycrystalline thin film photovoltaics. In: Burnham L (ed) Chapter 10 of renewable energy, Island press, Washington, D.C.

Chapter 4
Introduction to Wind Energy

Abstract This chapter gives an overview of wind energy, beginning with a study of wind as a resource that covers its properties and regional variations. It goes into more detail about the analysis of wind data, wind speed distribution, average wind speed, and statistical analysis utilizing the Weibull and Rayleigh distributions. As a whole it presents comprehensive overview of wind energy providing into its technological component's performance characteristics and resource potential.

Keywords Wind energy · Wind data · Power coefficients · Tip speed ratio · Wind power · Coefficient · Betz's Law

4.1 Wind—The Resource

Wind power is one of the three major renewable energy resources, alongside solar power and hydropower, that are being exploited on a large scale for global power generation. As an energy resource, wind is widely distributed and is capable of providing power in most parts of the world but it is both intermittent and unpredictable, making it difficult to rely solely on wind for electrical power. When used in conjunction with other forms of generation however, or in combination with energy storage, wind can make a valuable contribution to the global energy balance. According to the International Energy Agency, the contribution of wind energy to total electricity generation, worldwide rose from 342,203 GWh in 2010 to 1,864,068 GWh in 2021, making it the second most important renewable energy source after hydropower.

4.1.1 The Nature of Wind

As wind power has a cubic relationship with wind speed, learning the unique features of the wind resource is essential for all aspects of utilizing wind energy, from

© The Author(s), under exclusive license to Springer Nature Singapore Pte Ltd. 2024
K. Namrata et al., *Wind and Solar Energy Systems*, Energy Systems in Electrical Engineering, https://doi.org/10.1007/978-981-99-9710-7_4

finding suitable locations and forecasting the commercial sustainability of wind farm improvements to developing the wind turbines themselves and being able to understand their impact on electricity distribution systems and users.

From the perspective of energy stored in the wind, the most noticeable attribute of the wind resource is its inconstancy. The wind intensity is always profoundly fluctuating, both geologically also, temporally. Moreover, this diversity endures a very wide range of spatial and temporal dimensions. The cube of the available energy multiplies the significance of this fact.

For a large scale, spatial inconstancy portrays the way that there is a wide range of climatic zones on the planet, some a lot windier than others. These areas or zones are to a great extent directed by the latitude, which influences the measure of insolation. Inside any one climatic zone, there is a lot of fluctuation on a smaller scale, to a great extent directed by physical geology—the extent of ocean and land, the size of landmasses, and the presence of mountains or fields for instance. The kind of vegetation may likewise have a critical impact through its consequences for the assimilation or impression of solar-powered radiation, influencing surface temperatures, and moistness.

The influence of topography has a very remarkable impact on wind nature. On top of hills and mountains, the wind is felt more intensely than, say, in the lee of high terrain or sheltered valleys. There are obstacles like trees or buildings that decrease the wind speed considerably.

At a particular area, the instability for a wide scale implies that the wind may differ starting with one year and then onto the next, with significantly larger scale varieties over times of decades or more. These changes over the prolonged period are not well known and may make it hard to make precise forecasts of the financial feasibility of specific wind farm ventures, for example.

If the period is shorter than a year, the natural climatic changes are more foreseeable, but there are still significant fluctuations in shorter time scales, which are also not very reliable more than a few days but fairly well known. These "synoptic" changes are related to the section on climate frameworks. Contingent upon area, there may likewise be significant varieties with the hour of the day (diurnal varieties) which again are typically genuinely unsurprising. On these time scales, the consistency of the wind is significant for incorporating huge sums of wind power into the power organization, to permit the other producing plant providing the organization to be properly organized.

The layout and efficiency of the particular wind turbines, the quality of the electricity transmitted to the network, and their impact on customers can all be significantly impacted by turbulence in further shorter time frames of minutes to seconds or even less (Burton et al. 2021).

Using long- and short-term observations from Brookhaven, New York, Van der Hoven (1957) created a wind speed spectrum that displayed distinct peaks correlating to the synoptic, diurnal, and turbulent impacts mentioned in Fig. 4.1 (Van der Hoven 1957). The so-called spectral gap between diurnal and turbulent peaks, which

4.1 Wind—The Resource

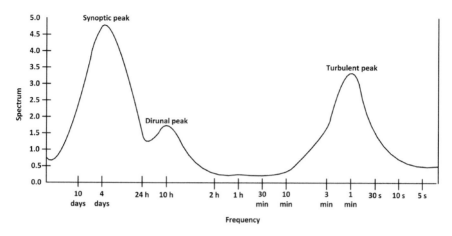

Fig. 4.1 Wind spectrum farm (Brookhaven) (Van der Hoven 1957)

demonstrates that the synoptic and diurnal changes may be viewed as completely different from the higher-frequency oscillations of turbulence, is of special importance. In the area between 2 h and 10 min, relatively little energy is available in the spectrum.

4.1.2 Geographical Variations in the Wind

Eventually, the winds are driven for the most part by the sun's energy, causing differential surface warming. The warming is generally extreme on part of the landmasses which are nearer to the equator; also, the warming phenomenon happens more during the day period, which implies that the locale of most prominent warming moves around the world's surface as it turns on its pivot. Owing to the earth's rotation, the resulting large motion of the wind is greatly impacted by Coriolis forces. The effect is a trend of the global circulation on a wide scale. Certain recognizable highlights of this, for example, the exchange winds and the "roaring forties" are notable.

The irregularities of our earth's surface, which comprises landmasses and oceans, guarantee that the minor instability on continental scales creates trouble for the global circulation format. In a highly dynamic and nonlinear manner, these differences interact and deliver a very messy outcome, which is the main cause of the uncertainty of the atmosphere in specific regions. Unmistakably, however, basic tendencies remain which lead to clear climatic contrasts between locales. These distinctions are tempered by more neighbourhood geographical and thermal impacts (Burton et al. 2021).

Hills and mountains result in higher wind intensity for local communities. This is partially an outcome of the boundary layer of the earth suggesting that wind speed typically rises above ground with height, and hilltops and mountain peaks will "project" into the higher levels of wind speed. It is additionally mostly a consequence of the speeding up of the wind stream over and around hills and mountains and channelling through passes or along valleys lined up with the stream. Similarly, topography may create zones of decreased wind strength, for example, sheltered valleys, areas on a mountain ridge's lee, or where airflow fashion leads to points of stagnation (Hartmann 2016).

A significant local difference can also arise due to temperature changes. Because of the unequal heating between land masses and sea, coastal areas are regularly getting intense wind flow. Despite the fact that the water is warmer than the land, a localized circulation forms wherein surface air moves from the ground to the ocean, with hot air flowing over the ocean and cool air sinking over the land. This process gets reversed as the land becomes hotter than the sea. Because the land warms and cools faster than the ocean surface, the processes of land and ocean breezes will often be opposing throughout a 24-h cycle. These effects played a big role in the early development of wind energy in California, where a sea fluctuation sends cold water to the shore not far from desert areas that warm up significantly throughout the day. An interceding mountain extends pipes the subsequent wind current through its passes, producing locally intense and stable airflow (which are well associated with fluctuations induced by air-conditioning loads in the nearby demand for power). The variation in the altitudes can also be the reason for the thermal imbalance. This allows chilly air from mountainous areas to drop to the fields below, producing particularly strong and well-defined "downslope" winds.

Warming of the air by the exchange of reasonable heat energy from the landmasses is moderately little. The transfer of latent heat of vaporization from the landmasses, which is later converted to sensible heat during precipitation, is the main effort to reduce environmental radiative loss from the environment. The addition of latent heat energy has a very unique structure compared to radiative cooling, with latitude peaking at approximately $140 \, Wm^2$ in the tropics and decreasing to almost zero in high latitudes. The latitudinal nature of the transition differences in environmental vitality reflects the latitudinal structure of evaporation. Environmental movements trade around $50 \, Wm^2$ from the tropical district and import around $85 \, Wm^2$ into the polar locales. One of the major climatic influences on the general flow of the air is this poleward movement of energy by the environment. Primary and secondary storm courses, as well as the worldwide distribution of mean sea level pressure (in millibars) for January, are all depicted below in Fig. 4.2, along with the basic characteristics of the world's winds (Meserve 1978).

4.2 Worldwide Status of Wind Power

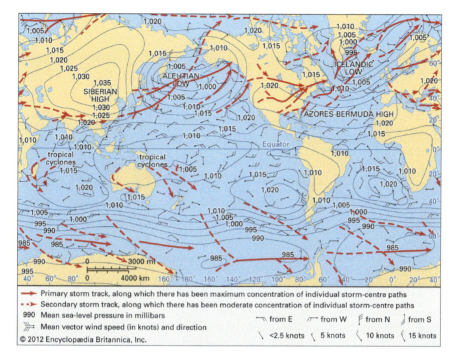

Fig. 4.2 Global winds scenario (*source* Wikipedia) (Meserve 1978)

4.2 Worldwide Status of Wind Power

4.2.1 Global Wind Power Statistics

As per IEA in 2022 wind electricity generation increased by a record 265 TWh (up 14%), reaching more than 2100 TWh as shown in Fig. 4.3 (Global Wind Energy Council 2022). This was the second-highest growth among all renewable power technologies, behind solar PV. However, to get on track with the net zero emissions by 2050 Scenario, which envisages approximately 7400 TWh of wind electricity generation in 2030, the average annual generation growth rate needs to increase to about 17%. Achieving this will require increasing annual capacity additions from about 75 GW in 2022 to 350 GW in 2030. So, more effort is required to enhance the installation capacity because still there are lots of wind energy both onshore and offshore those needs to be extracted.

Onshore wind market new installations decreased to 72.5 GW last year, although it was still the second-largest year ever. With more than 21 GW of grid-connected power, a record-breaking three times increase over the year before, the offshore wind sector experienced its best year yet in 2021. As a result of the astonishing increase in installations in China (offshore) and Vietnam, Asia Pacific has maintained its position as the region with the fastest-growing wind power industry, with a market

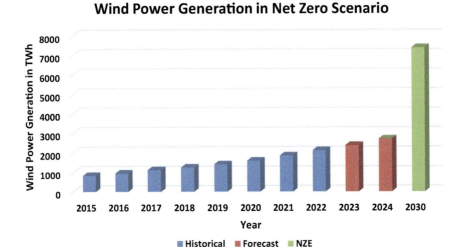

Fig. 4.3 Installed wind power capacity up to 2021 (in MW) (Global Wind Energy Council 2022)

share that is practically unchanged from that of 2020 (Global Wind Energy Council 2022).

Asia Pacific continues to take the lead in global wind power development accounting for 59% of the global new installations last year, followed by Europe (19%), North America (14%), Latin America (6%) and Africa and Middle East (2%) as shown in Fig. 4.4.

The five biggest markets by total installations by the end of 2021 stayed the same. These markets include China, the USA, Germany, India, and Spain, and collectively they accounted for 72% of all wind power installations worldwide, a 1% decrease

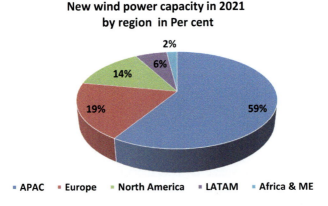

Fig. 4.4 New capacity 2021 installed by region (%) (Global Wind Energy Council 2022)

4.2 Worldwide Status of Wind Power

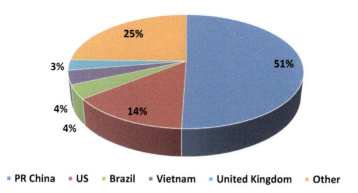

Fig. 4.5 Share of the top five countries for new wind power capacity 2021 (%) (Global Wind Energy Council 2022)

from 2020. Figure 4.5 (Global Wind Energy Council 2022) depicts the share of the top 5 countries for a new wind power capacity in 2021.

The total installed capacity of onshore wind increased by 72.5 GW in 2021, increasing the total installed capacity to 780 GW. Although new onshore installations in the continents of Europe, Latin America, and Africa and the Middle East experienced a record year, overall installations in 2021 are still 18% lower than in 2020. The two biggest wind power markets in the world, China and the USA had a slowdown in onshore wind growth, which was a major contributor to the fall (Global Wind Energy Council 2022).

The top five onshore wind markets included China (30.7 GW), the USA (12.7 GW), Brazil (3.8 GW), Vietnam (2.7 GW), and Sweden (2.1 GW). The amount of onshore wind capacity allotted for 2021 demonstrates that the nation is on track to meet these challenging objectives. More than 20 GW of onshore wind capacity was floated for auction last year as part of the restart of wind, renewable energy, and technology-neutral auctions in Europe.

However, due mostly to difficulties with permits, procurement was undersubscribed in several important onshore wind markets, including Germany, Italy, and Poland. The total amount of onshore wind capacity allocated in Europe for 2021 was just 10.3 GW. The details of the new installation of onshore wind power and total installation all over the world have been shown in Fig. 4.6 (Global Wind Energy Council 2022).

By the end of 2021, the total installed capacity of offshore wind energy had increased to 57.2 GW, a record-breaking achievement in the history of offshore wind. 21.1 GW of offshore wind energy became grid-linked globally in 2021. As projected by GWEC, China surpassed the UK by 2021's end as the country with the

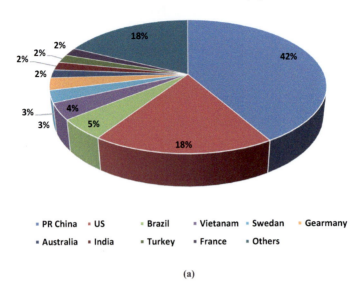

(a)

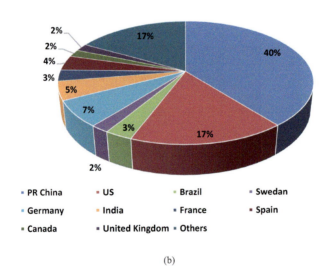

(b)

Fig. 4.6 **a** New installation onshore (%). **b** Total installations onshore (%) (Global Wind Energy Council 2022)

most aggregate installations. German, Dutch, and Danish markets round out the best five markets worldwide.

GWEC Market Intelligence expects the offshore wind market to proceed to develop at a quickened pace. More emphasis has to be given to extracting the wind power potential from offshore regions because there is a very small percentage of the contribution from the countries regarding offshore wind power installations as we can analyse from Fig. 4.7 (Global Wind Energy Council 2022).

Every country putting its efforts into enhancing the wind power contribution to the total renewable energy sector which is needed as per the necessity of the situation that the world is going on. The government also should make more policies regarding the enhancement of not only onshore wind power but also on the offshore potential wind. The detailed report of new installation and total installation of both onshore and offshore wind power has been shown in Tables 4.1 and 4.2 (Global Wind Energy Council 2022).

From the above-mentioned data, we can say that very few countries have succeeded in extracting wind energy from the offshore regions which is appreciable but others also should come forward with full pace to enhance this natural resource.

Despite 2021 having the second-highest installed capacity level in history, the CAGR for the following 5 years is 6.6%. According to current policy, 557 GW of additional capacity should be built during the next five years, according to GWEC Market Intelligence. Up to 2026, there will be more than 110 GW of fresh installations annually.

Onshore wind power will grow at a 6.1% CAGR during the subsequent five-year period and 93.3 GW is the estimated yearly average installation. 466 GW will likely be constructed overall in the years 22–2026. In the following five years, offshore wind's CAGR is 8.3%.

Considering that more than 21 GW of offshore wind capacity was installed last year, such a growth rate is highly encouraging. Over 90 GW of offshore capacity is anticipated to be installed globally between 2022 and 2026. The anticipated 18.1 GW yearly average offshore installations.

4.2.2 Indian Wind Power Statistics

India's rise in renewable energy is tremendous, and wind energy is the biggest powerful output to the issue of non-renewable fuel depletion, coal production, clean gas emissions, climate degradation, etc. Like a sustainable, nontoxic and inexpensive supply, wind energy directly prevents transportation and fuel dependence, which may contribute in favour of clean, green power.

Out of a total of 395 GW, renewable energy sources currently account for 38.5% of the total generating capacity. While wind makes up 10.2% of this at the moment, the Ministry of New and Renewable Energy (MNRE) estimates that 140 GW of wind energy capacity would be required by 2030 to meet the country's 2030 climate

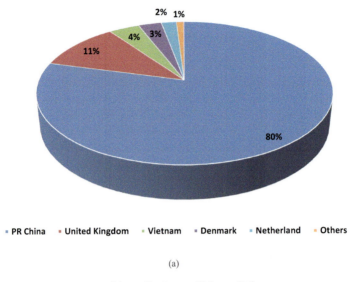

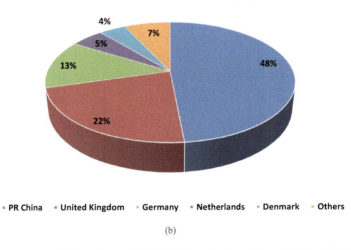

Fig. 4.7 a New installation offshore (%). **b** Total installations offshore (%) (Global Wind Energy Council 2022)

targets. With 40.1 GW of installed wind capacity as of January 2022, India is fourth globally.

The intrinsic power one of the largest sources of wind energy benefits of promoting rural jobs and uplifting the rural economy. In comparison, there is no water used in wind energy, unlike all other types of electricity, which in turn would become a finite asset. Overall, like power stability, as well as independence, are defined as

4.2 Worldwide Status of Wind Power

Table 4.1 Onshore new and total installation in 2021 (Global Wind Energy Council 2022)

S. No.	Country	New installations 2021 (MW)	Total installations 2021 (MW)
1	USA	12,747	134,354
2	Canda	677	14,255
3	Brazil	3830	21,580
4	Mexico	473	7262
5	Argentina	669	3287
6	Chile	615	3444
7	Other Americans	232	4051
8	Egypt	237	1702
9	Kenya	102	440
10	South Africa	668	3163
11	Other Africa	802	3780
12	China	30,670	310,629
13	India	1459	40,084
14	Australia	1746	9041
15	Pakistan	229	1516
16	Japan	211	4523
17	South Korea	64	1579
18	Vietnam	2717	3231
19	Philippines	0	427
20	Thailand	16	1554
21	Other Asia	240	2577
22	Germany	1925	56,814
23	France	1192	19,131
24	Sweden	2104	11,915
25	UK	328	14,065
26	Turkey	1400	10,681
27	Other Europe	7146	95,191

the key drivers, the future of India's wind energy is brilliant. The best benefit is a kind of renewables that power is abundant and does not cause CO_2 emissions either. Wind farms might be installed relatively easily, the wind farmland can be used for cultivation as well as having dual purposes, and relative to other sources of renewable energy, it is cost-effective.

The National of Wind Energy (NIWE) has progressed demonstrating procedures and returned to this examination according to the direction and mandates of MNRE and assessed the wind energy capacity as 302,251 MW at 100 m. These assessments have been shown due to the 2% area accessibility for every state aside from Northeastern states, Himalayan states, and the Andaman and Nicobar Islands, i.e.

Table 4.2 Offshore new and total installation in 2021 (Global Wind Energy Council 2022)

S. No.	Country	New installations 2021 (MW)	Total installations 2021 (MW)
1	UK	2317	12,522
2	Germany	0	7728
3	Belgium	0	2262
4	Denmark	605	2308
5	Netherlands	392	3003
6	Other Europe	4	331
7	China	16,900	27,680
8	South Korea	0	133
9	Other APAC	888	1167
10	USA	12	42

Tables 4.3 and 4.4 show the developed wind energy capacity at 80 and 100 m, respectively (http://www.inwea.org/wind-energy-in-india/wind-power-potential/).

The development using wind energy in India is surprising which shows the achievement of the strategies and the guidelines received throughout time by the Union and State governments. The endeavours of the wind energy sector have driven the nation to arrive at the achievement of 37 plus Gigawatts. Table 4.5 shows the overall analysis of installed wind power capacity (http://www.inwea.org/wind-energy-in-india/wind-power-installation/).

Following the success of the onshore wind power programme, since the existence of 7500 km the Indian government has agreed to promote the production of offshore wind energy in the area. As of October 2015, the service released the National Offshore Wind Policy in order to exploit offshore potential for wind energy in the Exclusive Economic Zone (EEZ) of India along its coastline. NIWE is located on the banks of both Gujarat and Tamil Nadu and is in the process of investigating the potential for offshore wind introductory examinations that have uncovered great capacity. For exact strong wind estimations, one light detection and ranging (LiDAR) has been introduced close to the Gujarat coast, which is producing information about the nature of offshore a windy year after Nov 2017.

Energized by the nature of seaward wind, a personal organization has additionally introduced Gujarat's Gulf of Kutch LiDAR for seaward wind asset estimations. Plans are brewing to introduce a greater amount of such types of gear in Gujarat and Tamil Nadu. Reviews to comprehend the oceanological, ocean state of the bed inside recognized areas of the Gujarati and Tamil Nadu coasts have likewise been arranged.

The biggest idea ever offered by NIWE serious offering measure to the foundation of Gujarat's Gulf of Khambhat's First Offshore Wind Project (FOWPI), which would have a maximum capacity of about 1 GW.

4.2 Worldwide Status of Wind Power

Table 4.3 Calculation of the 80 m level installable wind power potential (http://www.inwea.org/wind-energy-in-india/wind-power-potential/)

S. No.	States/UTs	Calculated potential (MW)	
		@ 50 m	@ 80 m[a,b]
1	Andaman and Nicobar	2	365
2	Andhra Pradesh	5394	14,497
3	Arunachal Pradesh[a]	201	236
4	Assam[a]	53	112
5	Bihar	–	144
6	Chhattisgarh[a]	23	314
7	Diu Daman	–	4
8	Gujarat	10,609	35,071
9	Haryana	–	93
10	Himachal Pradesh[a]	20	64
11	Jharkhand	–	91
12	Jammu & Kashmir[a]	5311	5685
13	Karnataka	8591	13,593
14	Kerala	790	837
15	Lakshadweep	16	16
16	Madhya Pradesh	920	2931
17	Maharashtra	5439	5961
18	Manipur[a]	7	56
19	Meghalaya[a]	44	82
20	Nagaland[a]	3	16
21	Orissa	910	1384
22	Pondicherry	–	120
23	Rajasthan	5005	5050
24	Sikkim[a]	98	98
25	Tamil Nadu	5374	14,152
26	Uttarakhand[a]	161	534
27	Uttar Pradesh[a]	137	1260
28	West Bengal[a]	22	22
Total		49,130	102,788

[a] The validation of wind potential by real measurements is still pending
[b] Mesoscale modelling is used for estimation (Indian Wind Atlas)
Source Indian Wind Energy Association (INWEA)

148   4   Introduction to Wind Energy

Table 4.4 Estimate of the 100 m level installable wind power potential (http://www.inwea.org/wind-energy-in-india/wind-power-potential/)

State	Rank I[a]	Rank II[a]	Rank III[a]	Total
Andaman & Nicobar	4.12	3.43	0.88	8.43
Andhra Pradesh	22,525.50	20,538.10	1165.00	44,228.60
Chhattisgarh	3.24	57.03	16.31	76.59
Goa	0.00	0.08	0.76	0.84
Gujarat	52,287.59	32,037.83	105.09	84,431.33
Karnataka	15,202.36	39,802.59	852.40	55,857.36
Kerala	332.63	1102.56	264.38	1699.56
Lakshadweep	3.50	3.40	0.77	7.67
Madhya Pradesh	2216.39	8258.55	8.93	10,483.88
Maharashtra	31,154.76	13,747.43	492.15	45,394.34
Odisha	1666.20	1267.06	160.22	3093.47
Puducherry	69.43	79.00	4.40	152.83
Rajasthan	15,414.91	3342.62	12.96	18,770.49
Tamil Nadu	11,251.48	22,153.34	394.82	33,799.65
Telangana	887.43	3347.52	9.34	4244.29
West Bengal	0.03	2.04	0.01	2.08
Total in MW	153,019.59	145,742.59	3489.31	302,251.49
Total in GW	153	146	3	302

[a] Rank I—Wasteland, Rank II—Cultivable Land, Rank III—Forest Land
Source Indian Wind Energy Association (INWEA)

Table 4.5 Installed wind power capacity by state and year (in MW) (as of 31.03.2020) (http://www.inwea.org/wind-energy-in-india/wind-power-installation/)

State	Up to 2019	2019–20
Tamil Nadu	8968.89	335.44
Karnataka	4694.9	95.70
Maharashtra	4794.18	206.20
Rajasthan	4299.65	00.00
Andhra Pradesh	4090.5	2.00
Madhya Pradesh	2519.9	0.0
Kerala	52.9	10.00
Gujarat	6073.05	1468.45
Telangana	128.1	00.00
Others	4.3	0.0
Total	35,626.37	2117.79
Cumulative Capacity	35,626.37	37,744.16

Source Indian Wind Energy Association (INWEA)

4.2.3 *Environmental Aspects*

The nation benefits from the use of wind energy since it reduces the side by product and expenses of fossil power sources and accordingly decreases influence leading to environmental change. There is likewise work furthermore, public vitality security advantages. The proprietors and landowners with propellers advantage by pay from sent out a force, and frequently by their utilization of their capacity. Improvement of system and scale economies have brought about the capital expenditure per unit limit of turbines wind diminishing essentially since 1990. Strong government approaches that perceive the advantages power of wind, for example, feed-in tariffs, as well as committed buys, uphold the development of establishments and production, so setting up a feasible company.

The most nearby effects of wind force might be summed up as below (Twidell and Weir 2015):

1. **Visual**: Turbines ought to be located in wide-open regions or coastal areas, with the goal that they be noticeable in plain view. Observe that the bigger the width, the increasingly slow "effortless" the tower and turbine blade tips rotate and become higher. Rotary engine shading may be picked as the more adequate, which is normally white. For the spectator, sighting can be hindered by close interference by mountains, big structures trees, etc. If visible from historic landmarks or in places of scenic beauty, turbines may be deemed counterproductive. Turbines might be viewed as inconvenient if perceptible from noteworthy destinations or in zones of grand excellence. Recreation programming is utilized to provide a powerful wind visual impression of farms among all perspectives before consent is conceded to build. Until permission to build is given, a modelling tool is used to provide a complex visual image from all the wind farm points of view.
2. **Noise pollution**: Audible noise from hardware, cutting edge tips, tower passing and so forth @ 250 m is anticipated to reach 40 dB which is normal; not from vibrations, infrasound perceptible or frequently recognized, yet combative. Such effects are not perceivable due to offshore turbines. Present-day devices are significantly calmer than quick improvements as makers look to react to open remarks and to enhance effectiveness: sound is regularly an indication of few effective power catches.
3. **Effect of birds and bats**: Normally sometimes (house windows); stay away from sitting close to supports or other areas where different insects are getting fed. Species fluctuate extensively in their conduct thus master examination is required previously allowing development. Flying creatures sometimes get themselves in trouble while moving across the wind blades and sometimes also get killed as can be seen in the Fig. 4.8.
4. **Agriculture**: Initially cattle or other animals should be taken care of, but as time passes they will get used to this noise as shown in Fig. 4.9. By and large, the past utilization of the location for creatures and yields proceeds with unbothered separated between 1 and 2% of the region utilized by the substations, tower foundation and uncovered paths.

Fig. 4.8 Effects of wind turbines of birds and bats (*source* Wikipedia)

Fig. 4.9 Wind turbines and cattle grazing (*source* Wikipedia)

5. **Electromagnetic interference**: Television and equipment using microwave signals may not work properly sometimes, also permissible interference affects the flight operators who can utilize particular "erasing" programmes for their system.
6. **Grid limitations**: As the demand for power increases, the grid connected to the wind power station needs to be upgraded and proper maintenance of the overall system has to be done periodically.

The influence of the wind energy system requires thought of numerous orders, including nature, style, social legacy, and public discernment. Wind industry engineers need to acquire nearby or public arranging authorization before introducing a breeze ranch, which may include the thought of all the above elements from autonomous specialists. As a result, the planning process for submission has been systematic and competent. All these processes are important, but they are costly and take much time to accomplish. However, the ultimate result is that wind power limits are expanding not only at the national level but also globally, carbon and different pollutants are lessening, the innovation is improving, and a large portion of the apparent unfavourable effects are diminishing per unit of produced yield.

4.3 Wind Energy Basics

Renewable power source assets present the enchantment answer for power demand requests by the electrical power market in the last many decades. Nowadays, the world is developing consideration wind power transformation frameworks utilized in producing electricity as one of the quickest developing power sector areas with a normal yearly development rate surpassing 25% over the previous decade. Even though wind energy conversion regions are accessible everywhere in the world, it is very significant to pick an appropriate area to diminish the general expense of the conversion of wind energy systems (WECS) (El-Ahmar et al. 2017). Guaranteeing that the WECS framework is working at greatest proficiency involves planning a wind turbine, especially for the particular region where it is to be introduced.

4.3.1 Power

"Wind power" or "wind energy" refers to the process of utilizing the wind to generate mechanical or electrical power. Turbines of wind change the kinetic power stored in the wind to mechanical power. This mechanical power may be utilized for particular tasks (such as grain mills or water pumping) or an alternator can convert it to electrical energy to run homes, businesses, schools, and other structures.

The wind is described by its speed, and direction, that affected by numerous factors, for example, geography and religious assets. A turbine of wind may produce a power source component with numerous sites-explicit variables and streamlined execution of the blade. High significance is the normal yearly pitch, wind speed, pretwist, attachment angle and area spread by the rotor; be that as it may, air density additionally influences the efficiency of wind turbine energy production and air density also gets affected with elevation and temperature variations.

Numerous atmosphere factors must be taken into thought while planning the breeze turbine, for example, wind velocity, the region that the turbine sweeps, rotor, air density, site temperature, and tower height. Choosing a wind farm ought to be founded on the atmosphere states of the specific regions. Wind power is legitimately relative to such factors.

The flow of wind creates wind power and the kinetic energy associated with such motion may be stated as follows (Umanand 2007):

$$\text{Kinetic energy} = \text{K.E.} = \frac{1}{2}mv^2, \tag{4.1}$$

$$\text{where} = \text{Air mass in Kg} = \text{Volume}(m^3).\text{Density}(kg/m^3) = Q \cdot \rho \tag{4.2}$$

$Q = $ Discharge, $v = $ Velocity of air mass in m/s.
Hence, the power equation can be obtained as:

$$\text{Power} = \text{Rate of change of energy} = \frac{dE}{dt}$$

$$= \frac{1}{2} \cdot \frac{d(m.v^2)}{dt} = \frac{1}{2} \cdot \frac{d(\rho \cdot Q \cdot v^2)}{dt} = \frac{1}{2} \cdot \rho \cdot \frac{dQ}{dt} \cdot v^2 \tag{4.3}$$

Here, $= \frac{dQ}{dt} = $ Rate of discharge $(m^3/s) = A \ (m^2) \cdot v \ (m/s)$,

where $A = $ Area of cross section of blade movement

$$\text{Power} = \frac{1}{2} \cdot \rho \cdot A \cdot v^3. \tag{4.4}$$

Normally area of blades is constant for a specific blade length so the air density also. Therefore that may be stated that wind energy fluctuates as a cube of wind velocity.

Example 4.1 Calculate the power of the wind generated by the wind turbine having a blade length of 50 m and wind flow at a speed of 20 m/s. The air density is 1.23 kg/m^3.

Solution: Given blade length $l = 50$ m, wind speed $v = 20$ m/s and air density $\rho = 1.23$ Area of the rotor $= \Pi \cdot \frac{D^2}{4} = \Pi \cdot \frac{100^2}{4} = 7850$ m^2.

So as per the Eq. (4.4), wind power $= \frac{1}{2} \cdot \rho \cdot A \cdot v^3 = \frac{1}{2} \cdot 1.23 \cdot 7850 \cdot 20^3 = 38.622$ kW.

4.3.2 Air Density

The properties of a wind turbine are commonly linked to the degree density of air. Air density is straightforwardly related to the power produced by the wind. These features are illustrated in Fig. 4.10 (El-Ahmar et al. 2017).

As per the gas law, the air pressure varies w.r.t. pressure and temperature as per the below equation

$$\rho = \frac{P}{RT}, \tag{4.5}$$

where P indicates the pressure, R and T denotes the gas constant and the absolute temperature.

At 1 atm (14.7 psi) and 60 °F the air density at sea level is 1.225 kg/m^3. It can be observed from the mentioned equation that the density of the air has an indirect relationship with temperature and a direct relation to pressure. Figures 4.11 and 4.12 show the change in wind power output concerning wind speed variation as pressure and temperature vary.

Moreover, the density limits and temperature decline while expanding height. Consequently, the variety in height outcomes changes the measure of power delivered because of variations in the air density measurements. The influence of the effects of pressure and temperature on air density may made simpler by the given expression, which is accurate up to a height of 6000 m (20,000 feet).

$$\rho = \rho_O - \left(1.194 \times 10^{-4} \cdot H_m\right), \tag{4.6}$$

where H_m is the location's altitude in metres, ρ_o is the standard figure for air pressure (1.225 kg/m^3) at water level.

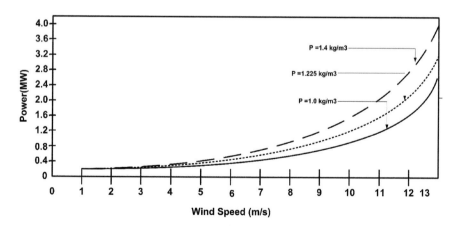

Fig. 4.10 Effect of air density on wind turbine power output (El-Ahmar et al. 2017)

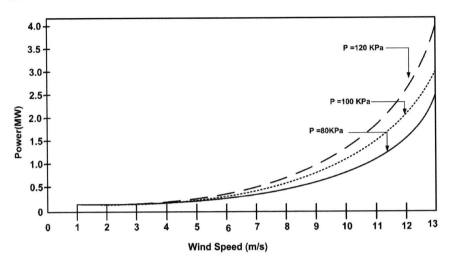

Fig. 4.11 Effect of air pressure on the power production of wind turbines (El-Ahmar et al. 2017)

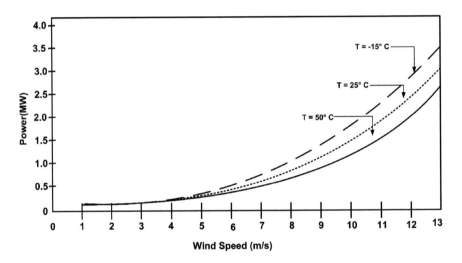

Fig. 4.12 Effect of temperature on wind turbine power output (El-Ahmar et al. 2017)

4.3.3 Swept Area

As per wind energy expression, the area swept by the rotor is directly varying with the generated wind power which is equal to:

$$A = \frac{\prod}{4} D^2, \qquad (4.7)$$

4.3 Wind Energy Basics

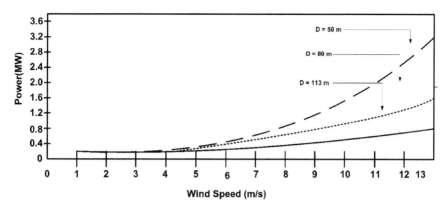

Fig. 4.13 Effect of surface area on wind turbine power output (El-Ahmar et al. 2017)

where D represents the rotor diameter.

With relation to the above expression, since the region that the rotor sweeps corresponds to the square of the diameter of the rotor, a gradual incremental change in rotor blade length results in a high rise in the power available to the turbine appears in Fig. 4.13 (El-Ahmar et al. 2017).

If the region covered by rotor blades and the diameter of the rotor is larger, incremental growth in wind energy will be extracted from the wind.

Example 4.2 Find the length of the wind turbine blade for producing 772.440 kW having an air density of 1.23 kg/m³ and wind speed of 10 m/s.

Solution: We have wind power $P = 772.440$ kW, air density $\rho = 1.23$ kg/m³, wind speed $= 10$ m/s.

So the area of the blade, $A = \frac{P}{\frac{1}{2} \cdot \rho \cdot v^3} = \frac{772.440}{\frac{1}{2} \cdot 1.23 \cdot 10^3} = 1256 \, m^2$.

Hence the length of the blade, $l = \sqrt{\frac{A}{\pi}} = \sqrt{\frac{1256}{\pi}} 20 \, m^2$.

4.3.4 Cube of Wind Speed

The wind output power has a direct relation with the wind speed which is a very necessary factor to forecast the information regarding wind power production. As per wind power output expression, power fluctuates with the cubic of wind velocity, thus more the intensity of a wind velocity results in larger power output.

Speed of wind can be of various types which are as follows (Umanand 2007):

- **Cut-in wind speed**—It is the slowest wind velocity that the rotor will activate and produce an amount of considerable power. Precisely this speed varies between 10–16 kph.

156   4 Introduction to Wind Energy

- **Rated wind speed**—It is the speed at which the rotor is designed to run and the wind turbine starts delivering the rated output power. This speed normally exists between 40–55 kph for several wind turbines. If the speed of wind lies between rated and cut-in limits the output power increases w.r.t. the wind. Most machines' performance increases above the rated velocity. Most manufacturers have diagrams that display describing their wind turbine performance differs based on wind velocity, called "power curves."
- **Cut-out wind speed**—At extremely extreme wind speeds, ordinarily somewhere in the range of 72 and 128 kph, the majority of windmills stop delivering electricity and turn off. The maximum shutting down at which the speed of wind happens is known as the speed of cutoff. This speed limit is a well-being highlight which shields the wind turbine from harm. There may be other different ways which cause the closing of working of wind turbines.

In certain machines, a programmed stop quickly is enacted by a wind velocity indicator. Certain devices spin otherwise "tune" the blades for spilling the wind. Others also use "plot twists" and utilize flaps that are brought on impulsively by high rotational rpm on the blades or hub or a spring-loaded mechanism that activates its mechanism and pushes the device angled towards the wind stream. When the windfall falls to a reasonable stage, daily windmill activity normally continues. So, it tends to be seen it is greater beneficial to place a conversion to a wind energy framework in a situation with normally extreme wind speed than in a place with less wind speed.

4.3.5 Tower Height Effect

One effective technique to put the wind turbines under more intense wind pressure is to place them at higher heights. Over the top of the earth inside the initial not many units of wind speed impressively affected by the contact that now the air brings about as it streams over the outside of the planet. Non-stick floors, similar to a quiet ocean, offer a limited quantity of opposition, and the distinction of speed compared with a range from the surface is just restricted. Winds, on the other hand, persist on unusual structures such as trees and big structures.

The following equation is frequently used to explain how the surface effects of the earth's corrosion affect wind velocity (El-Ahmar et al. 2017):

$$v = v_{\text{ref}} \left(\frac{\ln \frac{z}{z_o}}{\ln \frac{z_{\text{ref}}}{z_o}} \right), \tag{4.8}$$

where:

v is the wind speed at a height of z above sea level.

v_{ref} is reference velocity; specifically, a wind speed at height z_{ref}.

4.3 Wind Energy Basics

For the targeted speed v, z is the height above ground grade.

z_{ref} is reference height, i.e. height at which wind speed is v_{ref}.
z_o is grossness length of the wind's direction.

Figures 4.14 and 4.15 represent the effect of the height of the wind tower on wind direction and wind energy output with various stiffness lengths (0.1, 0.4, and 1.6).

The coarseness properties are shown in detail in Table 4.6. Normally 10-m height is taken as the base for an 8 m/s wind speed. The impact of tower height on wind energy can be observed properly from the graphs.

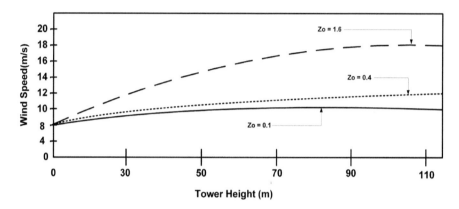

Fig. 4.14 Effect of the height of a tower on wind speed (El-Ahmar et al. 2017)

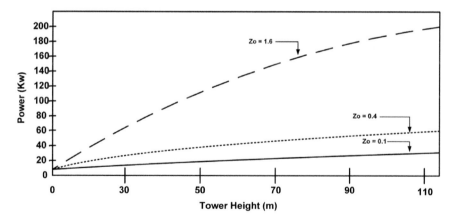

Fig. 4.15 Effect of structure height on wind power production (El-Ahmar et al. 2017)

Table 4.6 Roughness classifications (El-Ahmar et al. 2017)

Roughness class	Description	Roughness length (Zo)
0	Water surface	0.0002
1	A few windbreaks and some open spaces	0.03
2	Farmland with several windbreaks separated by more than a kilometre	0.1
3	Farmland and urban areas with plenty of windbreaks	0.4
4	Forest or dense urban areas	1.6

4.4 Analysis of Wind Data

4.4.1 Average Wind Speed

The sustained mean wind speed interval can be calculated by summation of a large no. of readings observed in that time interval and dividing by the number of observed readings. Numerous advanced information lumberjacks introduced in the course of the most recent couple of decades gathered normal wind speed information essentially for meteorological purposes, instead of surveying wind energy. The speed was recorded each hour and afterwards arrived at the midpoint of the day, which, thusly, was found the middle value throughout the year and the month. These steps were taken for the averaging (El-Ahmar et al. 2017):

$$V_{\text{avg}} = \frac{1}{n} \sum_{i=1}^{n} V_i, \tag{4.9}$$

where

V_i ith reading of wind speed
n number of readings.

The wind speed from month-to-month variety at a run-of-the-mill site throughout the year might be between 30 and 35% more than the yearly normal. As observed before, for surveying the power of wind, the root mean cube speed is what makes a difference. What might be compared to advanced information logging is as per the following:

$$V_{\text{rmc}} = \sqrt{\frac{1}{n} \sum_{i=1}^{n} V_i^3}. \tag{4.10}$$

The previous expression does not take into consideration the mass density of air, which is an important measurement (even though of 2nd-order) density of wind

4.4 Analysis of Wind Data 159

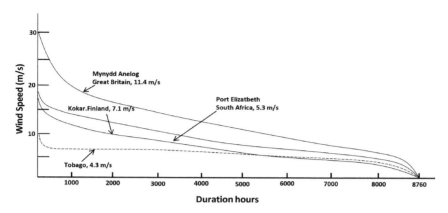

Fig. 4.16 Example of a velocity duration curve (Rohatgi and Nelson 1994). Reprinted with the Alternative Energy Institute's permission

power. So the best way to process power calculations, and wind speed information is, therefore, to digitize the annual average power density (Xu et al. 2011).

The average wind speed can be calculated graphically by taking the area under the curve of the velocity duration where the x-axis shows the total number of hours per year when the speed is equal to or greater than every specific number and the y-axis displays the wind speed. Figure 4.16 gives Rohatgi and Nelson (1994) an example of velocity length curves for different areas of the globe (Considering typical wind velocities ranging from around 4–11 m/s). An estimated understanding of the existence of the site's wind conditions is given by this form of Fig. 4.16. A calculation of the average wind speed is all of the curve's area beneath it (Xu et al. 2011).

When the period for collecting wind data is small, the calculation of the typical speed of the wind and the best is standard deviation, but the least-square approach is best when the period is long. Maximum wind speed and average speed of wind prediction is the simplest, it just has to be aware of the average speed of wind and the maximum speed of wind, and with the wind data, they may simply get off (Manwell et al. 2010).

4.4.2 Wind Speed Distribution

As wind speed has a cubic relationship with the power it is the very crucial variable required to calculate the capability of the specific area chosen for wind power installation. We know that the wind speed is always variable in any region. It is affected by the climate framework, the nearby land landscape and its height over the surface of the ground. Wind's speed differs continuously, minute, daily, month, and sometimes year. Hence, the yearly average speed should be found in the middle value of more than 10 yr or more. Hence the average wind speed taken over a long duration helps a lot for power extraction of a particular site. In any case, if the procedure for data estimation is too long then it would be expensive and most of the customers cannot entertain this. In such cases, the small period of collected data is matched with the data over time from a nearby site to foresee the drawn-out yearly wind speed at the site viable. This is called a means to link, quantify, and forecast (MCP) method. Since the sun, the seasons, and the wind all influence each other cycles get repeated for the most part every year. The wind site is generally defined by the speed information arrived at the midpoint of over-scheduled months. Often, the month-to-month information is accumulated throughout the year for quickness in detailing the general "windiness" of different areas. Wind speed varies throughout time, and it may represent the use of a distribution function for probabilities.

The histogram is commonly referred to as the popular representation of data on wind speed. In the histogram, it is usually obtained by dividing the complete amount of data into equal-sized bins, defined as classes. A class is defined by the middle of each bin's value. The equivalent frequency with the Δv width of each bin b_j is (Nayak and Mohanty 2018):

$$f_{r_j} = \frac{n_j}{n},\qquad(4.11)$$

where n_j represents the resulting data point counts underclass resulting from wind's speed; v_j and f_{r_j} relates the matching frequency to the class "j."

$$\sum_{j=1}^{N} n_j = n \qquad(4.12)$$

$$\sum_{j=1}^{N} f_{r_j} = 1 \qquad(4.13)$$

n indicates the total class number.

4.4.3 Wind Data Statistical Analysis

To assess the wind power capacity at a certain location and to measure the power generated by a windmill built there, statistical analysis may be used. Similar approaches have been pointed out by many researchers, including Rohatgi and Webber (1985), Justus (1978), and Justus and Mikhail (1994). If time arrangement estimated information is accessible at the ideal area and altitude, there might be little requirement for an information examination as far as likelihood appropriations and factual methods. That is, the recently portrayed procedures might be everything necessary. Then again, if the projection of estimated information starting with one area then onto the next is required, or when just synopsis information is accessible

4.4 Analysis of Wind Data

then the means of empirical illustrations for the wind speed probability distribution has specific advantages. There is a probability distribution concept for mathematical analysis that explains the possibility that a random variable (like wind speed) will have a specific value. As talked about forward, a possibility distribution is ordinarily portrayed either by a cumulative density function or a probability function density. Generally, two probability distributions are utilized for the analysis of wind data which are the Rayleigh and the Weibull, respectively. Mean wind speed is the sole characteristic required for the Rayleigh distribution. Weibull's distribution is based on two parameters and may, thus, best express a broader spectrum of various wind patterns. All the distributions of Weibull and Rayleigh are referred to as distributions that "skew" (Manwell et al. 2010).

(a) **Weibull Distribution**

This probability distribution is mostly utilized to represent wind speed data for the long term which is as follows:

$$f_w(v) = \frac{k}{c}\left(\frac{v}{c}\right)^{k-1} e^{\left[-\left(\frac{v}{c}\right)^k\right]}, \tag{4.14}$$

where k indicates the form component, c indicates the scalar component, and v is the wind speed

The shape and scalar parameter can be represented as:

$$k = \left[\frac{\sigma}{v_{avg}}\right]^{-1.086} \quad 1 \leq k \geq 10 \tag{4.15}$$

$$c = \frac{v_{avg}}{\Gamma\left(1 + \frac{1}{k}\right)}, \tag{4.16}$$

where v_{avg}, Γ and σ are the typical wind speed, gamma function and, correspondingly, the standard deviation which can be calculated by the following expressions:

$$V_{avg} = \frac{1}{n}\sum_{i=1}^{n} V_i \tag{4.17}$$

$$\Gamma(x) = \int_{0}^{\infty} t^{x-1} e^{-t} dt \tag{4.18}$$

$$\sigma = \sqrt{\frac{1}{n-1}\sum_{i=1}^{n}(v_i - v_{avg})^2}. \tag{4.19}$$

Figure 4.17 shows that as the rise in k value for a constant value of c the peak of the curve becomes sharper and sharper which indicates the slow variation of wind speed (Mohsin and Rao 2018).

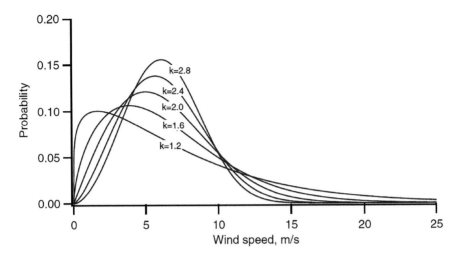

Fig. 4.17 Function of Weibull probability density example for $V_{avg} = 6$ m/s (Mohsin and Rao 2018)

(b) **Rayleigh Distribution**

This probability distribution is the easiest way to show the wind speed data because it considers only one parameter, i.e. the mean of wind's speed. It may be referred to the Weibull distribution as a specific example where the shape parameter is made equal to two, i.e. $k = 2$. The probability distribution function is as follows:

$$f_r = \frac{2v}{c^2} e^{\left[-\left(\frac{v}{c}\right)^2\right]}. \tag{4.20}$$

The Rayleigh probability density function for various, average wind speeds is shown in Fig. 4.18. The higher the mean wind speed value, as shown, gives a greater chance of increased wind speeds (Mohsin and Rao 2018).

4.5 Overview of Wind Turbines and Its Components

4.5.1 Introduction

Modern wind turbines are getting complex in design consisting of various components and subcomponents to ensure better performance. These turbines are accountable for the transformation of the rotational power extracted due to the wind to the rotary power, finally turning it into electric power. The basic structure of the wind turbine has been shown in Fig. 4.19, and the short details of the key components of the wind generator system are as follows (Bezzaoucha et al. 2018):

4.5 Overview of Wind Turbines and Its Components

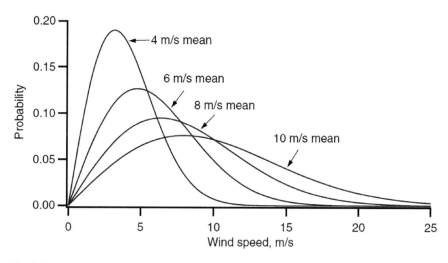

Fig. 4.18 Rayleigh distribution function example for $V_{avg} = 6$ m/s (Mohsin and Rao 2018)

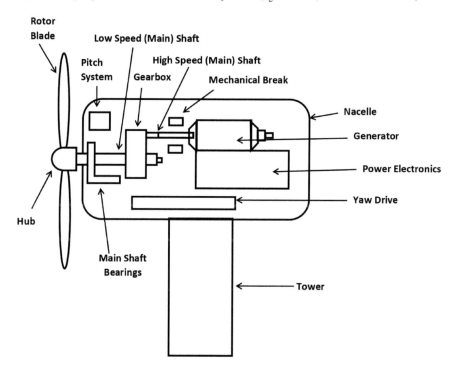

Fig. 4.19 Wind turbine components

1. **Rotor**: It is a wind turbine component that absorbs wind power. The rotor normally comprises two or more rotating wooden, fibreglass, or metal blades at a regulated speed based on the wind direction and the blades around a vertical or horizontal axis. The hub is a connection between the drive shaft and the blades mounted to the hub.

2. **Pitch system**: There are two types of pitch systems: hydraulic and electrical. It aims to adjust the blade angle to adapt the energy. To prevent the blade from rotating in excessive winds fast or too slow to produce energy, the blades are pitched or turned.

3. **Gearbox:** It is normally placed between the generator and the rotor. That increases the mechanical energy rate transfer compared to the generator and is among the biggest critical components in terms of maintenance. The slow-moving shaft is connected by gears, and the high-speed shaft, which raises the rotor from 30 to 60 rpm to 1200 to 1800 rpm range is the speed needed largely produced to generate power efficiently. Since the gearbox is an expensive and heavy feature, designers are looking into low-speed direct-drive generators that do not require a gearbox.

4. **Hydraulic system**: It governs the mechanical brake and regulates the pitch angle of the blade in the pitched framework. It also optimizes wind's energy output in the system of yaw.

5. **System of yaw**: Brakes, an engine, gears, and bearings make up the yaw mechanism. Its task is to monitor the tower's rotation so that the nacelle can face the wind correctly. If the turbines take advantage of forced or free yaw affects the system of yaw components. The alignment of the rotor is typically used to decide the yaw-style mechanism (either the tower's upwind or downwind). At least one yaw bearing is needed, and a yaw brake, a yaw drive (yaw bull and gear motor), and a yaw damper are optional. In the yaw drive, it is driven using a yaw engine. Unlike upwind turbines need rotation drive because the rotor is blown downstream by the wind.

6. **Generators**: It converts the mechanical energy received as an input to the electrical energy as an output. It's typically a commercially available magnetic field producer that generates AC electricity at 50 or 60 Hz.

7. **Electrical control**: It aims to control the generator output.

8. **Braking mechanism**: Normally used to bring the windmill to a halt in the event of heavy wind or crucial hardware failures. This braking system can be operated in mechanical, electrical or hydraulic mode.

9. **Power electronic components**: Generally, the converter establishes the synchronization between the voltage and frequency of the generator to the grid.

10. **Tower**: The tower category consists of the tower itself, its base, and probably the machine's self-erection mechanisms. the primary shaft connecting the blade to the base is known as the tower. It also lifts the rotor to a higher altitude, where stronger winds can be found. The steps for the tower are repair and examination of wind turbines with a horizontal axis.

4.5 Overview of Wind Turbines and Its Components

In addition to the main components, there are several other subcomponents like batteries, inverters, charge controllers, vanes, etc. that are needed to support the wind turbine mechanism for proper operation (Al-Shemmeri 2010).

4.5.2 Classification of Wind Turbines

Even though the contemporary wind turbine industry began in the 1970s, the wind turbine's fundamental anatomy had been defined much earlier. The utilization of wind caught by sails or blades to turn a shaft to which they are coupled, so providing a rotational motion that can later be utilized to generate an output valuable to civilization, is the basic principle upon which all stationary wind turbine systems are designed. Wind turbines were first employed to grind grain or pump water, but they were also utilized to generate power from the late nineteenth century forward (Ali 2017).

The selection and utilization of turbines is the key to harnessing wind for electricity generation. Windmills may revolve along either a vertical or horizontal plane. Wind turbines with horizontal and vertical axes, sometimes known as "eggbeaters" designs, also known as Rotors by Darrieus, are the most common and oldest configurations. Vertical turbines are still behind on the utility-scale due to the high expense of positioning the rotors at increased wind velocities at heights of 80–100 m upper floor. However, with newer entry into the market on a routine basis, usage for residential units is gaining appeal. Figure 4.20 depicts these two wind turbine topologies schematically.

4.5.2.1 Wind Turbines with a Vertical Axis

A wind turbine having an axis in a vertical shape is the oldest wind turbine ever observed. This sort of machine, with a vertical axis and fabric sails, was employed for milling and pumping in Iran and Afghanistan in the ninth century. The generator shaft is vertically positioned with the blades pointing up in the vertical axis wind turbine (VAWT), and the generator is situated on the ground or in a small tower. The overall benefit of a Vertical axis wind turbine is basic since those that do not need a pitch or yaw framework; and all of the accessories which would normally be in a cockpit that is on the ground, facilitating simpler servicing. The fast variations of the velocity vector horizontal-axis wind turbine (HAWT) may be directed to "seek the wind" adjusting the blade's rotational axis, resulting in a significant loss of energy which is not the issue with VAWT. The primary rotor shaft of a VAWT is oriented vertically. VAWT and the generator are situated on the ground or in a short tower. The three types of VAWT have been shown in Fig. 4.21 and have been described as follows:

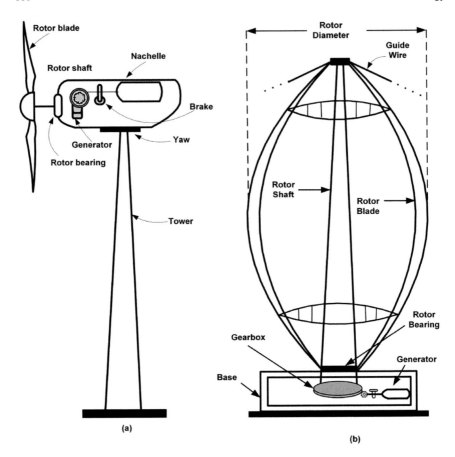

Fig. 4.20 Typical arrangements of wind turbines

1. **Vertical axis wind turbine made by Savonius**: It is a slide device with a high power coefficient of less than 0.2 when using cup-type blades. These turbines feature low-cost rotors and are simple in construction, but their ineffectiveness does not justify their use. In anemometers, these devices of the drag type are wind turbines having a pair (or above) of lobes. The wind flows through the two curved sheet airfoils and around them in a complicated pattern. Savonius rotors can be utilized for water pumping because of their high solidity, which creates a high initial torque. Twisted Savonius is a Savonius that has been modified to have an extended conical for smooth torque and scoops. That is frequently utilized as a windmill on a rooftop. The parallel turbine, which employs another kind of vertical axis, is a cross-flow or centrifugal fan.
2. **Darius Turbine**: These turbines have two or three aerodynamic blades that create forces due to aerodynamic lift. Darrieus VAWT has a lesser theoretical effectiveness of 0.554153 than HAWT which is 0.593. The blades are usually shaped like a

4.5 Overview of Wind Turbines and Its Components

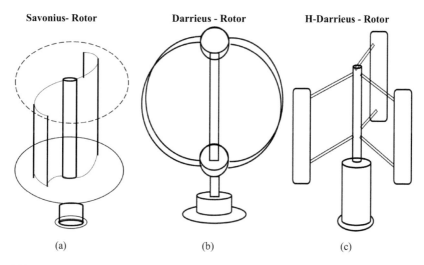

Fig. 4.21 Vertical axis wind turbine types. **a** Savonius Rotor. **b** Darrieus Rotor. **c** H-Darrieus Rotor

beater for eggs. Inverter and additional elements occupy the land, making operation easier. The drawback is unable to be used, these turbines are erected up above the surface, resulting in substantially lower wind speeds for the rotor. Because the blades rotate around a vertical axis, or the attack's angle varies over time, causing the torque to pulse. Structure collapses are caused by oscillating torque and centrifugal forces. Darrieus turbines, often turbines, sometimes called "egg-beater" turbines, are named after Georges Darrieus, a French scientist. Despite their efficiency, they generate a lot of cyclical stress on the tower and torque ripple, which affects dependability. Because of the low starting torque, they usually need an outside energy input or an extra Savonius rotor to get started. That is the torque ripple is minimized when more than three blades are used, resulting in increased rotor strength. By dividing the rotor area by the blade area, used to determine solidity. Guy wires are no longer used to support newer Darrieus-type turbines, which instead have an outer framework attached to leading bearings.

3. **The Giromill VAWT or H rotor**: It is a kind of the Darrieus turbine with flat, rather than curved, blades that are parallel to the rotational axis. This has become a popular arrangement among tiny VAWTs. The blades are shaped like airfoils. As the blades spin around the axis, the angle of attack varies, resulting in undesirable lift and drag. The cycle-turbine variant is self-starting and features variable pitch to prevent torque pulsation. The variable pitch has a strong initial a large, flat torque curve, and a higher performance coefficient, better performance-reduced blade bending strains are caused by decreased blade speed ratios in turbulent winds.

The following are the fundamental benefits of a wind turbine with a vertical axis in theory:

1. The generator, gearbox, and other components may be put on the ground, and the machine may not require a tower.
2. Against the wind, rotate the rotor, no yaw mechanism is required.

The following are the primary drawbacks:

1. Wind speeds are relatively few near the surface stage, therefore even if a tower is spared, wind speeds on the bottom half of the rotor will be quite low.
2. The vertical axis turbines' total efficiency isn't very noteworthy.

4.5.2.2 Wind Turbines with a Horizontal Axis

The primary rotor shaft and electrical generator of a horizontal-axis wind turbine are positioned at the top of a tower and are required to face the wind. Small wind turbines are pointed using a straightforward wind vane, whereas bigger turbines are pointed using a wind sensor and a servo motor. Most have a gearbox that accelerates the blades' slow rotation into one that is suitable for powering an electricity device. The (HAWT) horizontal-axis wind turbine aerodynamics are complicated. The flow of air at the turbine blades is different from airflow that is further apart. Air is deflected by the turbine as a result of how energy is extracted from the atmosphere. The HAWT can be classified based on the wind movement across the rotor blades as shown in Fig. 4.22 (Ali 2017):

1. **Horizontal upwind**: The wind strikes the blade before the tower because the generator shaft is horizontal. Blades of turbines are strengthened to protect strong gusts from pushing them into the tower, and they are placed directly ahead of the tower and sometimes slanted up a little amount.
2. **Horizontal downwind**: The wind strikes the tower first, then the blade, since the generator shaft is horizontal. Horizontal downwind does not require an extra system to maintain it aligned with the severe gusts of wind, the propellers may flex, reducing their swept area and hence their air resilience. Disruptions are not an issue with a horizontal downwind turbine (Fig. 4.22).

4.5.3 Aerodynamics of Rotor

The wind turbine's horizontal-axis blades had an aerofoil-shaped border, similar to that of an aircraft wing. Air moving across a wing's top surface must go beyond the wind flowing across the lowest ground due to the aerofoil's structure. As a result, the air pressure above the wing is lower than below it. As seen in Fig. 4.23, the differential pressure causes a force that pushes the plane's wing to generate lift, which is utilized to lift it into the air. A wind turbine blade uses the same force to deliver power to the rotor. In a moving stream of air, another pressure is created over an aerofoil or wing, which is known as drag. This is the aerofoil's resistance to wind, the power everybody feels when subjected to the wind, and it has the effect of slowing the aerofoil down.

4.5 Overview of Wind Turbines and Its Components

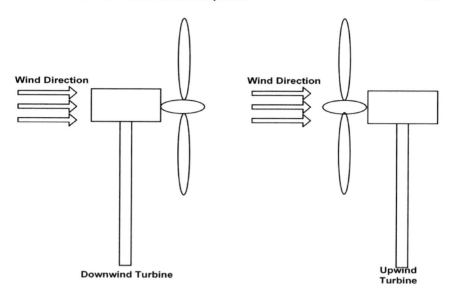

Fig. 4.22 Types of HAWT

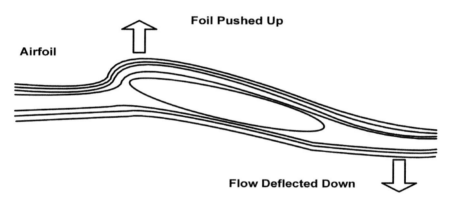

Fig. 4.23 Principle of airfoil

On an aerofoil, drag operates perpendicular to lift. A plane's lift is employed to propel it into the air. It has been observed that the moment an aerofoil is aimed straight towards the wind; it has the wind's smallest cross section and hence the least drag.

This, however, does not produce the greatest lift. Increasing the attack's stance, or raising the front of the aerofoil in relation to the wind direction, increases lift while simultaneously improving drag. If the striking angle is also raised high, the aerofoil's upper surface was being swept by air will eventually produce a reduced pressure zone, causing the airflow to become turbulent which leads the amount of lift to rapidly decrease, this phenomenon known as stalling in an aeroplane.

If a wind turbine blade is thought of as an aeroplane wing, then when the blade is stationary the lift that it experiences as the wind passes over it acts perpendicular to the blade, in the plane of the circle swept out by the blades, and generates a force about the shaft of the turbine rotor—a torque—causing it to turn. Since the whole propellers are geometrically equal contributing the force in the same rotating direction irrespective of their position, force of each blade's resistance will be experienced like a pressure along that turbine, pressing the rotor panel contrary to the pole in this case.

The condition becomes more challenging as the rotor turns. When a blade moves across the wind direction, the blade perceives the wind as flowing from a different direction than the original wind direction, a direction known as the apparent wind direction, which can be calculated by adding the wind and blade speed. This has the consequence of changing the real direction where both lift and drag are sensed, such that some of the lift seems like a force parallel to the wind turbine shaft, pressing against the tower, while some of the drag decreases the shaft's actual torque.

As a result, the optimum aerodynamic performance is achieved when the aerofoil blade is pointed in the direction of the apparent wind. Furthermore, as the blade revolves, the real speed of the blade rises from root to tip. This necessitates bending the blade aerofoil longitudinally. To get the optimum aerodynamic performance, a twist of $10°$–$20°$ is usually necessary. A wind turbine blade's ideal aerofoil shape is relatively thin. While this would extract the most energy from the wind, the blade must be strong sufficiently to resist the forces of lift and drag, as well as the gravitational force exerted on it due to its centrifugal forces and weight while rotating. This implies the blade must be broader than the ideal, with the highest thickness at the base, where the most strength is needed (Breeze 2016).

4.5.4 Transmission System

Wind turbines usually contain a rotor with huge blades that are rotated by the wind. The wind's kinetic energy is transformed into rotational mechanical energy by the rotor blades. To generate electrical power, mechanical energy is normally used to operate one or more generators. As a result, wind turbines have a power transmission system that performs and converts mechanical energy into electrical energy. The wind turbine's power transmission is occasionally known as the "power train." The drive train is the section of a power transmission system that connects the rotor of a wind turbine to the generator.

The rotating speed of the wind turbine rotor must frequently be raised to meet the generator's requirements which is achieved by the gearbox that is situated between the generator and the rotor. So, the gearbox is the important element of the drive train, converting the wind turbine rotor's low-speed, high-torque input into a lower-torque, higher-speed output for the generator. Designing gearboxes for wind turbines raises serious difficulties because of the dimensions, complexity, and variability of the forces experienced by the rotor and drive train.

4.5 Overview of Wind Turbines and Its Components

Some manufacturers handle this issue by excluding the gear stage from their power transmission systems. In such setups, the rotor of a wind turbine powers a low-speed generator directly. Even if there are no worries regarding gearbox dependability, the absence of a gear stage frequently causes additional issues. To produce equal amounts of electricity, low-speed generators in direct-drive wind turbines are often bigger than their high and medium-speed counterparts in geared systems. In addition to financial issues, the bigger size poses transportation, assembly, and maintenance problems. Due to the competing challenges of traditional drive trains and direct-drive machines, interest in medium-speed solutions has grown. An integrated gearbox and a medium-speed generator led to the development of the system, i.e. "hybrid systems" which is sometimes known as multibird having a design which is light and compact with fewer rotating elements.

The typical power transmission system consists of a gearbox and generator. The gearbox consists of a gearbox housing and a gearbox output member. The generator consists of the following components which are as follows:

1. A generator housing with a drive-end side linked to the gearbox housing and a non-drive-end side.
2. The generator housing is followed by a stator.
3. The rotor has a shaft connected to the gearbox output and a body that revolves around the shaft.
4. A non-drive-end shield connected to the generator housing's non-drive-end side.
5. A spindle trying to extend in the axial direction from the non-drive-end shield.
6. There should be at least one generator bearing between the rotor shaft and the spindle.

The rotor shaft and gearbox output member are supported by the generator bearing(s) which refers to a bearing that is shared by the generator and gearbox is required for their operation. The generator bearing(s) are used as pivoted support for the gearbox output member and rotor shaft.

Wind turbines are generally classified into two categories based on their rotating speed: fixed-speed wind turbines (FSWT) and variable-speed wind turbines (VSWT). The speed of FSWT is determined by a variety of elements including gear ratio, generator type, number of poles, and frequency of grid. In the FSWT, a squirrel cage induction generator is usually linked to the grid. Between the generator and the grid, a capacitor bank and a soft starter are also used to decrease imaginary power requirements. Due to a tip speed ratio, the FSWT has a disadvantage of limited power extraction capacity which can be solved by the VSWT because it is more sophisticated and so more expensive. The rotor speed should be disconnected in VSWT since the generator frequency is proportional to its rotational speed. In VSWT, an induction or synchronous generator is used in conjunction with a complete power–frequency converter, or a DFIG is used in conjunction with a partial power–frequency converter. A gearbox between the turbine rotor and the generator can be used to set the generator speed to a desired level. As a result, there are two types of power transmission methods from the turbine rotor to the generator: mechanical power transmission and hydraulic power transmission (Mahato and Ghoshal 2019).

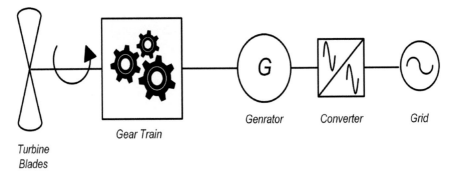

Fig. 4.24 Mechanical power transmission

1. **Mechanical power transmission**: Mechanical transmission, often known as gear train transmission, is created by appropriately installing gears on a frame. Because the provision to decrease power fluctuation is low, this transmission is not suited for offshore wind turbine applications. When the input wind speed is high enough owing to a storm, cyclone, or other event, it is unable to store the excess power created by the turbine rotor. As a result, a large quantity of energy might be lost. When the input wind velocity is low enough, it is incapable of balancing the power requirement. Figure 4.24 shows a schematic diagram of a wind turbine and its gear train power transmission.
2. **Hydraulic power transmission**: These power transmission systems are of two kinds: closed loop and open loop. Both kinds are utilized in wind turbines, and they differ in a variety of aspects, including the transmission medium (oil and saltwater), equipment (fixed and variable displacement pumps, fixed and variable displacement hydraulic motors), and so on. Onshore wind turbines utilize a closed loop and offshore wind turbines use an open-loop hydraulic transmission system. The closed-loop system is more expensive due to the usage of variable displacement hydraulic pumps, motors, and other hydraulic equipment. In addition, the closed-loop hydraulic system complicates the control approach for onshore wind turbines. Figure 4.25 shows a schematic representation of a wind turbine having a closed-loop power transmission system.

A variable displacement pump is directly linked to the turbine rotor in an open-loop hydraulic system of an offshore wind turbine and pressured sea water can be utilized as the power transmission means. After hydroelectric power generation, the heat transfer fluid (sea water) can also be utilized as a cooling medium in the district cooling system by passing it via a heat exchanger, as illustrated in Fig. 4.26.

The use of an open-loop power hydraulic system in an offshore wind turbine has shown several limits due to the medium of transmission (sea water), temperature differences, and environmental factors (humidity).

4.5 Overview of Wind Turbines and Its Components

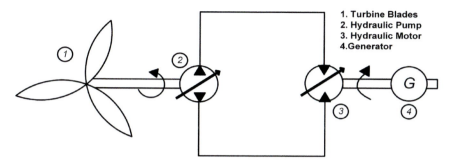

Fig. 4.25 Closed-loop hydraulic power transmission

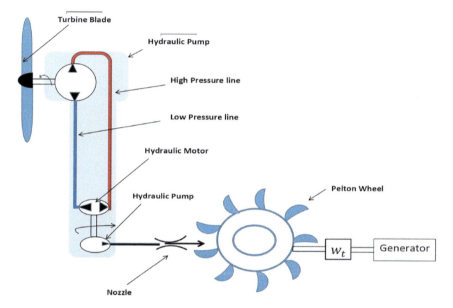

Fig. 4.26 Open-loop hydraulic power transmission

4.5.5 Generator

The generator does the work of converting mechanical energy from spinning into electrical energy. Throughout time, several generator types have been used in wind energy systems. Examples include the synchronous generator (SG) (wound rotor and permanent magnet), doubly fed induction generator (DFIG), and squirrel cage induction generator (SCIG), with power ratings ranging from a few kilowatts to several megawatts. The SCIG is made simply and robustly. It is reasonably priced and takes little upkeep. There are still direct grid-connected wind energy systems on the market today. These turbines all run at a set speed and use SCIGs. A tapped stator winding may be modified to modify the pole pairs in two-speed SCIGs so that they are commercially available. Figure 27a is an illustration of a 2 MW two-speed SCIG with 6/4 poles that is rated at 1000/1500 rpm. The SCIGs are also used in wind energy systems with varying speeds. The biggest SCIG wind energy systems are now 3.5 MW in size and are found in offshore wind farms. The current workhorse of the wind energy sector is the DFIG. The generator's stator is directly linked to the grid, while the rotor connects through a power converter system with a lower power capacity. For the majority of wind speed circumstances, the DFIG normally works roughly 30% above and below synchronous speed. Moreover, it permits grid- and generator-side reactive power regulation. Since the reduced-capacity converter is less costly and takes up less space, the DFIG WECS is a popular product on the market right now. Figure 27b depicts a huge DFIG as an example. Direct-drive wind turbines are well suited for the synchronous generator. In wind energy systems with a maximum power rating of up to 7.5 MW, wound rotor synchronous generators (WRSGs) and permanent magnet synchronous generators (PMSGs) are employed. Compared to wound rotor generators, permanent magnet generators are more efficient and have a greater power density. A shift towards direct-drive turbines with PMSG is suggested by recent developments. Several manufacturers have created SG turbines with gearbox drive trains even though the majority of SG-based turbines are directly driven. Examples of 4-pole and 120-pole PMSGs are shown in Fig. 27c, d respectively. Figure 4.27e depicts a multipole WRSG for a direct-drive WECS (Wu et al. 2011).

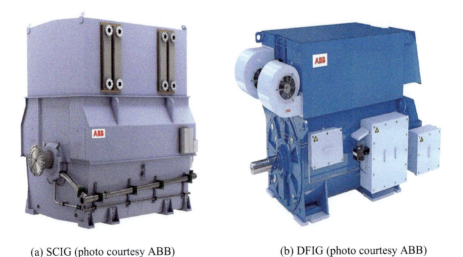

(a) SCIG (photo courtesy ABB) (b) DFIG (photo courtesy ABB)

Fig. 4.27 **a** SCIG (*photo courtesy* ABB). **b** DFIG (*photo courtesy* ABB). **c** Low-pole number PMSG (*photo courtesy* ABB). **d** High-pole number PMSG (*photo courtesy* Avantis). **e** WRSG High-pole number (*photo courtesy* Enercon GmbH)

4.5 Overview of Wind Turbines and Its Components

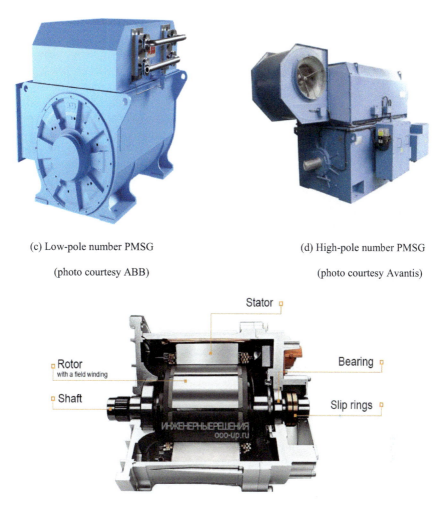

(c) Low-pole number PMSG

(photo courtesy ABB)

(d) High-pole number PMSG

(photo courtesy Avantis)

(e) WRSG High-pole number (photo courtesy Enercon GmbH)

Fig. 4.27 (continued)

4.5.6 Power Electronics Interface

The power source transforms rotational energy towards electric energy, which is then delivered into the electrical system via an electronic power link. The electrical/electronic connection must fulfil the both grid's and the generator's demands simply and affordably to maintain results because it is located in the generator for the wind turbine and the electricity system. This connection guarantees that the turbine's rotating speed is constantly modified on the generator side for extracting the optimum power from the wind going to follow the highest monitoring position. Inside the power system, regardless of wind speed, the electrical/electronic connection should meet system standards, such as the ability to manage the regulation of voltage, frequency, and active and reactive power.

Since the 1980s, the use of power electronics in wind turbine systems has steadily increased, having brought considerable shaft performance enhancements, not just by lowering physical pressure/load and improving power production, but also by allowing windmills will function as functional, programmable elements in the electrical grid and assist the electricity system in the same way that traditional power plants do. Elements can now tolerate larger ratings for both voltage and current, lowering energy waste and increasing device reliability.

Over the years, the most often utilized applications involving wind turbines, and power electronic connections are as follows:

- Soft starters are a basic and inexpensive electrical component for power that was first employed in windmills using SCIG in the 1980s to minimize a surge of current while grid interconnections, hence limiting grid disruptions. If gentle starting is not present, surge currents may rise to seven to eight times the authorized current or more, causing significant grid voltage fluctuations.
- An electrical capacitor bank device which offers reactive power to wind turbines' asynchronous provider banks of mechanically switchable capacitors has long been the simplest and most cost-effective technique to reduce the imaginary energy needed via generators of asynchronous connected to the system. Wind turbine producers may support the dynamic compensation for the entire load, in which a fixed set's capacitor is constantly either linked or unplugged based on the generator's standard power of reactance requirement over a predefined timeframe. Because an asynchronous generator's reactive power requirement is greatly influenced by wind velocity might affect the capacitor banks activated by a large quantity of changing occasions.
- A frequency converter, which has been used since 2000 in wind turbines, is a system that allows two electrical networks with different frequencies to be connected. It enables wind turbines to modify and manage generator frequency and voltage, improving their capacity to perform and function as active power system components. A conventional frequency converter, often known as an adjustable speed drive, is made up of the following components:

  1. Alternating current (AC) is converted to direct current using a rectifier (DC)).
  2. Capacitors.
  3. Inverters transform DC to AC.

Various back-to-back, multilayer, and resonant converter topologies have been studied in recent years to see if they may be employed in windmills. The converter for back-to-back is extremely important in the present design of wind turbines. It represents the present status and can thus be utilized to compare alternative converter topologies. The multilevel matrix converters have been the more severe back-to-back converter competitors and are thus suggested for further research.

4.5 Overview of Wind Turbines and Its Components   177

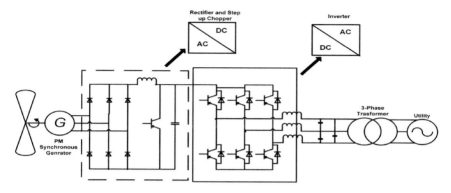

Fig. 4.28 Power electronic layout of a synchronous generator connected wind turbine system

The basic power electronic topologies for some generators have been discussed as:

A. Synchronous generator power electronics layout: The basic power electronic interface of a permanent magnet synchronous generator has been shown in Fig. 4.28. A diode bridge rectifies the wind turbine's three-phase varying output, i.e. voltage and frequency.

DC's side voltage of a diode converter fluctuates as the synchronous generator's speed changes. Using a step-up chopper modify the voltage to the rectifier to keep the inverter's DC link voltage constant. The system's generator and rectifier therefore treated as a prime source of current when seen to the inverter from the DC inputs. A huge capacitor filters the diode output that has been rectified bridge to create a smooth DC pulse. The DC signal is then reversed turning it into a three-phase, 60 Hz pulse using semiconductor switches. The voltage may be scaled to this waveform level needed by the AC system of the utility with a transformer. Because an electrically powered DC connection decouples the grid-powered generator, the PE interface offers tunable input to the wind energy grid. The grid-connected power converter allows for quick handling of power, both reactive and active. The disadvantage is that the system becomes more complicated, necessitating the use of more delicate power electronic equipment.

B. DFIG power electronics layout: An IGBT-based AC-DC-AC PWM converters are needed for the DFIG's power electronics. The stator is linked whereas the rotor is connected directly to the 60 Hz grid, supplied at a different using the AC-DC-AC converter, frequency. A typical DFIG electrical layout using a DC link to power a back-to-back rectifier and inverter exchanges has been shown in Fig. 4.29. The rectifier's AC side is linked to the induction machine's rotor by slip rings, and the inverter's output is associated with the utility system. Variable-speed fixed frequency is the classification for this system. The DFIG arrangement is typically seen in wind turbines with capacities larger than 1 MW. Modular multilevel converter designs are another PE-based architecture. These designs

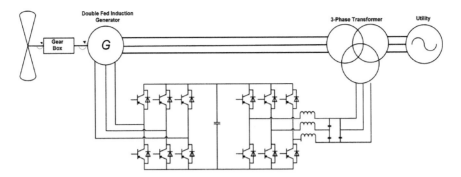

Fig. 4.29 Power electronic layout of DFIG connected wind turbine system

have the potential to increase efficiency and are currently in development, but they have yet to be applied in the market.

4.5.7 Control System

A windmill is developed to provide an optimum energy intake throughout variable wind velocities. Wind turbines are all built to withstand a certain highest wind velocity, known as the speed of survival. Commercial wind turbines have a survival speed of 144 km/h (89 mph) to 72 m/s (259 km/h, 161 mph) in speed. The more typical rate of survival speed is 60 m/s (216 km/h, 134 miles/h). There are three ways of functioning for wind turbines:

- Operation at lower-than-rated wind speeds.
- Operation at or near the average wind velocity (usually at nameplate capacity).
- Operation at higher-than-rated wind speeds.

The force has to be limited if the specified wind velocity is surpassed. A few methods to do this using the control system. A basic control system of a wind turbine comprises sensors, a system, and actuators comprised of devices and programmes that evaluate sensor input data and create actuator output signals. Generally, two control systems have been described as follows.

4.5.7.1 Control in a Closed Loop

A closed-loop microcontroller is often an application device that modifies the turbine's operating condition to maintain it on an operating curve or characteristic that has been predetermined.

1. **Generator Torque Control**: Variable-speed machines are used in today's huge windmills. Generator torque is reduced when the wind speed is less than the rated

4.5 Overview of Wind Turbines and Its Components

wind speed applied to adjust the rotational speed so that even more strength as you can is captured. When the tip speed ratio is maintained fixed at its very highest point, the peak strength is collected (often 6 or 7). It implies that when the wind's velocity rises, the speed of the rotor must rise in synch. The rotor speed is controlled by the variation in the dynamic feature collected through the propellers and the generator's applied torque. The rotor speeds when the torque of the generator is low and slows down when the generator torque is high. When the wind's velocity falls down the wind rating speed, control of the generator's torque kicks in, and the pitch of the propeller is normally kept at the fixed inclination that collects the greatest energy, which is relatively parallel to the wind. The generating torque is reduced when the wind speed exceeds the rated wind speed usually retained constant while the blade pitch remains active.

The fluctuating variable induction generator is a specific example in which the torque/speed correlation may be changed by actively controlling the impedance connected to the windings of the rotor. This is feasible to maintain continuous above-rated torque using closed-loop regulation according to present measurements, effectively permitting variable speed operation in this range.

2. **Control of Pitch**: The more popular way of controlling the aerodynamic power of the turbine rotor is pitch control. Pitch control will have a considerable influence on all aerodynamic loads on the rotor. The pitch method regulates the incline of blades in relation to the rotating plane. Smaller turbines lack a pitch mechanism, relying rather to stall in limit rotational speeds in strong wind speeds. Controlling the pitch of a turbine allows it to catch power with little wind speeds while maintaining a steady quantity of energy at higher wind speeds.

A system that regulates the angle of the blades can adjust the angle of pitch in numerous ways. The wind's velocity and energy output are monitored continuously by the control strategy, which changes the pitch of the blades as needed. When the wind velocity is higher unlike the valued wind velocity, the blades are pitched sufficiently to modify the angle of attack and cause stalling.

The turbine should just strive to create more power at all times below maximum wind speeds, therefore there is usually no requirement to change the angle of pitch. Because inertia-based loads under the maximum wind speeds are often smaller than those higher valued, pitch control is not required to regulate them. The ideal pitch angle for aerodynamic efficiency in fixed-speed turbines, on the other hand, changes little with wind speed. As a result, in reaction to a substantial power output signal or anemometer average, the pitch angle on certain turbines is slowly altered by a few degrees below rated. The control of pitch is a highly efficient way of managing the stresses and power generated by the rotor's aerodynamics above-rated wind speed so that technical restrictions are not surpassed. However, to be able to accomplish effective management, controls for pitch must react quickly to changing situations. Because it converses so aggressively, control action interacts substantially with the turbine dynamics and demands an expert layout.

To get the most power, the pitch parameter must be at its highest value. At periods of high wind is below maximum wind velocity. As a result, whenever the wind velocity exceeds the specified value, either a rise or fall torque will decrease as a consequence of a change in pitch angle. Lowering torque by tilting the leading edge towards the wind increases the pitch angle, which increases the angle of attack and, therefore, the lift. This is generally termed as pitching towards the feather while reduced pitch, or shifting the downwind leading edge, decreased lift is achieved by increasing torque when the angle of attack approaches stall begins to diminish and drag rises is termed as the stall pitches.

The pitch shaft is close to the hub in full-span pitch control and is used in the majority of pitch-controlled turbines. Aerodynamic control can also be achieved using a pitch mainly a blade's ends, or by modifying the aerodynamic qualities with air jets, ailerons, flaps, or certain instruments; however, this is not usual. In heavy winds, these tactics will cause the majority of the blade to stall (Liebst 2012). It can be challenging to suit an appropriate into the outboard actuator section of the propeller if just the tips of the blades are pointed, and maintenance can be challenging. The pitching control for the wind turbine has been given in the Fig. 4.30.

The most popular pitching systems are:

- Hydraulic system, where the blades are acted on by piston-driven cylinders operated by a pump for oil.
- Pinion and rack, where the rack has a circular shape and engages with the pinions having three or two propellers. There is a motor and a worm gear for driving the rack.
- Each blade is controlled by its motor. This is the most prevalent means of pitch control.

3. **Yaw Control**: Each HAWT must include a way to adjust the machine's orientation when the wind direction varies. Yaw motion has always been free in downwind machines. Like a weather vane, the turbine chases the wind. The

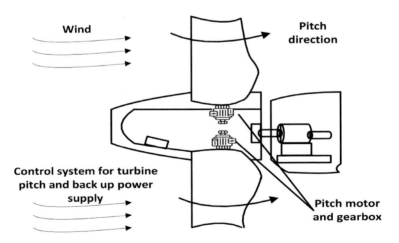

Fig. 4.30 Pitch control system for wind turbine

4.5 Overview of Wind Turbines and Its Components

propellers are often a couple of degrees coned lower in order for the yaw to operate well. Dampers for yaw are occasionally employed in loose yaw wind devices to control the rate of yaw and consequently the gyroscopic stresses on the propellers. Active yaw control is almost always included in upwind turbines. When correctly aligned, this normally comprises keeping the turbine fixed in yaw, there is a yaw motor, gearing, and brake. Towers housing torsional forces must be manageable for turbines with active yaw stresses that the yaw system will generate.

The turbines in a yaw system face the wind or align the rotational axis such that it is at 90° to the wind's direction. Passive yaw is used in small turbines, and there are two kinds: (1) a tail vane that orients the rotational plane, and (2) a downstream turbine, where blades are turned by the wind passing through the nacelle. Nearly total major unit-scale turbines have active yaw and are upwind turbines. Because it uses an electrical structural motor with a wind direction monitoring control system, active yaw is more costly. The nacelle structure and its equipment link them from the tower's nacelle house the yaw motor.

The yaw system also incorporates brakes that lock the yaw position. A mean valued incline between the route of the wind and the rotational axis is caused by poor control techniques or faulty yaw drive operation. As a result, energy output is reduced and non-symmetrical loads are increased. Big wind turbines in the modern era are carefully controlled to head the route of the wind, which is sensed on the rear side of the nacelle. The energy output is maximized and non-symmetrical loads are reduced by decreasing the yaw angle (the mismatch between the turbine and wind heading orientations). The wind direction varies quickly, and the turbine would not match the direction exactly and would be yawed somewhat. Yawing can reduce turbine production considerably at low to moderate winds, with wind direction fluctuations of 30° being frequent, and turbine reaction times to changes in wind direction being long. The wind direction generally varies less for higher wind speeds (Ciri et al. 2018). Figure 4.31 describes a diagram depiction of adjusting pitch and yaw.

The transverse plane motion of the complete term "yaw" refers to a wind turbine. Yaw management maintains the turbines pointing all the time, into the wind, increasing the useful rotors' size and, as a consequence, power. The turbine might be out of alignment with the approaching wind, which causes energy generation waste since the wind's direction might quickly shift.

4. **Stall Control**: Stalling minimizes generated and raises the angle at which the drag comparative wind attacks the propellers. Stalling is easy as it can be executed reactively (it rises when the wind speeds up), and it maximizes the blade cross-sectional area towards the wind and hence the normal drag. When a blade of the turbine is entirely stalled and halted, the blade's flat side points directly towards the wind.

In comparison to its pitch-regulated version, the turbine must run closer to stall in order to achieve stagnation-controlled at moderate flow speed, which leads to a reduced aerodynamic performance than valued. This drawback in a variable-speed

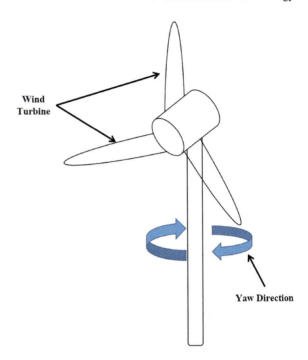

Fig. 4.31 Yaw adjustment

turbine may be overcome by adjusting the rotation speed under the base speed to maintain the highest capacity factor.

The rotor speed must be controlled for the turbines to halt, instead of quickening in strong winds. At greater wind speed, the attack angle of a fixed-speed HAWT automatically rises as the speed of the blades accelerates. If the wind speed picks up, the blade might stall is a natural approach. Previously, HAWTs were generally successfully employed using this strategy. However, it was discovered that the blade pitch degree seemed to increase several of these blades' audible noise level pairs (Fig. 4.32).

Figure 4.32 shows that at higher speeds the stall-regulated wind turbine power gets reduced while pitch-regulated regulated becomes constant (Kerho 2012).

The frequency of the power supply limits the rotor speed in a fixed-speed turbine given that the speed stays under the pulling force. The speed of a turbine with variable speed is kept constant by varying the producer acceleration to meet the torque of aerodynamics. In heavy winds, an adjustable rotor can slow down the rotor to put it into the stall. In low winds, hence, the turbines may run moving away from the stagnation spot, resulting in increased aerodynamic efficiency. When a gust strikes the turbine, however, the load speed must not only increase to meet the torque due to wind, but it must accelerate further to prevent the rotor from stalling. One of the key benefits of functioning at varied speeds is that it allows for extremely gentle adjustment of power and torque over the specified levels.

4.5 Overview of Wind Turbines and Its Components

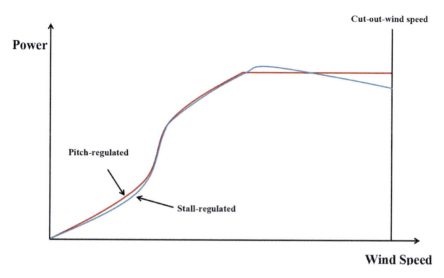

Fig. 4.32 Stall control (Kerho 2012)

4.5.7.2 Administrative Regulation

Administrative regulation may be thought of as the process of moving a turbine from one operational condition to another. The working stages might involve:

- Keep an eye out while the turbine is ready to run if external circumstances allow.
- Beginning.
- Energy generation.
- Switch-off.
- Shutdown (if fault occurs).

Other states could be imagined, or some of these states could be subdivided even further. The administrative regulation will control of sequence in addition to selecting when to begin a changeover from one state to another. The control of sequence to the launch of a wind turbine with a set speed and pitch control may consist of steps which are as follows:

- Powering the pitch actuator.
- Shaft brake release.
- Pitch position the Ramp requirement up to a certain initial pitch at a certain rate.
- Wait till the rotor speed reaches a minimal threshold.
- Use speed pitch control in a circuit.
- Increase the desire for speed until it reaches the fixed speed.
- Wait till the speed is near to the intended speed for a certain amount of time.
- Generator contractors should be connected.
- Apply power pitch control in a closed loop.
- Increase the power consumption until it reaches the rated level.

Before going on to the next position, the administrative regulation should ensure that every phase has been effectively finished. The administrative regulation must switch to switch-off intervals if any step is not finished within a particular amount of time or if any defects are identified.

4.6 Power Coefficients and Characteristics

4.6.1 Introduction

A wind turbine can be referred to as a device which has the potential to transform the kinetic energy available in the wind into mechanical energy and finally into electrical energy but it is in contrast to the "windmill," behaviour that generates the mechanical power from the wind power. Like electrical generators, wind turbines are associated with some power networks. These organizations incorporate battery-charging circuits; private-scale power frameworks separated or island organizations, and huge utility grids. The most commonly available wind turbines are typically very limited having the capacity of 10 kW or less. Wind turbines having a rating of up to 30 kW can be installed to harness wind energy not only in rural but in urban regions also. Furthermore, they can be utilized for several various applications in areas where there are no power sources or where continuity of electrical power cannot be assured.

Many agriculture firms or small and medium-sized organizations can benefit from this effort. Larger turbines generally have a total generating capacity in the range of 1.5–5 MW. These huge-sized turbines are utilized basically in bigger utility frameworks, from the start generally in Europe and the United States and all the more as of late in China and India. Figure 4.33 shows a modern wind turbine, i.e. General Electric 1.5 MW in a wind farm set up, linked to a transmission network.

Wind turbines of larger capacity and photovoltaic boards have progressed so much that the technology has developed drastically. Be that as it may, the innovation of small-sized wind turbines has not yet developed similarly, with the end goal that they have lower effectiveness. Two significant reasons are recognized as being liable for lower productivity. Firstly, the plan of bigger wind turbines cannot just be downscaled to wind turbines of small capacity because of the distinction in wind conditions. Large varieties of wind speed, which are regular at a tallness of 15 m, because of fast changes in the turbine's rate of rotation.

To see how windmills are utilized, it is beneficial to quickly recognize a portion of the principal realities fundamental to their activity. In new wind turbines, the real change measure utilizes the fundamental streamlined power lift required to produce a total benefit force on a pivoting shaft, coming about first in the generation of mechanical force and afterwards its change to electrical power in a generator. Wind turbines, in contrast to most different generators, can deliver power because of the quickly accessible resource only.

4.6 Power Coefficients and Characteristics

Fig. 4.33 Wind turbine used nowadays in utilities (*source* General electric)

Wind cannot be stored and utilized as per our requirements like other energy sources. Hence the wind power is always varying and cannot be transported efficiently. (The most one can do is to restrict the generation beneath the wind may deliver.) Any framework with which a windmill is associated might, here and there, consider this fluctuation. In a bigger electrical network, a windmill operates to lessen the absolute electrical burden and hence brings about a decline in either the quantity traditional producers are utilizing or in the fuel utilization the ones that are moving quickly. There is the possibility that a smaller network is equipped with backup generators, energy storage, and other advanced regulator systems. A further truth is that the wind cannot be transported it must be utilized over where it is blowing. Truly, an item, for example, the windmill ground the wheat, and afterwards moved to its place of utilization. Today, the plausibility of passing on electrical power through electrical cables repays somewhat for the wind transportation issue. Later on, hydrogen-based frameworks may add to this chance.

4.6.2 Tip Speed Ratio

The Tip speed ratio normally is written in short as TSR has a greater role in wind turbine manufacturing. It is generally defined as the ratio of the outer blade tip speed of the turbine to the unperturbed wind speed and denoted by the symbol "λ" which is given as:

$$\lambda = \frac{r \cdot \Omega}{v}, \tag{4.21}$$

where r is the radius of the outer blade Ω is the angular or rotational frequency and v is wind speed.

The wind turbine rotor will rotate fast if this ratio is more at a specific wind speed. Wind turbine output is influenced by the number of propellers that comprise a rotor and the total size that they encircle. The wind must flow smoothly over the blades for a lift-type rotor to work effectively. Spacing between blades should be adequate to eliminate vibration such that one blade does not experience the disrupted, poorer airflow induced by the blade that went before it. The majority of the wind volume will go undisturbed through the gap area among the blades of the wind turbine rotor. On the other hand, if the rotor turns excessively fast, the rotor's sharp edges will seem like a strong opposition to the flow of wind. Hence, wind turbines are planned with ideal tip speed proportions to extricate however much force out of the breeze as could reasonably be expected (Parker and Leftwich 2016).

The number of blades in the wind turbine rotor determines the ideal tip speed ratio. To capture the most power from the wind, the wind turbine rotor must revolve more quickly as the number of blades decreases. The ideal tip speed ratio for a two-bladed rotor is around six, for a three-bladed rotor it's about five, and for a four-bladed rotor, it's about three.

High rotational speeds are necessary for generating electricity. Wind turbines of the lift type have overall tip speed ratios of about 10, while drag-type ratios are around 1. The lift-type wind turbine is the most practical for this use, considering the high rotational speed specifications of electric generators.

Example 4.3 Calculate the tip speed ratio of the wind turbine having a blade radius of 10 m and rotating at 1 rotation per second at a wind speed of 15 m/s.

Solution: Frequency, $f = 1$ [rotation/s], $\omega = 2\pi f = 2\pi$, Radius $r = 10$.
So Wind speed, $v = wr = 2\pi \cdot 10 = 20\pi$ m/s.
Hence tip speed ratio, $\lambda = \frac{\omega r}{v} = \frac{20\pi}{15} = \frac{62.83}{15} = 4$.

4.6.3 Wind Power Coefficient and Betz's Law

The wind power coefficient represents the performance of a wind turbine. It is defined as the ratio of the total mechanical power extracted at the output to the total wind power available at the input side. It is a dimensionless quantity and expressed as (Twidell and Weir 2015):

4.6 Power Coefficients and Characteristics

$$C_p = \frac{\text{Total output power extracted}}{\text{Total power in the wind}} \tag{4.22}$$

The maximum value of the power of coefficient can be calculated from the expressions maximum power extracted and the power of the wind which are already derived in earlier sections as follows:

$$\text{The total maximum power extracted} = P_{\text{max}} = \frac{8}{27}\rho A v_i^3. \tag{4.23}$$

$$\text{Power available in the wind} = P_{\text{total}} = \frac{1}{2}\rho A v_i^3. \tag{4.24}$$

So, the maximum value of the power coefficient

$$= C_{p_{\text{max}}} = \frac{\frac{8}{27}\rho A v_i^3}{\frac{1}{2}\rho A v_i^3} = \frac{16}{27} = 0.593. \tag{4.25}$$

The power coefficient's greatest value is referred to as Betz's bound.

According to Betz's law, which was proved nearly a century ago by the German physicist Albert Betz, no turbine can capture more than 59.3% of the power available in the wind and he also verified theoretically that the maximum power that can be derived from the wind or some free flow, regardless of the turbine size (Ragheb 2021).

Example 4.4 A wind turbine with a rotor diameter of 20 m operates in an area where the wind speed is consistently 8 m/s. The air density (ρ) in the area is 1.225 kg/m³. Calculate the maximum power extracted from the wind.

Solution: Given rotor diameter $D = 20$ m and Wind speed $v = 8$ m/s.

So, the swept area $A = \pi \frac{D^2}{4} = \pi \frac{20^2}{4} = 314$ m².

Using the above data the actual power $P = 0.5 \cdot \rho.A.v^3 = 0.5 \cdot 1.23 \cdot 314 \cdot 8^3 = 98872.32$ W.

As we know, the theoretical value of the wind power coefficient is 59.3%. The maximum wind power can be calculated as:

Maximum power extracted from the wind, $P_{\text{max}} = C_{p\,\text{max}} \cdot P = 0.593 \cdot (98872.32) = 58334.66$ W.

So, the maximum wind power extracted from the turbine is 58.33 kW.

Practically, the maximum power coefficient can have lower values due to the following factors:

- Rotates the wake left by the rotor.
- Restricted blade count and related tip losses.
- Aerodynamic drag from nonzero.

It is seen that the more the value of the power coefficient the better be power extraction from the wind turbine. As the power coefficient value will increase the wind power extraction efficiency will also increase which make the system more economical. The derived power often varies with the difference in the power coefficient. The reason behind this is the shift in their measured velocity and rotor diameter. It is analysed that the susceptibility of power to power coefficient thus improves with the rise in rotor diameter. That's why it is advised to keep the power coefficient values in a limited range for bigger machines.

The maximum power efficiency of an ideal wind turbine in terms of percentage can be calculated by taking the ratio of the maximum power extracted by the wind turbine to the total power available in the wind at the input side of the turbine which is based on the Betz's law and equals to maximum power coefficient, i.e. $C_{p_{max}}$ So the maximum power efficiency can be expressed as:

$$\eta_{max} = C_{p_{max}} \times 100 = 0.593 \times 100 = 59.3\%. \tag{4.26}$$

For a practical situation, the actual efficiency of the turbine is always less than the maximum efficiency. A commercial wind turbine has a maximum power coefficient of about 0.4 in practical service. This can be referred to as having an efficiency of 68% relative to the Betz criterion. Practically the actual efficiency of a wind turbine is approximately 60% of the maximum efficiency. Hence the actual efficiency will be:

$$\eta_{actual} = 0.6 \times \eta_{max} = 0.6 \times 0.593 \approx 0.35 \approx 35\%. \tag{4.27}$$

So, we can say that the actual efficiency of a wind turbine is 35% which is not very impressive, so extra efforts need to be carried out to improve this wind turbine efficiency so that maximum electrical output can be generated and optimum utilization of wind energy can be achieved.

4.6.4 Power Coefficient Versus Tip Speed Ratio Curve

A wind turbine's output can be characterized by the way the three key indicators, strength, torque and thrust, differ with wind speed. The strength determines the amount of energy collected by the rotor, the torque produced determines the size of the gearbox, and whatever engine is powered by the rotor must be balanced. The thrust of the rotor affects the structural architecture of the tower tremendously. Normally it is helpful to indicate the behaviour of the wind energy system with the help of non-dimensional characteristic performance curves. The real performance can be obtained by these curves easily without considering too much about how the turbine is worked, e.g. at fixed rotational speed or some variable rotor speed. Expecting that the streamlined exhibition of the rotor's sharp edges doesn't break down the non-dimensional streamlined execution of the rotor will rely on the tip speed ratio and, if suitable, the pitch setting of the rotor blades. It is common, subsequently, to show the power, torque and thrust coefficients as elements of the tip speed ratio.

4.6 Power Coefficients and Characteristics

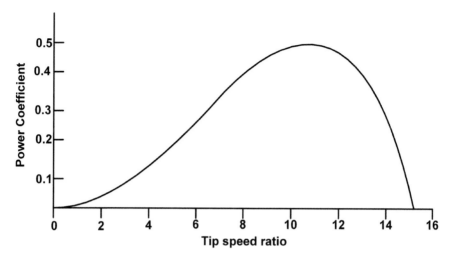

Fig. 4.34 Sample C_P–λ curve for a high tip speed ratio wind turbine (Kooning et al. 2013)

After the blade has been designed for efficient operation at a given tip speed ratio, it is important to evaluate the rotor's output overall predicted tip speed ratios. This can be achieved using various methods. The aerodynamic requirements on each blade segment must be calculated for each tip speed ratio. From these, the overall efficiency of the rotor can be analysed. The outcomes are normally introduced as a graph of power coefficient versus tip speed ratio, called a C_P–λ curve, as appeared in Fig. 4.34.

C_P–λ curves can be utilized in wind turbine configuration to decide the rotor power for various wind and rotor speed values. They give a fast response regarding the maximum rotor power coefficient and ideal tip speed ratio. This performance curve must be utilized carefully so that the result should be optimum. The information for such a relationship can be found in turbine tests or different turbine models. In either case, the outcomes rely upon the lift and drag coefficients of the airfoils, which may differ as an element of the stream conditions. Varieties in airfoil lift and drag coefficients rely upon the airfoil and the Reynolds numbers being thought of, at the same time, airfoils can have unique conduct when the Reynolds number changes by as meagre as a factor of 2 (Kooning et al. 2013).

The other key parameter to consider at this point is solidity, defined as the total blade area divided by the area swept. By modifying the blade chord, the solidity may also have been affected. The effect of solidity on the C_P–λ curve can be observed as shown in Fig. 4.35.

- A long, flat curve is generated by low solidity, which indicates that the power coefficient value will not vary considerably over a large value of tip speed ratio, but since the drag losses are more, the maximum value of the power coefficient is less (drag losses approximately vary with the cube of the tip speed ratio).

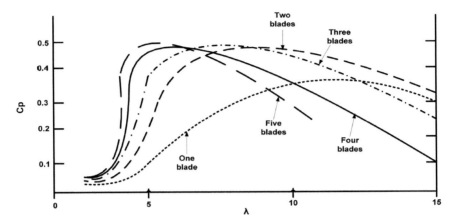

Fig. 4.35 Solidity effect on the C_P–λ curve

- A narrow performance curve having a sharp peak caused due to the high solidity value makes the turbine too sensitive as the tip speed ratio varies. The power coefficient has a lower maximum value if the solidity is increased too much. Stall losses are responsible for the decline in the maximum value of the power coefficient.
- A better result for solidity has all the earmarks of being accomplished with three rotor blades, however, two blades may be an adequate option because although the greatest value of C_p is a little bring down the spread of the pinnacle is more extensive and that may bring about a bigger wind energy catch.

A performance map of five different types of turbines is shown in Fig. 4.36 (Manwell et al. 2010). For the ratio of blade tip speed to wind speed, the performance or power coefficient varies, with the peak value being the amount cited for a turbine reference. The two-bladed propellers, the Darrieus and the Savonius, have peak efficiencies of over 30%, while the American Multiblade and the Dutch windmills range at around 15%.

These efficiencies suggest that the American Multiblade is not sufficient for power generation, although it is almost the best and most efficient for water pumping.

Review Questions

(1) What variation of wind speed is ideal for the production of wind energy?
(2) Which elements prompted the rapid growth of wind energy?
(3) Indicate the direction of the global wind circulation using a diagram. What factors govern the velocities and directions of the world's winds?
(4) Describe the process through which regional winds are produced.
(5) What are the determinants of the distribution of wind energy over the earth's surface?
(6) What does gust mean to you?

4.6 Power Coefficients and Characteristics

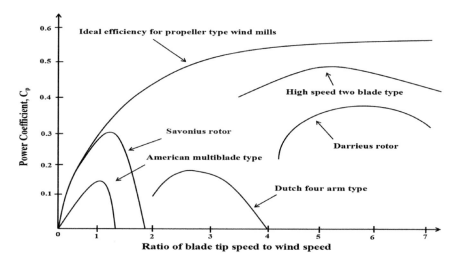

Fig. 4.36 C_P–λ curve for different wind machines

(7) What criteria can be applied to determine wind speed? What height is considered the norm for measuring wind speed?

(8) What are the benefits of displaying the wind information as a wind rose?

(9) Provide an explanation of the wind speed fluctuation with height from the ground using a diagram. Describe the meanings of the following terms: Ekman layer, planetary boundary, surface layer, and wind shear.

(10) What locations are more suitable for erecting wind turbines?

(11) Describe the main uses of wind energy.

(12) Explain the words solidity, pitch angle, chord, drag and lift forces, and free and relative wind velocities, with the aid of a figure.

(13) What does "connected" and "separated flow" mean to you? Show a connected and halted flow using a diagram. What negative characteristics of halted flow exist? How is the functionality that has stagnated enhanced?

(14) Create a formula to describe the energy in the wind.

(15) Determine the formulation for power collected from the wind using the Betz model of a wind turbine. What are the conditions and greatest theoretical power that can be extracted?

(16) Get the conditions for such operation as well as the expression for the greatest axial thrust that a wind turbine may experience.

(17) Describe how a wind turbine's power production varies depending on the rotor's tip speed.

(18) Draw a HAWT schematic and describe the roles of its constituent parts.

(19) Describe the different HAWT blade designs and their associated attributes.

(20) What do you mean when you say a rotor is teetering? What scenarios call for it?

192   4   Introduction to Wind Energy

(21) What do you mean by a machine being upwind, downwind, or having its yaw fixed?

(22) Draw a VAWT's schematic and describe the roles of its constituent parts.

(23) Describe the different VAWT blade designs and their associated attributes.

(24) What are the differences between machines of the drag and lift types?

(25) What impact does solidity have on a wind turbine's performance?

(26) What similarities and differences do HAWT and VAWT share?

(27) Discuss the properties of a wind turbine's power vs wind speed using a diagram.

(28) Compare the differences between a stalled-regulated and pitch-regulated wind turbine's performances.

(29) Explain the purposes of the different WECS blocks using a block diagram.

(30) Analyse the appropriateness of different generator types for the production of wind energy.

(31) Talk about the different driving systems used in wind turbines.

(32) Describe the major characteristics of hybrid wind-diesel producing systems. Additionally include the different diesel unit operating scheduling schedules.

(33) Comments about the effects of wind energy on the environment.

Problems

(1) For a two-blade HAWT, the following information was gathered: At a typical height of 10 m, the average free wind speed is 8 m/s or $\alpha = 0.13$ m/s. The air density is 1.226 kg/m^3, the hub height above the ground is 80 m, and the rotor diameter is 60 m. Find

    (a) Wind energy is accessible for use.
    (b) Energy generated by the turbines
    (c) Force axial on the turbines
    (d) Force axial on the turbine
        i. when the most power is derived, and
        ii. When no power is extracted and the blade stalls completely.

Ans: 1.995 MW, 1.1219 MW, $1.429 * 10^5$ N, $1.693 * 10^5$ N, Zero

(2) A HAWT has the following data: speed of wind $= 10$ m/s at 1 atm and 15 °C, Diameter of rotor $= 120$ m, Speed of rotor $= 40$ rpm. Calculate the maximum possible torque produced at the shaft.

Ans: 0.98

(3) A two-blade HAWT is installed at a location with a free wind velocity of 20 m/s. The rotor diameter is 30 m. What rotational speed should be maintained to produce maximum output?

Ans: 80 rpm

(4) A multi-blade windmill lifts 1.03 m^3/h of water through a head of 28 m when the wind speed is 3.3 m/s. Calculate the power coefficient if the rotor diameter is 4.5 m, given that the transmission efficiency $= 0.95$ and pump efficiency $= 0.7$.

4.6 Power Coefficients and Characteristics 193

Ans: 0.343

Objective Type Questions

(1) The range of wind speed suitable for wind power generators is

- (a) 5–25 m/s
- (b) 0–5 m/s
- (c) 50–75 m/s
- (d) 25–50 m/s

(2) Favourable winds for small-scale applications exist

- (a) On 75% of the earth's surface
- (b) Everywhere on the earth's surface
- (c) On 25% of the earth's surface
- (d) On 50% of the earth's surface

(3) Compared to conventional sources of power such as thermal plants, the cost of wind power is

- (a) 10 times
- (b) 100 times
- (c) 1/10 times
- (d) Comparable

(4) The energy payback period of wind generation is

- (a) 2 years
- (b) 1 year
- (c) 4 years
- (d) 3 years

(5) There is little wind in the

- (a) South pole region
- (b) North pole region
- (c) $\pm 5°$ around the equator
- (d) Tropical region

(6) When solar radiation falls on the earth's surface, the temperature of

- (a) Land mass rises slower than water mass
- (b) Land mass rises faster than water mass
- (c) Only land mass increases and water remains at a fixed temperature
- (d) Land mass and water mass rise uniformly

(7) Stalled flow occurs when the value of the incident angle is

- (a) 180°

194　　　　　　　　　　　　　　　　　　　　　　　　4　Introduction to Wind Energy

(b) 0°
(c) Beyond 16°
(d) In the range from 0 to 16° (approximately)

(8) A wind turbine extracts maximum power from the wind when the downstream wind speed reduces to

(a) Half that of upstream wind
(b) One-third of upstream wind
(c) Zero
(d) Two-thirds of that upstream wind

(9) If the speed of a wind stream remains unchanged while passing through the rotor,

(a) Zero power will be generated
(b) A large power will be generated
(c) The speed of the rotor will be very high
(d) The flow is known as stalled flow

(10) As per Betz criterion, the maximum energy extractable by an ideal wind turbine is

(a) 39% of that available in wind
(b) 29% of that available in wind
(c) 59% of that available in wind
(d) 49% of that available in wind

(11) The maximum axial thrust occurs when interference factor α is

(a) 0.33
(b) 0
(c) 1.0
(d) 0.5

(12) A two-blade wind turbine produces maximum power when the tip speed ratio is equal to

(a) 2π
(b) 1π
(c) 0.593
(d) 3π

(13) The wind turbine rotor has a low value of solidity

(a) Runs faster
(b) Runs slower
(c) Has low efficiency
(d) Produces high torque

(14) Stall regulation is used with turbines

    (a) Having diameters of more than 25 m
    (b) Having diameters less than 25 m
    (c) Having rotors with pitch control
    (d) Having rotors of large solidity

(15) Grid-connected wind generators usually have maximum penetration of:

    (a) 20–30%
    (b) 10–20%
    (c) 40–50%
    (d) 30–40%

(16) As per the size of the installed capacity of wind power generation in the world, India ranks

    (a) Second
    (b) First
    (c) Fourth
    (d) Third

(17) Potential sites for wind generation are those having average energy densities of

    (a) $50–100$ w/m^2
    (b) $25–50$ w/m^2
    (c) 200 w/m^2 or more
    (d) $100–200$ w/m^2

Answer

1. a	2. a	3. d	4. b	5. c	6. b	7. d	8. b	9. a
10. c	11. d	12. a	13. a	14. b	15. b	16. c	17. c	

References

Ali MH (2017) Wind energy systems: solutions for power quality and stabilization. Routledge
Al-Shemmeri T (2010) Wind turbines, p 83
Bezzaoucha FS, Sahnoun M, Benslimane SM (2018) Failure causes based wind turbine components classification and failure propagation: for proactive maintenance implementation. In: 2018 international conference on wind energy and applications in Algeria. ICWEAA 2018. https://doi.org/10.1109/ICWEAA.2018.8605082
Breeze P (2016) Rotors and blades. In: Wind power generator, pp 29–40
Burton T, Jenkins N, Bossanyi E, Sharpe D, Graham M (2021) Wind energy handbook (3rd edn). Wind Energy Handb, pp 1–952. https://doi.org/10.1002/9781119992714

Ciri U, Rotea MA, Leonardi S (2018) Effect of the turbine scale on yaw control. Wind Energy 21:1395–1405. https://doi.org/10.1002/WE.2262

De Kooning JDM, Gevaert L, Van De Vyver J, Vandoorn TL, Vandevelde L (2013) Online estimation of the power coefficient versus tip-speed ratio curve of wind turbines. IECON Proc 1792–1797. https://doi.org/10.1109/IECON.2013.6699403

El-Ahmar MH, El-Sayed AHM, Hemeida AM (2017) Evaluation of factors affecting wind turbine output power. In: 2017 19th international middle-east power systems conference MEPCON 2017—Proceedings of 2018-February, pp 1471–1476. https://doi.org/10.1109/MEPCON.2017. 8301377

Global Wind Energy Council (2022) GWEC Global Wind Report 2022. Glob Wind Energy Counc 102:140

Hartmann DL (2016) Atmospheric general circulation and climate. Glob Phys Climatol 159–193. https://doi.org/10.1016/B978-0-12-328531-7.00006-2

Meserve JM (1978) U. S. Navy marine climatic atlas of the world. North Atlantic Ocean, vol 1

Justus CG (1978) Use of power laws in analyzing natural phenomena. Transactions of the ASAE 21(2):366–374

Justus CG, Mikhail A (1994) Height variation of wind speed and wind turbulence statistics. J Wind Eng Ind Aerodyn 52(3):305–315

Kerho M (2012) Adaptive airfoil dynamic stall control. J Aircraft 44:1350–1360. https://doi.org/ 10.2514/1.27050

Liebst BS (2012) Pitch control system for large-scale wind turbines. J Energy 7:182–192. https:// doi.org/10.2514/3.48074

Mahato AC, Ghoshal SK (2019) Various power transmission strategies in wind turbine: an overview. Int J Dyn Control 7:1149–1156. https://doi.org/10.1007/S40435-019-00543-8/FIGURES/12

Manwell JF, McGowan JG, Rogers AL (2010) Wind energy explained: theory, design and application. https://doi.org/10.1002/9781119994367

Mohsin M, Rao KVS (2018) Estimation of weibull distribution parameters and wind power density for wind farm site at Akal at Jaisalmer in Rajasthan. In: 3rd international innovative applications of computational intelligence on power, energy and controls with their impact on humanity. CIPECH 2018, pp 14–19. https://doi.org/10.1109/CIPECH.2018.8724170

Nayak AK, Mohanty KB (2018) Analysis of wind characteristics using ARMA Weibull distribution. In: 2018 national power engineering conference, NPEC 2018

Parker CM, Leftwich MC (2016) The effect of tip speed ratio on a vertical axis wind turbine at high Reynolds numbers. Exp Fluids 57:1–11. https://doi.org/10.1007/S00348-016-2155-3/FIG URES/9

Ragheb M (2021) Wind energy conversion theory. Betz equation

Rohatgi VK, Nelson TA (1994) A comparison of two wind spectra. J Wind Eng Ind Aerodyn 51(2):137–146

Rohatgi VK, Webber JD (1985) Comparison of measured wind spectra with some theoretical models. J Wind Eng Ind Aerody 20(2):195–212

Twidell J, Weir T (2015) Renewable energy resources

Umanand L (2007) Wind generation, history of wind-mills

Van der Hoven I (1957) Power spectra of horizontal wind speed in the frequency range from 0.0007 to 900 cycles per hour. J Meteorol 14(2):160–164

Wind Power Installation. http://www.inwea.org/wind-energy-in-india/wind-power-installation/

Wind Power Potential. http://www.inwea.org/wind-energy-in-india/wind-power-potential/

Wu B, Lang Y, Zargari N, Kouro S (2011) Power conversion and control of wind energy systems. Power Convers Control Wind Energy Syst. https://doi.org/10.1002/9781118029008

Xu C, Yan Y, Liu D, Zheng Y, Li C (2011) Study of different anemometer time intervals influence on wind speed probabilistic distribution parameters. In: 2011 international conference on electrical and control engineering ICECE 2011—Proceedings, pp 1744–1747. https://doi.org/10.1109/ ICECENG.2011.6058397

Chapter 5
Wind Energy Conversion System

Abstract Detailed analysis of wind energy conversion systems (WECS) has been thoroughly discussed in this chapter. Followed by wind turbine topologies, including designs with horizontal or vertical axis layouts architectural styles, and constant or variable speed designs. To maximize energy extraction from wind turbines and ensure efficient power conversion, the last section deals with converter control strategies and maximum power point tracking control.

Keywords Wind energy conversion system (WECS) · Soft-starter · Stall and pitch aerodynamic power control

5.1 Introduction

Due to the rising costs, depleting reserves, and unfavourable environmental effects of fossil fuels during the past 20 years, renewable energy sources have gained a lot of attention. Some such renewable energy sources are now more competitive on the market thanks to technology developments, cost reductions, and governmental subsidies. One of the sources of renewable energy that is developing the quickest among them is wind energy (Derbel and Zhu 2019).

For centuries, people have utilized wind energy to sail the oceans, pump water, and process crops. The creation of a 12 kW DC windmill generator in the late nineteenth century is when windmills were first used to produce electricity. Yet it wasn't until the 1980s that technology advanced enough to create power reliably and efficiently. A range of wind power technologies has been created during the past 20 years, improving conversion efficiency and lowering production costs for wind energy. Wind turbines now range in size from a few kilowatts to many megawatts. A massive size of wind turbines has been placed offshore and onshore to generate more energy and lessen their effect on land usage and the environment (Ali 2017).

This chapter gives a summary of wind energy conversion systems (WECS) and the technologies that support them. The chapter's goal is to give background information on several topics relevant to this innovative technology and market trends, including installed capacity, rate of growth, and costs.

© The Author(s), under exclusive license to Springer Nature Singapore Pte Ltd. 2024
K. Namrata et al., *Wind and Solar Energy Systems*, Energy Systems in Electrical Engineering, https://doi.org/10.1007/978-981-99-9710-7_5

5.2 Overview of Wind Turbine Topologies

Wind energy generation represents one of the most cost-effective and environmentally sustainable means of producing electricity from renewable sources, and it has been highlighted as the one with the fastest-evolving technology. Wind energy conversion systems transform the kinetic energy of the wind into electricity or other forms of energy. Wind power generation has increased dramatically over the past 10 years and is now recognized as a reliable and economical way to generate electricity (Yaramasu and Wu 2016). The kinetic energy (K.E.) of the wind is converted into mechanical or electrical energy by all wind energy systems. Wind turbines come in a wide range of sizes, but they all function in the same manner, no matter how big they are. The arrangement is the same overall. Each system consists of a nacelle (enclosure), which houses the drive system, a rotor (blades) and a generator, which changes the wind energy into shaft energy for rotation (Iov and Blaabjerg 2009).

The kinetic energy (K.E.) that powers the wind is applied to propel the blades. This mechanical energy is subsequently used to move the drive train which is then changed into electrical energy in the generators, which is subsequently transmitted to utility companies or home power networks for routine use or stored in batteries. Figure 5.1 represents a brief process flow of the WECS (https://ieeexplore.ieee.org/book/6047595).

The installed capacity, efficiency, and aesthetics of wind turbines have all considerably increased since the 1980s when the very first commercialized wind turbines were built. Despite various approaches that have been explored in search of the optimal turbine design, there has been a substantial amount of standardization in the last ten years. Today, the great majority of commercial turbines have three uniformly spaced blades on a horizontal axis. They are connected to a rotor, which uses a gearbox to transmit power to a generator. A box called a nacelle houses the gearbox and generator. Certain turbine models are not equipped with a gearbox in favour of direct driving. The electrical power is then transferred from the tower to a transformer, and then it is sent to the utility.

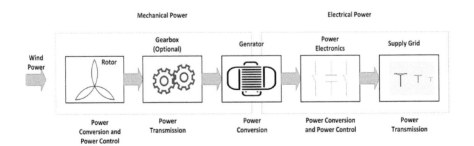

Fig. 5.1 WECS layout

Wind turbines may operate at speeds between, 3–4 m/s to around 25 m/s, or 90 km/h (56 mph) or gust 9 or 10. Nowadays a large number of turbine types make use of the ongoing wind variations by controlling pitch, which is done by turning or yawing the whole rotor as the wind direction varies and by running at varying speeds. The capacity to operate at variable speeds boosts the turbine's ability to integrate with the functioning of the electrical grid and respond to changing wind speeds. Advanced control systems allow for precise adjustment of the turbine's operation and power production.

Existing wind technology can function properly in a number of environments, including severe climates, deserts, and both low and high wind speeds. Wind farms, which are arrangements of turbines that operate with high reliability and often get along with the environment, are accepted by the community (Lee and Lee 2017). Emerging turbine designs employ light materials to reduce their mass and are streamlined (Mousavi et al. 2009).

The main design criteria for present wind technology are dependability, noise elimination, grid synchronization, aerodynamic performance, optimum efficiency, and improved quality at low wind speeds, and offshore development. In the same manner, wind turbine size and height have risen. The generators of the most recent turbines are 100 times bigger than those from 1980. The diameters of their rotors have increased eight-fold during the same time span. The China State Shipbuilding Corporation (CSSC) Haizhuang's new H260, which has a rotor diameter of 128 m and an output power of 18 Megawatt, is the largest turbine currently in operation.

However, in recent years, the size of land-based turbines has been established in the 1.5–3 MW range. Due to this, thousands of the same-design turbines have been produced in series, allowing for the elimination of early issues and an increase in dependability. Examples of continuous advancements in turbine construction include the production of blades utilizing different composite material combinations, particularly to keep their weight to a minimum, modifications to the drive train system to minimize loads and improve stability, and improved control systems, in part to make sure better grid integration.

5.2.1 Wind Turbine Architectures

A key component of wind energy conversion systems is the wind turbine. Several kinds of wind turbines have been created over time and the installation costs of turbines in the mid-twentieth century have been shown in Table 5.1 (Breeze 2015) and the comparison of on-land wind energy systems and offshore wind energy systems' total installation costs has been shown in Fig. 5.2. An overview of wind turbine technology, including fixed-speed and variable-speed turbines with horizontal and vertical axes, is given in this section.

Table 5.1 Mid-2000s installation costs for both small and big wind turbines (Breeze 2015)

Parameters	Wind turbine (small)	Large wind turbine (large)
Output power (rated)	50 kW	1.7 MW
Turbine cost	$110,000	$2074,000
Installation	$55,000	$782,000
Total installed cost	$165,000	$2856,000
Total cost per kW installed	$3300	$1680

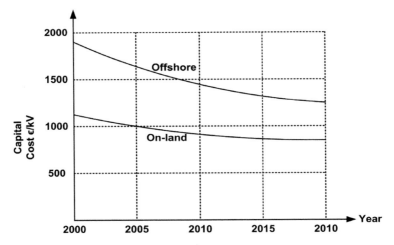

Fig. 5.2 On-land wind energy systems and offshore wind energy systems' total installation costs are compared (Breeze 2015)

5.2.2 Fixed and Variable-Speed Wind Turbines

When compared to the latest cutting-edge technology, wind energy producers have remarkably evolved. In recent years, wind turbines have become more popular among generating technologies as a result of these advancements. At that time, there were a few different generators that were in use as wind energy converters. Synchronous generators were the popular form of generator in the previous years, but thanks to technological advancements, other types of generators are currently becoming more and more popular in the WECS field (Nouh and Mohamed 2014). Induction generators, in particular doubly fed induction generators, are becoming increasingly widely used in the field of clean energy. Regarding the commercial use of wind turbines, the direct-driven wind turbine (WT) with a permanent magnet synchronous generator (PMSG) has advanced the fastest. This is due to its simple construction, cheap cost of maintenance, high conversion efficiency, and high dependability. Currently, there are two main wind turbine types (Qazi and Mustafa 2016).

5.2 Overview of Wind Turbine Topologies

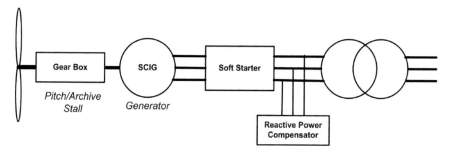

Fig. 5.3 Fixed speed squirrel cage induction generator

A. Fixed-Speed Wind Turbine System

Fixed-speed WECS are comparatively simple systems that drive either a wound rotor induction generator (WRIG) or a squirrel cage induction generator (SCIG) through a gearbox and shaft (Chen 2005). The generator's rotor speed and slip vary in direct proportion to the quantity of electricity produced. Due to the frequent little rotor speed change (between 1 and 2%), this WECS is also known as a constant or fixed-speed WECS. Because no power electrical equipment is employed, this form of WT has a minimal capital cost from a design and production standpoint. These are designed to operate as efficiently as possible at any given wind speed. The energy collected does not yield the highest grade of energy for the wind speed which is either higher or lower than the specified wind speed.

Fixed-speed WECS is structurally uncomplicated, dependable, steady, and quite good. Maintenance and purchasing of electrical parts are less expensive. On the other hand, the negative effects of WECS are mechanical stress and inadequate power quality control. Figure 5.3 displays a fixed-speed wind turbine with a squirrel cage induction generator (SCIG).

B. Variable-Speed Wind Turbine System

The inherent problems of fixed-speed WECS have emerged as becoming increasingly evident, especially in areas with a fairly poor grid structure, as the WECS's scope grows and wind power's entry into a power network expands. Variable-speed concepts are frequently used in emerging WECS innovation to address these issues and meet grid code connection requirements.

With the proliferation of power electronic converters, generally utilized for connecting wind turbines and the power supply, variable-speed wind power energy is gaining popularity. The main advantages of variable-speed WECS are improved power collection, enhanced efficiency, better quality of power with fewer flickers, less mechanical stress, less fatigue, and reduced acoustic noise. Power converters are also used in wind turbines to give high-potential control capabilities for both large wind farms and wind turbines to fulfil the exacting standards set by the power grid authorities. The capacity to adjust active and reactive power (frequency and voltage regulation), the quick reaction in situations where transient and dynamic

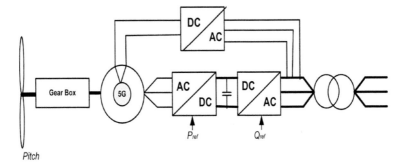

Fig. 5.4 DFIG-based variable-speed wind turbine system

power is present, the effect on network stability, and better power quality are the key characteristics of variable-speed WECS (Bhowmik et al. 1999).

To separate the generator's electrical frequency from the grid frequency, which is continually changing due to the wind speed, power electronic converters might be placed in between the generator and the power grid. To enable the varied functioning of the wind turbines, power converters separate the mechanical frequency of the rotor from the grid system frequency.

For variable-speed WECS, both conventional synchronous generators and the double-fed induction generator (DFIG) idea can be employed. Figure 5.4 displays the wind turbine built on a DFIG. Due to its compact construction and cheapest price of power converters, better controllability, the requirement for minimal space, enhanced stability of the power system, and capacity to generate reactive power, the DFIG development seems to be more suitable for variable wind speed performance.

Asynchronous generators with modest rotational speeds (5–30 rpm) can also be used instead of induction generators. As shown in Fig. 5.5, a synchronous generator with large poles and permanent magnets or constant excitation can be specially connected to a turbine without a gearbox (direct drive wind turbine) (Morimoto et al. 2005). Additionally, the power converters that are utilized for linking WTG to the grid network have a rating that is equal to the generator's power rating. Because there are so many poles and the generator turns slowly, it must transmit a lot of torque. As a result, it is often large and heavy, which inevitably has an impact on the growth and dimensions of the nacelle. This type of power converter transmits the entire generator's power. The development and application of permanent magnet generators for direct-driven systems have been adversely affected by the decline in the price of permanent magnets and their availability on the market (Morimoto et al. 2005).

The following are some key distinctions between current big wind turbines' variable-speed operation and earlier, more traditional fixed speed operation:

1. The grid voltage may not be regulated with fixed-speed operations, and reactive power management is limited.

5.2 Overview of Wind Turbine Topologies

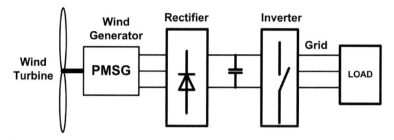

Fig. 5.5 Variable-speed wind turbine systems (PMSG)

2. Increased energy capture may be possible with variable-speed operating below-rated power.
3. Even within a very narrow speed range, variable-speed capability overrated power can significantly reduce loads, alleviate pitch system workload, and lessen output power fluctuation.

5.2.3 Horizontal and Vertical Axis Wind Turbine

According to the direction of their spin axis, wind turbines may be divided into horizontal-axis wind turbines (HAWT) and vertical-axis wind turbines (VAWT), as shown in Fig. 5.6.

The spin axis of horizontal wind turbines is oriented parallel to the ground, as shown in Fig. 5.6a. In order to obtain better wind conditions and offer enough area for the rotor blade to spin, the tower raises the nacelle. The generator, gearbox, and, in certain configurations, power converters are all housed in the nacelle, which also supports the rotor hub that carries the rotor blades. The three-bladed, upwind design used by the industry standard HAWT is positioned in front of the nacelle. Yet, there are additional downwind designs with blades towards the back that are used in real-world settings. Wind farms may also have turbines with one, two, or more than three blades.

The spinning axis orientation is perpendicular to the ground in vertical-axis wind turbines. Curved, vertically mounted airfoils are used in the turbine rotor. As shown in Fig. 5.6b, the gearbox and generator are typically installed at the turbine's base on the ground. The VAWT's rotor blades come in a range of styles with various blade counts and shapes. One of the common designs is that seen in the image. Guide wires are often required by the VAWT to maintain the rotor shaft's fixed position and reduce any mechanical vibrations.

Due to the blade design and access to greater winds, the HAWT has a better wind energy conversion efficiency, but its installation is more expensive and requires a larger tower to support the nacelle's substantial weight. Contrarily, the ground-level gearbox and generator installation of the VAWT has the benefits of less installation costs and simple maintenance, but its wind energy conversion efficiency is not good

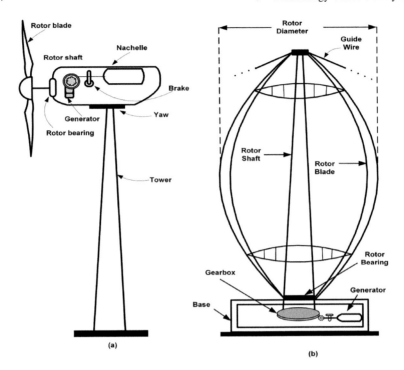

Fig. 5.6 a Horizontal axis wind turbine. b Vertical axis wind turbine

due to the less amount of wind on the lower side of the blades and the finite blades aerodynamic functioning. The shaft of the rotor is also large, which makes it susceptible to mechanical vibrations. These drawbacks make it difficult to use vertical-axis turbines for the practical conversion of significant amounts of wind energy. Today's wind industry is fully influenced by HAWT, specifically in commercial wind farms.

Advantages

Hawt

- Improved wind energy conversion efficiency.
- Due to the tall tower, have an availability of stronger winds.
- Pitch and stall control at greater wind speeds.

Vawt

- Less installation cost and better maintenance because of the generator- and ground-level gearbox.
- Best fit for rooftops (stronger wind without tower requirement).
- Not fully dependent on wind direction.

5.2 Overview of Wind Turbine Topologies

Disadvantages

Hawt

- Longer cable running from the tower's top to the ground.
- More expensive installation, a heavier tower to accommodate the nacelle's massive weight.
- Orientation required (yaw control) VAWT.

Vawt

- Less wind energy conversion efficiency.
- Susceptible to mechanical vibrations and significant torque swings.
- Limited alternatives for power regulation in high wind conditions.

5.2.4 Stall and Pitch Aerodynamic Power Control

Turbine blades are aerodynamically developed to catch the optimum power from the wind during typical operation with a wind speed limit of 3–15 m/s. At high wind speeds of around 15–25 m/s, the turbine's aerodynamic power management is needed to avoid harm to it (Tan and Islam 2004). Pitch and stall controls are the two methods used the most frequently to manage aerodynamic forces on turbine blades.

The design of the wind turbine blades results in air turbulence on the surface not facing the wind when the speed of the wind surpasses the wind speed of about 15 m/s. The most fundamental control method is this one. Turbulence minimizes the lift force on the blade, which reduces the gathered power and guards against damage to the turbine. As there are no sensors, mechanical controllers, or actuators, power management using passive stalling is cost-effective and dependable (Kumar and Das 2014). The primary problem of this technique is that power conversion efficiency is reduced at low wind speeds. WECS that are small to medium-sized frequently use the passive stall.

Large wind turbines typically employ pitch control. To maximize wind production, the pitch angle needs to be fixed at its highest value in normal operating conditions with wind speeds between 3 and 15 m/s. When the wind speed exceeds the rated value, the blade is moved away from the wind to reduce the gathered power. The blades' longitudinal axis is rotated, changing the pitch angle, using an electromechanical mechanism situated in the rotor hub and linked to a gear at each blade base. Therefore the power that the turbine can capture is maintained near to its rated value (Tan and Islam 2004).

When the speed of the wind reaches its upper limiting value of approximately 25 m/s, the blades are fully pitched or feathered, this prevents any power from being captured. By using this technique, you can prevent wind damage to the turbine and its supporting structure. As soon as the blades become fully pitched and the turbine is in resting mode, a mechanical brake locks the rotor into place. The pitch mechanism's added complexity and expense as well as the slow pitch-control dynamics' power variations during high wind gusts are the main drawbacks of pitch control.

Another way to handle aerodynamic power is active stall control, which is essentially a pitch-control system with the blade attack angle tilted into the wind, producing stall (turbulence on the rear of the blade). The active stall mechanism can reduce the maximum collected power during strong wind gusts and improve power conversion efficiency during slow winds in contrast to the passive stall technique. However, it is a complicated system just like pitch-controlled WECS. In the medium- to large-size WECS, active stall techniques are typically employed.

5.3 Generators for Wind Turbines

A generator is an electromechanical unit in a wind turbine that transforms mechanical power into electrical power. As a result, it is the very essential energy conversion element, in addition to the wind turbine and rotor for the conversion of wind power into electrical power (Wu et al. 2011).

A. Induction Generators

In the early 70s and 80s, the earliest commercial wind turbines relied heavily on induction generators for electrical energy generation. With this generator, the 3 phases of a grid energy supply are provided by a typical stator with three coils. The rotor, on the other hand, is either a wrapped rotor or a squirrel cage, a more durable construction. The rotor coil in both cases, forms a closed-loop system. Figure 5.7 shows the schematic diagram of Induction generator connected wind turbine (Ramos et al. 2019).

A magnetic field is created in the stator windings when any induction generator is linked to the grid. This magnetic field causes a severe current to develop in the rotor windings. This causes the rotor to start spinning because it works against the magnetic field of the stator. When the rotor speed is just below the grid frequency, an optimal state with a low induced torque is reached. This happens as the rotor speed increases. The slip is the difference in speed between the rotational speed of the stator and the grid frequency of the rotating magnetic field in the stator windings.

Due to the slip issue, the induction machine has non-synchronism behaviour with the grid frequency; for this reason, it is referred to as an asynchronous generator. When a grid-connected induction generator receives torque from a wind turbine rotor, the additional rotational energy causes the generator to spin faster than the grid frequency's rotating magnetic field in the stator. The generator starts to produce a current in the stator windings, which is then fed back again into the grid system as a consequence of a negative slip. Usually, 15% more electricity is produced at its peak than the frequency of the grid (Fig. 5.7).

Because of their relative potential over traditional generators synchronous generators (SGs) and induction generators (IGs) are becoming more popular. Brushless

5.3 Generators for Wind Turbines

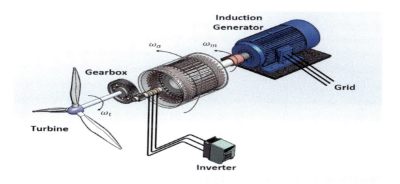

Fig. 5.7 Induction generator connected wind turbine (Ramos et al. 2019)

and durable design, cheap cost, maintenance and operating simplicity, fault self-protection, strong dynamic responsiveness, and the capacity to generate power at varied speeds are among these qualities. The other attribute allows the induction generator to operate in a standalone or isolated mode to provide far-flung and isolated places where grid extension is not financially viable, in conjunction with the synchronous generator to meet growing regional power needs, and in grid-connected mode to support the grid's real power requirement by integrating power from various sites.

As this type of generator depends on the grid for generator excitation, there is a grid-reactive power loss, which is its main disadvantage. Additionally, since the generator can only run at or close to grid frequency, the operation at variable speed over a wide range cannot be done. The induction generator is also less efficient than the default choice. This type of generator is currently not being used in large commercial wind turbines since present grid rules and regulations require wind turbines to be self-sufficient rather than relying on the grid for assistance. It has also the problem of requiring reactive power. Induction generators have been used in constant-speed WECS for a long time, where pitch control or active stall control is required for power limiting and protection, and a soft starter is utilized to prevent transients when the generator is linked to the grid.

B. Synchronous Generators

A synchronous generator is an alternative to an induction generator which can be seen in Fig. 5.8. This generates an alternating current that changes as per the rotational speed. In the simplest form of using this generator for wind energy, the turbine rotor is directly attached to the generator via a gearbox, which causes an alternating current to be produced at the grid frequency when the turbine rotates at its rated speed. It does, however, necessitate that the turbine and generator rotate at this fixed frequency alone, restricting flexibility. The fluctuations in rotor speed are generally due to the varying nature of wind sent to the grid as frequency varies from the normal value.

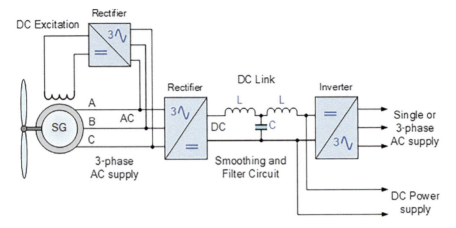

Fig. 5.8 Synchronous generator connected to wind turbine (Singh et al. 2010)

By adjusting a tiny percentage of the generator's output, the rotor functions as an electromagnet that is powered to generate magnetic fields from an external direct-current source. The use of permanent magnets instead of magnetic coils in the rotor of wind turbine generators is an alternative to this that is becoming increasingly popular and is easy to handle and need not any excitation, also they are lighter than traditional synchronous generators. However, they need powerful magnets that include up to thirty per cent by weight of the rare earth element neodymium.

The synchronous generator, like the traditional induction generator, is limited to operating at or near its synchronous speed. Otherwise, power will be sent into the system at an incorrect frequency. Power electronics are the best method for converting the generator's output from AC to DC at grid frequency. With this configuration, the generator may operate at any speed while still supplying power to the grid at the proper frequency. A variable-speed generator is built around this (Singh et al. 2010).

C. Doubly Fed Induction Generators

A more effective and resilient induction generator type utilized in larger wind turbines is the doubly fed induction generator (DFIG). A typical induction generator's rotor is a closed-loop coil, whereas the stator is directly connected to the grid. In contrast, the three-phase windings in the rotor of a doubly fed induction generator are connected to the grid via power electronic DC/AC converters. This makes it possible for the rotor windings to generate a magnetic field, which combines with the magnetic field present in the stator windings to produce torque (Boubzizi et al. 2018).

Both the two field's strength and their phasor angle affect how much torque is generated. By changing this, the wind turbine may run at different rotational speeds that correspond to around 63% of the frequency at which the grid operates. A design of a DFIG is presented in Fig. 5.9. These generators are now most popular in present wind turbines that have 5 MW capacity.

5.3 Generators for Wind Turbines

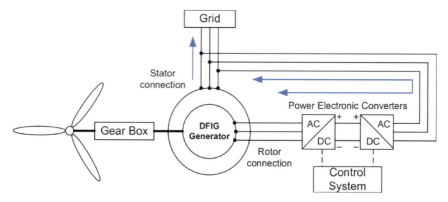

Fig. 5.9 Doubly fed induction generator connected wind turbine (Boubzizi et al. 2018)

D. Permanent Magnet Synchronous Generators

The permanent magnet generator is one form of electrical equipment that is increasingly being utilized in wind turbine applications. The schematic diagram of the permanent magnet synchronous generator connected wind turbine has been shown in Fig. 5.10. Nowadays, most small wind turbine generators up to at least 10 kW employ this generator as their first option; they may also be found in bigger wind turbines. These generators do not require field windings or a source of current to the field because the magnetic field is supplied by permanent magnets. In one illustration, a cylindrical cast aluminium rotor has the magnets built right in. There is no need for a commutator, slip rings, or brushes because the power is supplied by a fixed armature. The permanent magnet generator is highly robust due to the machine's straightforward construction.

The permanent magnet generator is one form of electrical equipment that is increasingly being utilized in wind turbine applications. Nowadays, most small wind turbine generators up to at least 10 kW employ this generator as their first option; they may also be found in bigger wind turbines. These generators do not require field windings or a source of current to the field because the magnetic field is supplied by permanent magnets. In one illustration, a cylindrical cast aluminium rotor has the magnets built right in. There is no need for a commutator, slip rings, or brushes because the power is supplied by a fixed armature. The permanent magnet generator is highly robust due to the machine's straightforward construction (Fig. 5.10).

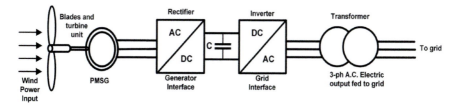

Fig. 5.10 Permanent magnet synchronous generator connected wind turbine

5.4 Power Electronics in Wind Energy

5.4.1 Soft Starters

Due to its primary role in assisting the wind turbine to begin gently with less mechanical stress and inrush current, the AC voltage controller is sometimes referred to as a "soft starter" in WECS. The AC voltage controller is often bypassed (short-circuited) once the system is turned on, eliminating the controller's power losses. The SCR is often used as a switching device by the AC voltage controllers in wind energy conversion systems. The controller's output voltage may be changed from zero to its supply voltage using SCRs control with the delay angle variation for the, which effectively lowers the system's beginning current.

(a) **Single-Phase AC Voltage Controller**

Figure 5.11 depicts the basic circuit diagram of a 1-ϕ AC voltage controller. It consists of two SCR thyristors linked antiparallel to each other between the power source and the load. Figure 5.12 shows the voltage controller's workings, the thyristors' gating configuration, and the output voltage and current waveforms that result (Manias 2017).

The gate signal waveforms, supposing a resistive load i_{g1} and i_{g2}, output current i_o and Fig. 5.12a show the output voltage v_o with the delay angle $\alpha = \pi/3$ of the controller. The thyristor T_1 is switched on at $\omega t = \alpha = \pi/3$ by i_{g1} in the +ve half cycle of the power supply and is turned off π when its current reaches zero. The thyristor T_2 is turned on at $\omega t = (\alpha + \pi) = 4\pi/3$ and turned off 2π during the negative half cycle.

The output voltage's V_o rms value with a resistive load can be calculated using

$$V_o = \left(\frac{1}{\pi} \int_{\alpha}^{\pi} \left(\sqrt{2} V_s \sin \omega t \right)^2 d(\omega t) \right)^{1/2} = V_s \left(1 - \frac{\alpha}{\pi} + \frac{\sin 2\alpha}{2\pi} \right)^{1/2} \quad (5.1)$$

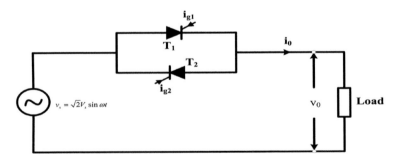

Fig. 5.11 1-ϕ AC voltage controller

5.4 Power Electronics in Wind Energy

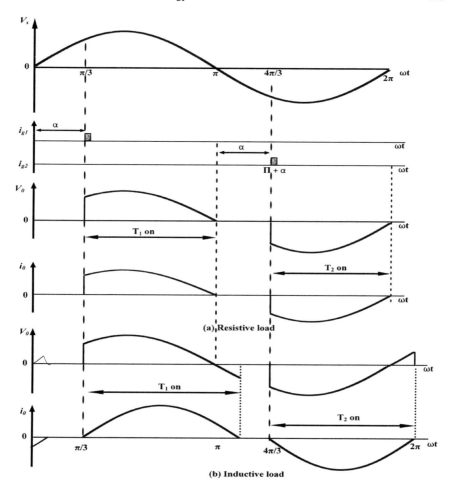

Fig. 5.12 1-ϕ AC voltage controller waveforms

The voltage controller's use with an RL load is shown in Fig. 5.12b and $\alpha = \pi/3$. The thyristor T_1 is turned on at $\omega t = \pi/3$, however, will remain on even if the supply voltage v_s drops to zero at $\omega t = \pi$. This is because the current flowing through the lagging inductive load, which is not equal to zero at $\omega t = \pi$. Thyristor T_1 continues to operate until its current reaches zero, at which time the load inductance's stored energy is completely released. Then T_1 is turned off.

It should be kept in mind that when the load power factor angle φ, which is determined by $\varphi = \tan^{-1}(\omega L/R)$, V_o is no longer adjustable since the controller's output voltage V_o will be equal to input voltage V_s. Consider Fig. 5.13, where the delay angle is equal to $\pi/6$ and the load power factor angle φ is equal to $\pi/3$. A particular amount of time occurs when the thyristor T_1 conducts during the positive half cycle of the supply voltage. Whenever the signal for the gate T_2 arrives at

$\omega t = \pi + \alpha$, due to the inductive load's continued positive load current, T_1 will continue to conduct and T_2 won't switch on. If the gate current i_{g2} for T_2 is still there, T_2 won't switch on until i_o reaches zero and becomes negative. T_1 is reverse-biased and, as a result, turned off when T_2 is switched on. Both T_1 and T_2 perform 180° alternately, according to the fundamental-frequency cycle, and as a result, the supply voltage V_s and the output voltage V_o are equal.

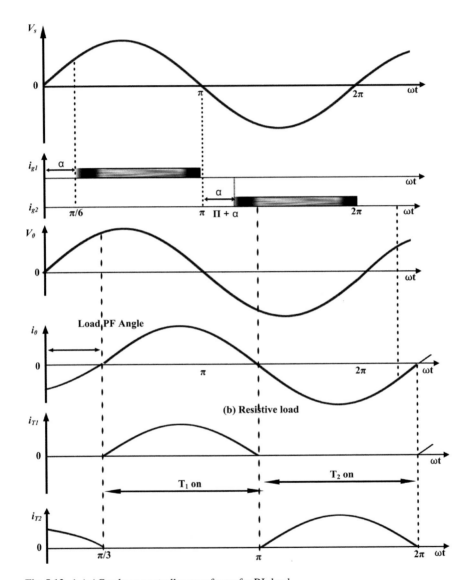

Fig. 5.13 1-ϕ AC voltage controller waveforms for RL load

5.4 Power Electronics in Wind Energy

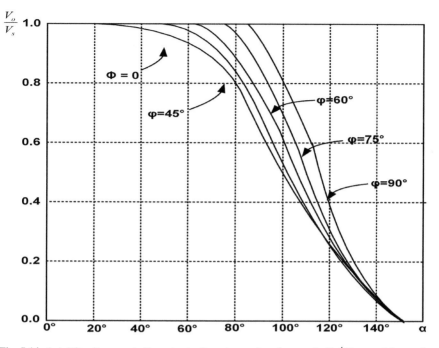

Fig. 5.14 1-φ AC voltage controller output voltage to supply voltage ratio V_o/V_s w.r.t. delay angle α

Moreover, it should be remembered that with an inductive load, continuous gating with prolonged duration, useful examples are i_{g1} and i_{g2} in Fig. 5.13. Short-duration gate signals will cause the controller to malfunction. For example, T_2 won't switch on during the supply voltage's negative cycle if i_{g2} is a short gating pulse, like the one seen in the figure with a solid block.

The output voltage V_o of the controller's rms value may be computed using the following formula, assuming a pure inductive load:

$$V_o = \begin{cases} V_s & \text{for } 0 \le \alpha < \pi/2 \\ V_s \left(2 - \dfrac{2\alpha}{\pi} + \dfrac{\sin 2\alpha}{\pi} \right)^{1/2} & \text{for } \pi/2 \le \alpha \le \pi \end{cases}, \quad (5.2)$$

where v_o is equal to v_s for $0 \le \alpha < \pi/2$.

Equations (5.1) and (5.2), describe the connection between the voltage ratio V_o/V_s and angle of delay α with a pure resistive ($\varphi = 0°$) and pure inductive ($\varphi = 90°$) load as shown in Fig. 5.14. Another curve in the load power factor angle of the figure for φ, 45°, 60°, and 75° are obtained by computer simulation.

(b) Three-Phase AC Voltage Controller

Figure 5.15 depicts the 3-φ AC voltage controller connected to 3-φ Y-connected load. It is made up of 6 thyristors in 3 pairs linked in series between the 3-φ supply and

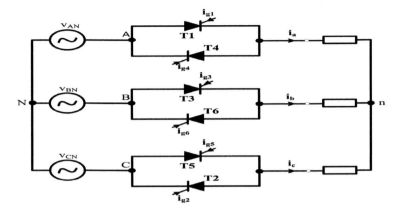

Fig. 5.15 Three-phase AC voltage controller with Y-connected load

the load. Figure 5.16, shows the input voltage waveforms, thyristors gating currents, and phases a load voltage V_{an} illustrates the controller's working concept.

Let us take the controller as a 3-ϕ balanced resistive load in star form. The line-to-line voltage V_{AB} is supplied to the phase-a and phase-b load resistors while thyristors T_6 and T_1 are switched on. The phase voltage v_{an} is equal to $v_{AB}/2$ as illustrated in Fig. 5.16 since T_5 and T_2 in phase-c are both off. Thyristors T_1 and T_2 conduct for period II, resulting in $v_{an} = v_{AC}/2$. Thyristors T_2 and T_3 are switched on during period III, but none of the phase-a thyristors are, resulting in $v_{an} = 0$. The v_{an} voltage in the next half cycle may be obtained using the same approach. Similarly to that, it is possible to derive all waveforms of load voltage for the v_{bn} and v_{cn}.

Wave shapes for the load $v_{ab} = v_{an} - v_{bn}$, $v_{bc} = v_{bn} - v_{cn}$ and $v_{ca} = v_{cn} - v_{an}$ may be used to calculate load line-to-line voltages respectively. Figure 5.17 shows the waveforms for v_{an} and v_{ab} angles gradually shift from $2\pi/3$ zero with the delay. One can see that when the delay angle (α) decreases, the load's phase voltage (v_{an}) and line-to-line voltage (v_{ab}) both increases and eventually equal the power supply's phase voltage (v_{AN}) and line-to-line voltage (v_{AB}) at $\alpha = 0$, respectively.

The 3-ϕ AC voltage controller's functioning may be divided into three operating modes based on the delay angle α; Mode I for $\pi/2 \le \alpha < 5\pi/6$, which includes intervals where either none or two thyristors conduct each phase; Mode II for $\pi/3 \le \alpha < \pi/2$, whereby each phase is initiated by turning on two thyristors; and Mode III for $0 \le \alpha < \pi/3$, when two or three thyristors are operating simultaneously. The 3-ϕ AC voltage controller's lies in the range from $0°$ to $5\pi/6$ ($150°$) as opposed to the 1-ϕ AC voltage controller's range of zero to π for the delay angle α beyond which ($5\pi/6 < \alpha \le \pi$) the controller's output voltage is held at zero. Thus, it is unnecessary to increase the delay angle beyond $5\pi/6$. Figure 5.18 depicts the typical v_{an} waveforms for the voltage controller operating in the 3 modes, the location of α is $2\pi/3$($120°$) in Mode I, $5\pi/12$ ($75°$) in Mode II, and $\pi/6$ ($30°$) in Mode III, respectively.

5.4 Power Electronics in Wind Energy 215

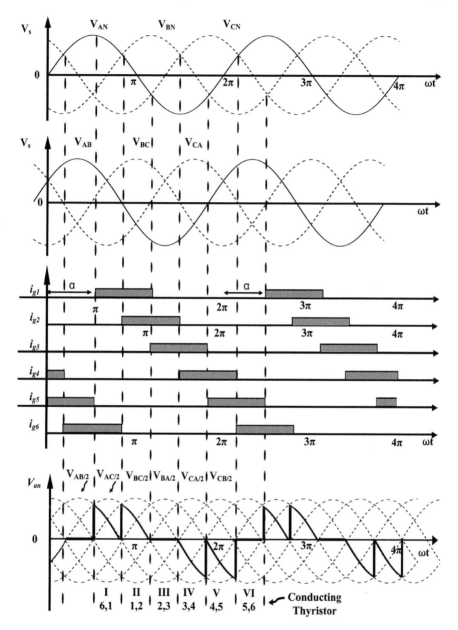

Fig. 5.16 Waveforms of three-phase AC voltage controller with a Y-connected resistive load

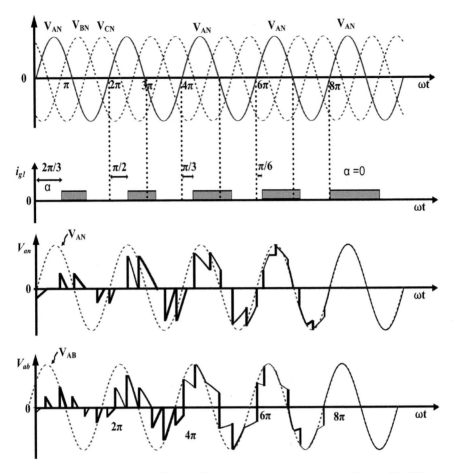

Fig. 5.17 3-ϕ AC voltage controller waveforms connected to a resistive load and with different delay angles. $\alpha = 2\Pi/3$ (1st cycle), $\Pi/2$ (second cycle), $\Pi/3$ (3rd cycle), $\Pi/6$ (fourth cycle), and 0 (fifth cycle)

We may determine the rms value of the load phase voltage by observing the waveforms in Fig. 5.18 as follows:

$$V_{an}\left(\frac{1}{\pi}\left(\int_{\alpha}^{5\pi/6}\left(\frac{\sqrt{6}}{2}V_s\sin(\omega t + \pi/6)\right)^2 d(\omega t)\right.\right.$$

$$\left.\left. + \int_{\alpha+\pi/3}^{7\pi/6}\left(\frac{\sqrt{6}}{2}V_s\sin(\omega t - \pi/6)\right)^2 d(\omega t)\right)\right)^{1/2}$$

5.4 Power Electronics in Wind Energy

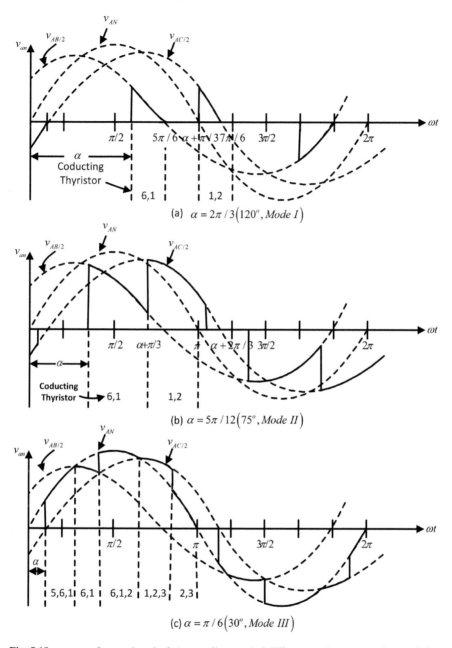

Fig. 5.18 v_{an} waveforms when the 3-ϕ controller runs in 3 different modes connected to a resistive load

$$= V_s \left(\frac{5}{4} - \frac{3\alpha}{2\pi} + \frac{3\sin(2\alpha + \pi/3)}{4\pi} \right)^{1/2} \quad \text{for Mode I } (\pi/2 \leq \alpha < 5\pi/6)$$

$$(5.3)$$

$$V_{an} = \left(\frac{1}{\pi} \left(\int_{\alpha}^{\alpha+\pi/3} \left(\frac{\sqrt{6}}{2} V_s \sin(\omega t + \pi/6) \right)^2 d(\omega t) \right. \right.$$

$$\left. \left. + \int_{\alpha+\pi/3}^{\alpha+2\pi/3} \left(\frac{\sqrt{6}}{2} V_s \sin(\omega t - \pi/6) \right)^2 d(\omega t) \right) \right)^{1/2}$$

$$= V_s \left(\frac{1}{2} + \frac{3\sqrt{3}}{4\pi} + \sin(2\alpha + \pi/6) \right)^{1/2} \quad \text{for Mode II } (\pi/3 \leq \alpha < \pi/2) \quad (5.4)$$

and

$$\left(\frac{1}{\pi} \left(\int_{\alpha}^{\pi/3} (\sqrt{2} V \sin_s \omega t)^2 d(\omega t) + \int_{\pi/3}^{\alpha+\pi/3} \left(\frac{\sqrt{6}}{2} V_s \sin(\omega t + \pi/6) \right)^2 d(\omega t) \right. \right.$$

$$+ \int_{\alpha+\pi/3}^{2\pi/3} \left(\sqrt{2} V_s \sin \omega t \right)^2 d(\omega t)$$

$$\left. \left. + \int_{2\pi/3}^{\alpha+2\pi/3} \left(\frac{\sqrt{6}}{2} V_s \sin(\omega t - \pi/6) \right)^2 d(\omega t) + \int_{\alpha+2\pi/3}^{\pi} \left(\sqrt{2} V_s \sin \omega t \right)^2 d(\omega t) \right) \right)^{1/2}$$

$$= V_s \left(1 - \frac{3\alpha}{2\pi} + \frac{3\sin 2\alpha}{4\pi} \right)^{1/2} \quad \text{for Mode III } (0 \leq \alpha < \pi/3), \quad (5.5)$$

where V_s represents the rms value of the phase voltage of the power supply can be expressed by

$$v_{an} = \sqrt{2} V_s \sin \omega t; \quad v_{bn} = \sqrt{2} V_s \sin(\omega t - 2\pi/3)$$
$$\text{and } v_{cn} = \sqrt{2} V_s \sin(\omega t + 2\pi/3) \quad (5.6)$$

The 3-ɸ AC voltage controller with inductive load has a difficult analysis because, unlike the 1-ɸ AC voltage controller, the thyristor's conduction does not stop when the supply voltage reaches zero and finally becomes −ve. A useful method for getting the waveforms of load voltage and current is computer simulation. When the voltage controller runs with a 3-ɸ, Y-connected RL load with a power factor equal to 0.9 at various delay angles, waveforms for phase-a load voltage v_{an}, line-to-line voltage v_{ab}, and load current i_a are shown in Fig. 5.19. Due to the filtering effect of the load

5.4 Power Electronics in Wind Energy

inductance, the phase-*a* load current i_a waveform is significantly smoother than its phase voltage v_{an}. According to the figure, the load power factor angle φ has a value of 25.8°.

Using a purely inductive load ($\varphi = \pi/2$) the three-phase AC voltage controller's load voltage v_{an} rms value is provided by

$$V_{an} = \begin{cases} V_s & \text{for } 0 \leq \alpha < \pi/2 \\ V_s \left(\frac{5}{2} - \frac{3\alpha}{\pi} + \frac{3\sin(2\alpha)}{2\pi} \right)^{1/2} & \text{for } \pi/2 \leq \alpha < 2\pi/3 \\ V_s \left(\frac{5}{2} - \frac{3\alpha}{\pi} + \frac{3\sin(2\alpha + 5\pi/3)}{2\pi} \right)^{1/2} & \text{for } 2\pi/3 \leq \alpha < 5\pi/6 \end{cases} \quad (5.7)$$

The connection between V_{an}/V_s is shown in Fig. 5.20 taking the load power factor angle as a reference parameter based on Eqs. (5.3)–(5.7). The other curves are produced by computer simulations for $\varphi = 45°$, $60°$, and $75°$.

The load voltage V_{an} has the same value as the supply voltage V_s and, as a result, cannot be modified longer adjustable when the delay angle α is lower than the load power factor angle φ which is the same occurrence as that covered in the 1-ϕ AC voltage controller.

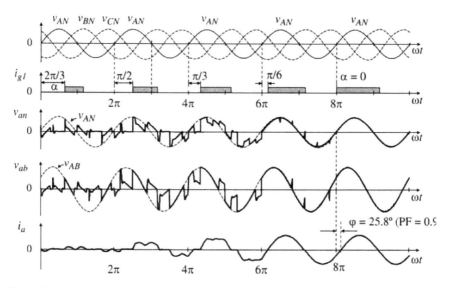

Fig. 5.19 3-ϕ AC voltage controller waveforms connected to RL load ($\cos \varphi = 0.9$) and different values of α, $\alpha = 2\Pi/3$ (1st cycle), $\Pi/2$ (2nd cycle), $\Pi/3$ (3rd cycle), $\Pi/6$ (4th cycle), and 0 (5th cycle)

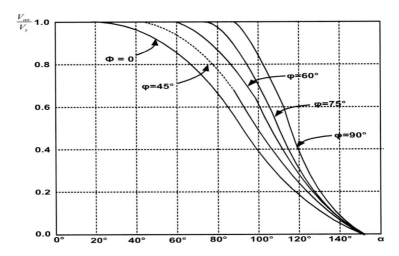

Fig. 5.20 V_{an}/V_s w.r.t. α(3-ϕ AC voltage controller)

5.4.2 Capacitor Bank

A capacitor bank is just a group of numerous capacitors having the same rating, as the name suggests. Depending on the required rating, capacitor banks can be linked in either series or parallel. Similar to a single capacitor, capacitor banks are utilized to retain and regulate the electrical energy flow. The energy storage capacity of a single device will rise when the number of capacitors in a bank is increased (Perera and Elphick 2023).

Today's high-tech society is extremely energy-hungry. Energy needs to be conveniently stored electrically to satisfy this requirement. High electrical energy charges may be stored in capacitors, which can also be used to control the energy's flow as required.

The following are some typical applications of capacitor banks:

- **Shunt Capacitor**: A shunt is a device that makes a low-resistance route around another point in the circuit so that electricity can flow through it. Capacitors are employed in electrical distortion bypass applications to send a high range of frequency noise to earth rather than allowing it to spread in the system and particularly to the load. Shunt capacitor banks mostly help to increase the electrical supply's quality and, as a result, the effectiveness of the power systems. A typical photograph of the shunt capacitor has been shown in Fig. 5.21.
- **Power-Factor Correction**: Capacitor banks are used in transformers and electric motors to rectify power-factor lag or phase shift in AC power sources. The power factor of an AC power system is calculated by dividing the "actual power," or the power utilized by the load, by the "apparent power," or the power provided to the load. To put it another way, the power factor is the ratio of the useful work carried

5.4 Power Electronics in Wind Energy

Fig. 5.21 Shunt capacitor bank (*source **Vishay Intertechnology***)

out by a circuit to the greatest amount of useful work possible at the specified voltage and ampere.

Capacitor banks are used in the distribution of electric power to correct power factors. These banks are required to neutralize inductive loading caused by components like electric motors and transmission lines, which causes the load to seem to be mostly resistive. Basically, power-factor correction capacitors make the system able to carry more current. The load of a system can be increased without changing the apparent power by adding capacitive banks. To improve the ripple-current capacity of a DC power supply or to raise the total quantity of stored energy, banks may be utilized in the DC power supply.

- **Energy Storage**: Capacitive banks store electric energy when they are coupled to a charging circuit and release that energy when they are discharged, just like individual capacitors. Electronic equipment frequently employs capacitors to sustain power while batteries are being changed. Due to space constraints, current consumer electronics like mobile phones require a large storage capacity in a relatively compact volume. This is a problem since a rise in capacitance usually implies an increment in plate area.

New materials that raise the permittivity "k" of the dielectric material between the plates of the capacitor have caused capacitive banks to become smaller. The calculations show that increasing the dielectric strength is another method of raising capacitance. The dielectric relative permittivity separating the plates is the ""k" element. "k" is equivalent to one or unity in free space. "k" is higher than one for every other media. Devices suitable for these applications typically include film and electrolyte capacitors.

Shunt capacitors are the most popular method of reactive power correction in wind turbines with induction generators because they are inexpensive. On higher voltage levels, such as in substations, shunt capacitors are also frequently utilized.

Shunt capacitors in wind turbines with directly linked induction generators are connected in banks and are switched off and on using contactors. The lifespan of the contactors is constrained by the surge current consumed by the capacitors during switching in. Another drawback is that the capacitors' effectiveness is diminished at lower voltages due to a sharp decline in reactive power capability.

Thyristor-switched capacitors are another extensively used similar technology. This circuit connects a pair of bidirectional thyristors, a tiny inductor, and a capacitor bank in series. To reduce switching transients and inrush currents, an inductor is used.

The capacitors are connected in a delta pattern for a three-phase system. Individual capacitor bank switches are used to manage reactive power. As a result, control is attained gradually. Because switching capacitors causes transients, the switching must be done with the fewest possible transients. After an integral number of half cycles, the switch is turned off at the current zero. The restricted fluctuation of only a few capacitance values is a clear drawback because reactive power fluctuates continually. The rating and number of parallel associated units in this design dictate the output characteristic, which is discontinuous. As a result, the voltage assistance offered will be irregular.

Capacitor banks have historically been used as a cost-effective and very straightforward method of reactive power adjustment for wind turbines with directly attached induction generators. No-load compensation is most frequently utilized to prevent overvoltage issues, particularly on islands. No-load compensation refers to compensation that is intended to balance the consumption of reactive power in a condition when there is no load and the generator torque is zero. Figure 5.22 depicts a typical example of a wind turbine with a capacitor bank compensation. Two banks of capacitors are used in the illustration to balance out the induction generator's no-load usage. The wind turbine controls these capacitors. The third capacitor, which is mounted on the main switch's grid side, is intended to make up for the step-up transformer's reactive power consumption (https://www.arrow.com/en/research-and-eve nts/articles/capacitor-banks-benefit-an-energy-focused-world).

5.4 Power Electronics in Wind Energy

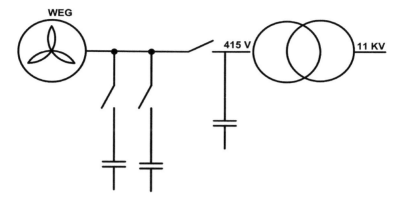

Fig. 5.22 Reactive power compensation using capacitor banks

5.4.3 Rectifiers and Inverters

(a) **Boost Converters**

In synchronous generator-based WECS, the DC/DC boost converter is among the common converters. The converter is often positioned in the power conversion system between the diode rectifier and the inverter. The two primary purposes of the boost converter are to track the maximum amount of wind energy and to increase DC voltage to an optimum level for the inverter (Ziogas 1980). The 2nd feature makes it easier to harness the full potential of the wind at all wind speeds. A single-channel boost converter is frequently used for low and medium-power wind energy systems that range in power from a few kW to hundreds of kW (Vaseee and Sankerram 2013).

One switching device may not be able to manage the current and voltage ratings in high-power megawatt wind energy systems. One approach may be to link many switching devices in parallel or series. To ensure an equal distribution of current or voltage among the parallel or series devices, further precautions must be taken. Paralleling or cascading power converters is a viable alternative to connecting the switching devices in series or parallel.

Multichannel interleaved boost converters are frequently employed in low-voltage (e.g. 690 V) wind energy systems to handle large currents in the system. The gating signals for each of the parallel converters can be interleaved (phase-shifted) to create an interleaved boost converter. The interleaved converter's enhanced equivalent switching frequency is one of its key advantages over the single converter. The equivalent switching frequency of the converter can be twice the device switching frequency for a two-channel converter, or three times for a three-channel converter. Increased equivalent switching frequency in the interleaved converter provides several benefits over the single-channel converter, including reduced input current ripple and output voltage ripple, quicker dynamic response, and enhanced power handling capacity.

An IGBT, rather than a MOSFET, which is frequently used in low-power switch-mode power supply, serves as the switching device in the interleaved boost converter

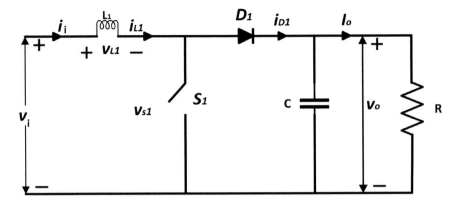

Fig. 5.23 Single-channel boost converter

in the WECS. The IGBT runs at low switching frequencies of a few hundred hertz to a few kilohertz to minimize switching losses, whereas the MOSFET frequently operates at considerably higher switching frequencies.

I. Single-Channel Boost Converter

Power converters known as boost converters have output DC voltages that are greater than their input DC voltages. Figure 5.23 represents the basic circuit for a single-channel boost converter. It is made up of a filter capacitor C, a DC inductor L_1, a switch S_1, and a diode D_1. The following study is predicated on the assumptions that (1) all converter components are perfect (have no power or voltage losses), (2) the output filter capacitor size is very large, and the converter's output voltage is ripple-free.

Diode D_1 is reverse-biased and the output is disconnected from the input when switch S_1 is switched on. The input supplies energy to the inductor L_1. Diode D_1 is forward-biased while the switch is off, allowing the energy held in L_1 to be delivered to the load via the diode. The converter's output voltage v_o is higher than its input voltage v_i in this situation because the output voltage v_0 is the result of adding the input voltage v_i and the inductor voltage v_{Ll}.

The functioning of the converter can be split into two operational modes: continuous current mode (CCM) and discontinuous current mode (DCM), depending on the continuity of the DC inductor current i_{L1}. The inductor current i_{L1} never goes completely to zero when a boost converter is running in CCM. The typical waveforms of the currents and voltages in this mode of the boost converter can be observed in Fig. 5.24.

The inductor voltage integral v_{L1} over time T_s must be zero for the converter to operate in a steady state. This suggests that the inductor L_1's average voltage over time T_s is zero. The area A_1 in Fig. 5.24 must equal the area A_2, according to its graphical interpretation, implying that

$$V_i t_{\text{on}} = (V_o - V_i) t_{\text{off}} \tag{5.8}$$

5.4 Power Electronics in Wind Energy

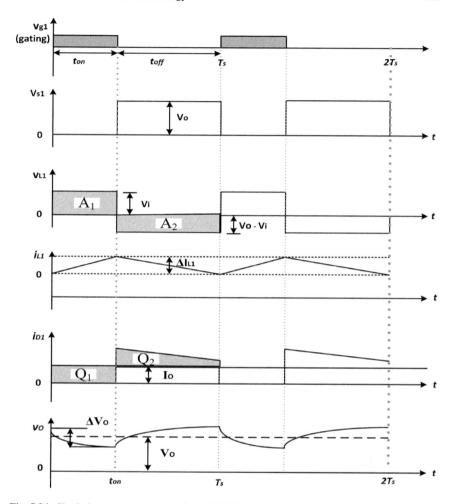

Fig. 5.24 Single-boost converter waveforms (CCM)

From which

$$\frac{V_o}{V_i} = \frac{1}{1-D} \text{ for } 0 \leq D < 1, \tag{5.9}$$

where D indicates the converter duty ratio, defined by $D = t_{on}/T_s$; T_s is the switching period; and t_{on} and t_{off} are the turn-on and turn-off times of the switch S, respectively.

The above formula shows that the converter's output voltage is always greater than its input voltage.

$V_i I_i = V_o I_o$, from which it is possible to determine the connection between the converter's input current I_i and output current I_o.

$$\frac{I_o}{I_i} = 1 - D \text{ for } 0 \leq D < 1 \tag{5.10}$$

As the inductor current varies linearly over time, the differential equation $\Delta v_{L1} = L_1(\Delta i_{L1}/\Delta t)$ may be used in place of the differential equation $v_{L1} = L_1(di_{L1}/dt)$. For the t_{off} period in Fig. 5.24, the inductor ripple current can be expressed by

$$\Delta i_{L1} = \frac{\Delta v_{L1}}{L_1} \Delta t = \frac{(V_o - V_i)}{L_1} t_{off} = D(1 - D)\frac{V_o T_s}{L_1} \tag{5.11}$$

The ripple current reaches its maximum value for the single-channel boost converter at duty cycle $(D) = 0.5$, at that point the $\Delta I_{L1,max}$ will be

$$\Delta I_{L1,max} = \frac{V_o T_s}{4L_1} \tag{5.12}$$

The load current and inductor L_1 current are both low while the converter is operating under conditions of light load. It's possible that the inductor didn't have enough energy during the t_{on} period to keep its current flowing during the t_{off} phase. Hence, the inductor current i_{L1} approaches zero before the completion of the t_{off} period and becomes discontinuous. Hence, the converter functions in discontinuous current mode. The following equation describes the inductor current at the CCM–DMM boundary:

$$I_{LB} = D(1 - D)\frac{V_o T_s}{2L_1} = \frac{\Delta i_{L1}}{2} \tag{5.13}$$

At $D = 0.5$, the inductor boundary current reaches its maximum value, which may be determined by

$$I_{LB,max} = \frac{V_o T_s}{8L_1} \tag{5.14}$$

The boundary output current may be calculated using

$$I_{oB} = D(1 - D)^2 \frac{V_o T_s}{2L_1} = (1 - D)\frac{\Delta i_{L1}}{2} \tag{5.15}$$

and its highest value is at $D = 1/3$, and it may be calculated by the following:

$$I_{oB,max} = \frac{2}{27}\frac{V_o T_s}{L_1} \tag{5.16}$$

We may examine the waveform of the current i_{D1} in the diode D_1 as shown in Fig. 5.24 to determine the output voltage ripple in the single-channel boost converter.

5.4 Power Electronics in Wind Energy

The massive output capacitor C, assuming it absorbs all of the ripple current components in D_1, is discharged to the load during the t_{on} period when the diode is turned off and charged during the t_{off} period when D_1 is switched on. The shaded regions in Q_1 during t_{on} and Q_2 during t_{off} should reflect equal amounts of charges. In that case, the peak-to-peak ripple voltage may be determined by the following:

$$\Delta V_o = \frac{Q_1}{C} = \frac{I_o t_{on}}{C} = \frac{V_o D T_s}{RC} \qquad (5.17)$$

From which

$$\frac{\Delta V_o}{V_o} = \frac{D T_s}{RC} \qquad (5.18)$$

As the duty cycle D is increased, the ripple voltage ΔV_o rises for a specified load resistance R and filter capacitor C. Figure 5.25 represents typical waveforms for the converter while it is in a discontinuous current mode of operation.

The DCM's converter operates in the same way as the CCM's converter during the t_{on} time. As time goes on, the current i_{L1} rises, and L_1 stores energy. The energy accumulated in L_1 during the t_{on} period is released at the completion of the $K_1 T_s$ period when the inductor current i_{L1} drops to zero during the t_{off} period. Due to the fact that the average voltage across the inductor throughout the switching time T_s is zero, area A_1 in Fig. 5.25 must equal area A_2, which means that,

$$V_i t_{on} = (V_o - V_i) K_1 T_s \qquad (5.19)$$

From which

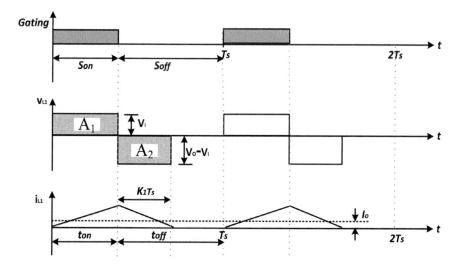

Fig. 5.25 Single-channel boost converter waveforms operating in a DCM

$$\frac{V_o}{V_i} = \frac{K_1 + D}{K_1} \quad \text{for } 0 \leq D < 1 \tag{5.20}$$

where K_1 can be calculated by the following:

$$K_1 = \frac{2L_1}{K_1 T_s D} I_o \tag{5.21}$$

II. Two-Channel Interleaved Boost Converter

A two-channel interleaved boost converter's architecture can be observed in Fig. 5.26. The circuit has two parallel converter channels. The first channel has an inductor called L_1, a switch called S_1, and a diode called D_1 while the second channel has L_2, S_2, and D_2. While they are effectively coupled in parallel, the two converter channels work in an interleaved fashion. At the output, they both use the same filter capacitor, C. The two channels' parameters are believed to be the same.

In Fig. 5.27, the converter's gating configuration and inductor current waveforms are displayed. The gating signals v_{g1} and v_{g2} for S_1 and S_2 in the interleaving structure are the same but shifted by $360°/N = 180°$, where N is the number of parallel converter channels. Each converter channel operation and waveforms are identical to those for the single-channel converter; hence they are not repeated here. The total input current i_i, which is the sum of the two inductor currents i_{L1} and i_{L2}, should be carefully considered. The following are the properties of the input current i_i:

- Input current (I_i) has an average DC component that is twice as large as the individual inductor ($I_i = I_{L1} + I_{L2}$). Just half of the load's total power is handled by each channel of the parallel converters since their input and output voltages are identical.

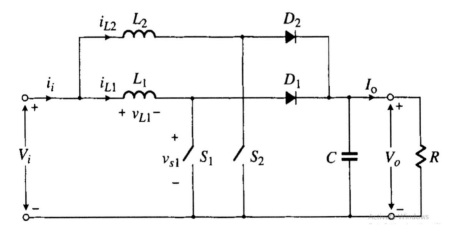

Fig. 5.26 Two-channel interleaved boost converter

5.4 Power Electronics in Wind Energy

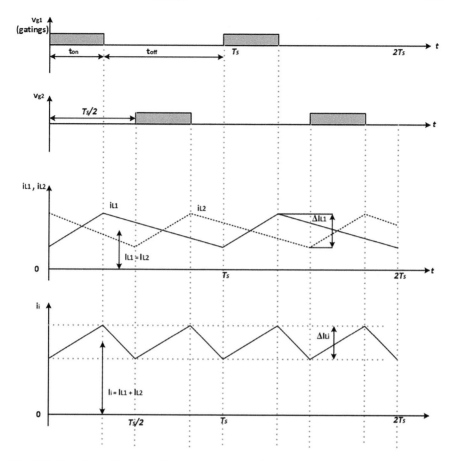

Fig. 5.27 Two-channel interleaved converter waveforms for the investigation of input current ripple ($D < 0.5$)

- Due to the interleaved technique's implementation, the peak-to-peak input current ripple ΔI_i is less than that in the individual channels. This aids in lowering the input filter's level.
- The input ripple current has a frequency that is doubled that of the individual channels. To put it another way, the interleaved converter has an equivalent switching frequency that is twice as high as each channel's. The output capacitor C's capacitance can be decreased for a specific output voltage ripple.

Input Current Ripple

It's important to observe that the ripples in the two inductor currents, i_{L1} and i_{L2}, cancel each other out and do not appear in i_i, i.e. $\Delta I_i = 0$, when the duty cycle D rises from about 0.35–0.5 in Fig. 5.27. When $D > 0.5$, the ripple current starts to rise. As a result, the input current ripple analysis may be done for the following 2 cases.

Case 5–1: $0 < D \leq 0.5$. The waveforms in Fig. 5.27 may be used to evaluate the ripple in the two-channel converter's input current, ΔI_i. The analysis may be carried out more easily for the time period t_{on}, during which the total input current i_i rises monotonically. The converter's input current ripple may be determined using

$$\Delta I_i = (K_1 - K_2)t_{on} = (K_1 - K_2)DT_s, \tag{5.22}$$

where $K_1 = di_{L1}/dt$ and $K_2 = di_{L2}/dt$, which are the inductor current slopes i_{L1} and i_{L2} during the charging and discharging processes, respectively. Thus,

$$\Delta I_i = \left(\frac{V_i}{L} - \frac{V_o - V_i}{L}\right)\left(1 - \frac{V_i}{V_o}\right)T_s = (1 - 2D)D\frac{V_o T_s}{L}, \tag{5.23}$$

where L represents each converter channel's inductance, i.e. $L = L_1 = L_2$.

The above equation may be differentiated with regard to V_i, to find the maximum input current ripple $\Delta I_{i,\,max}$:

$$\frac{\partial \Delta I_i}{\partial V_i} = \left(\frac{2V_i - V_o}{L}\right)\left(1 - \frac{V_i}{V_o}\right)T_s = 0 \tag{5.24}$$

From which

$$V_i = \frac{3}{4}V_o \text{ and } D = 0.25 \text{ for } \Delta I_i = \Delta I_{i,max} \tag{5.25}$$

The maximum current ripple can be obtained by substitution of Eq. (5.25) into Eq. (5.24):

$$\Delta I_{i,max} = \frac{V_o T_s}{8L} \tag{5.26}$$

Compare the aforementioned equation to Eq. (5.12), which states that $D \leq 0.5$, the maximum current ripple in the two-channel boost converter is half that of the single-channel converter.

Case 5–2: $0.5 < D < 1$. In Fig. 5.28, when the duty cycle D is 0.65, the waveforms for the analysis of the input current ripple in the two-channel converter are displayed. The analysis should be performed for the t_{off} time, during which the input current i_i monotonically drops.

The input ripple current can be calculated as follows:

$$\Delta I_i = (K_2 - K_1)t_{off} = (K_2 - K_1)(1 - D)T_s$$
$$= \left(\frac{V_o - V_i}{L} - \frac{V_i}{L}\right)\left(\frac{V_i}{V_o}\right)T_s = (2D - 1)(1 - D)\frac{V_o T_s}{L} \tag{5.27}$$

5.4 Power Electronics in Wind Energy

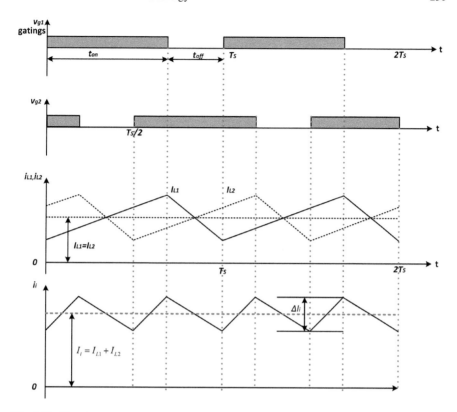

Fig. 5.28 Waveforms for analysis of input ripple current in a two-channel interleaved converter ($D > 0.5$)

The maximum input current ripple $\Delta I_{L1,\max}$, which is the same as that provided in Eq. (5.26) for the t_{off} period, may be calculated using the same technique.

Based on the above analysis the input current ripple ΔI_i for the single ($N = 1$) and two-channel ($N = 2$) converters w.r.t D has been shown in Fig. 5.29. As per Eq. (5.11), the ΔI_i has been normalized to the $\Delta I_{L1,\max}$ in the single-channel converter. It is proven that the two-channel converter's input ripple current magnitude is significantly smaller than that of the single-channel converter. In particular, the ripple current for the two-channel converter drops to zero at $D = 0.5$, but for the single-channel converter, it reaches its highest value. For the N-channel boost converters, an analysis between the ratios of the input ripple current ΔI_i to the channel ripple current ΔI_L with D has been represented in Fig. 5.30.

Output Voltage Ripple The decrease in output ripple is one advantage of interleaved converters, as was before highlighted. The two-channel converter's two parallel converters share the same output capacitor C, which makes the analysis a bit complex.

The output voltage ripples can be calculated using computer modelling methods. According to the presumption that the converter works in continuous-current mode,

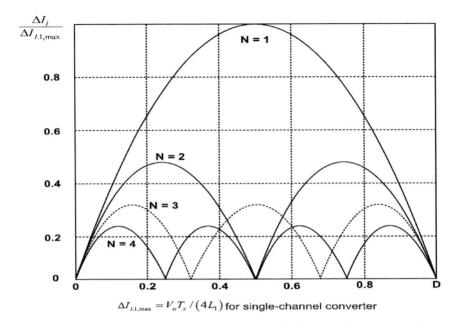

Fig. 5.29 For N-channel interleaved boost converters normalized input ripple current w.r.t. D

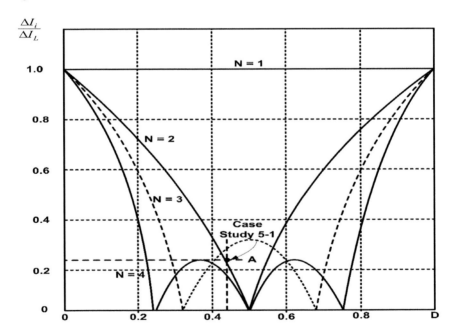

Fig. 5.30 For N-channel boost converters $\Delta I_i/\Delta I_L$ w.r.t. D

5.4 Power Electronics in Wind Energy

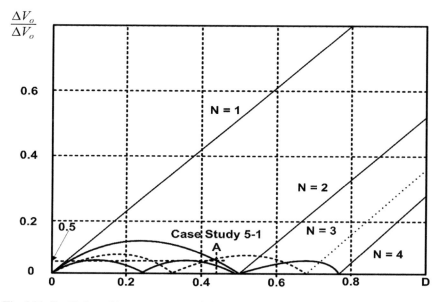

Fig. 5.31 For N-channel boost converters relative output ripple voltages $[xT_s/(RC)]$ with D

the findings for the relative output ripple voltage for the two-channel converter are shown in Fig. 5.31. The image also includes the single-stage converter's relative output ripple voltage, which was determined using Eq. (5.13). Compared to the single-channel converter, the two-channel converter generates significantly less voltage fluctuation.

III. Multichannel Interleaved Boost Converters

The boost converter layout for a three-channel ($N = 3$) interleaved by $360°/N = 120°$ is shown in Fig. 5.32a. In its simplest form, it consists of three single-channel converters that are linked in parallel and work via interleaving. Except for a $T_S/3$ time delay between the converters, the gate signals are identical for the converters. Figure 5.32b indicates the waveforms for the total input current, i_L, and the inductor currents, i_{L1}, i_{L2}, and i_{L3}. The input current, i_i, frequency is seen to be 3 times more than that of each converter (Dewangan and Vadhera 2022).

The input current ripple in the 3-channel converter may be determined and supplied by using the same method described for the two-channel converters.

$$\Delta I_i = \begin{cases} (1-3D)D\frac{V_oT_s}{L} & \text{for } 0 \leq D < \frac{1}{3} \\ \left(3D(1-D) - \frac{2}{3}\right)\frac{V_oT_s}{L} & \text{for } \frac{1}{3} \leq D < \frac{2}{3} \\ (3D-2)(1-D)\frac{V_oT_s}{L} & \text{for } \frac{2}{3} \leq D < 1 \end{cases} \quad (5.28)$$

where L indicates converter channel inductance, that is, $L = L_1 = L_2 = L_3$.

Figures 5.29 and 5.31, respectively, represent the input current ripple ΔI_i and output voltage ripple ΔV_o for the three- and four-channel interleaved converters.

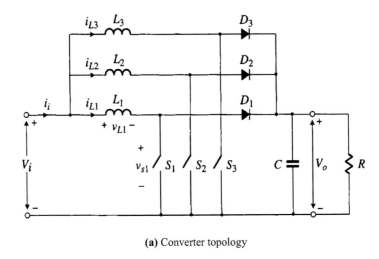

(a) Converter topology

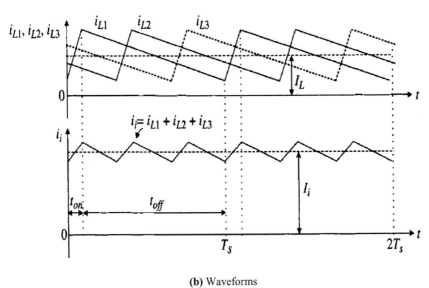

(b) Waveforms

Fig. 5.32 a Converter topology. b Waveforms

Compared to the single- and two-channel converters, these current and voltage ripples are even less. As a consequence, the input and output filters' size and price may be further decreased.

The highest power rating for each converter channel in real-world wind energy conversion systems ranges from 500 to 600 kW. Three interleaved converters are needed for a 1.5 MW WECS.

5.4 Power Electronics in Wind Energy

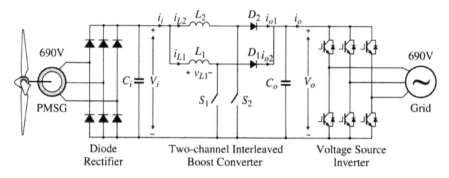

Fig. 5.33 A WECS with a two-channel interleaved boost converter

Case Study 5-1—PMSG Wind Energy System with a Two-Channel Interleaved Boost Converter. A 1.2 MW/690 V permanent magnet synchronous generator (PMSG) based wind energy conversion system is shown in Fig. 5.33, where a two-channel interleaved boost converter and a PWM voltage source inverter are employed. The boost converter provides two main functions: (1) to boost its input DC voltage V_i to a higher DC voltage V_o at its output, and (2) to perform maximum power point tracking (MPPT) such that the system can deliver its maximum possible power captured by the turbine to the grid at any wind speed (Wang and Chang 2004). The main function of the inverter is to keep its input DC voltage, which is the output of the boost converter, at a fixed value and also to control the reactive power to the grid.

At a given wind speed, the generator operates at a speed of 0.7 pu and delivers 412 kW (0.73 pu) power to the grid. The input voltage of the boost converter V_i is 680 V. The output voltage of the boost converter V_o is kept at a constant value of 1200 V by the inverter, which is the voltage required to deliver power to the grid of 690 V (refer to Sect. 4.7 for details). The inductance of the two-channel converter is $L = L_1 = L_2 = 270$ μH and the capacitance of the output filter capacitor C_o is 300 μF. The boost converter operates at a switching frequency of 2 kHz. Assuming that all the converters are ideal without power losses, investigate the input current ripple and output voltage ripple of the boost converter.

Assuming that the boost converter operates in the continuous-current mode, the duty cycle of each converter channel is

$$D = D_1 = D_2 = 1 - \frac{V_i}{V} = 1 - \frac{680}{1200} = 0.4333 \quad (5.29)$$

The ripple current in the two inductors can be calculated by the following:

$$I_L = \Delta I_{L1} = \Delta I_{L2} = D(1-D)\frac{V_o T_s}{L} = 545.6 \, \text{A} \quad (5.30)$$

The boundary inductor current relates to the inductor ripple current by

$$I_{\text{LB}} = I_{\text{LB1}} = I_{\text{LB2}} = \Delta I_L/2 = 272.8\,\text{A} \tag{5.31}$$

The boundary output current for each channel

$$I_{\text{oB}} = I_{\text{oB1}} = I_{\text{oB2}} = (1 - D)I_{\text{LB}} = 154.6\,\text{A} \tag{5.32}$$

The total average output current of the interleaved converter is

$$I_{\text{o}} = P_{\text{o}}/V_{\text{o}} = 343.3\,\text{A} \tag{5.33}$$

from which the average output current of each channel is

$$I_{\text{o1}} = I_{\text{o2}} = I_{\text{o}}/2 = 171.7\,\text{A} \tag{5.34}$$

Since $I_{\text{o1}} = I_{\text{o2}} > I_{\text{oB}}$ the converter operates in the continuous current mode. The total input current of the boost converter is given by

$$I_i = \begin{cases} P_i/V_i = P_{\text{o}}/V_i = 412 \times 10^3/680 = 605.8\,\text{A} \\ I_{\text{o}}/(1 - D) = 343.4/(1 - 0.4333) = 605.8\,\text{A} \end{cases} \tag{5.35}$$

The percentage inductor ripple current in each channel can be found from

$$\frac{\Delta I_{L1}}{I_{L1}} = \frac{\Delta I_{L2}}{I_{L2}} = \frac{\Delta I_L}{I_i/2} = \frac{545.6}{605.8/2} = 180.1\% \tag{5.36}$$

The total input current ripple

$$\Delta I_i = (1 - 2D)D\frac{V_{\text{o}}T_{\text{s}}}{L} = 128.4\,\text{A} \tag{5.37}$$

The total input current ripple can be found from

$$\frac{\Delta I_i}{I_i} = \frac{128.4}{606} = 21.2\% \tag{5.38}$$

which is much lower than the inductor ripple current of 180.1% in each of the channels.

The ratio of the total input ripple current ΔI_i, to the inductor ripple current ΔI_L of each channel is

$$\frac{\Delta I_i}{I_L} = \frac{128.4}{545.6} = 23.5\% \tag{5.39}$$

which is verified by Point A in Fig. 4.19.

5.4 Power Electronics in Wind Energy

To determine the output ripple voltage of the interleaved boost converter, it is assumed that the effect of the inverter operation on the DC voltage ripple is neglected. The load of the boost converter, which is the inverter, can be modelled by an equivalent resistance given by

$$R_{eq} = \frac{V_o}{I_o} = 3.495\ \Omega \qquad (5.40)$$

Making use of Fig. 4.20, the percentage output ripple voltage can be obtained by

$$\frac{\Delta V_o}{V_o} = 0.05\left(\frac{T_s}{R_{eq}C_o}\right) = 0.05 \times \frac{1/2000}{3.495 \times 300 \times 10^{-6}} = 2.38\% \qquad (5.41)$$

If the operation of the two-channel converters were not interleaved, switches S_1 and S_2 would be turned on and off simultaneously. The output ripple voltage would then be

$$\frac{\Delta V_o}{V_o} = D\left(\frac{T_s}{R_{eq}C_o}\right) = 0.433 \times \frac{1/2000}{3.495 \times 300 \times 10^{-6}} = 20.6\% \qquad (5.42)$$

which is around 8.6 times higher than that of the interleaved boost converter.

In summary, the multichannel interleaved converter produces a much lower input current ripple and output voltage ripple in comparison to the single-channel boost converter.

(b) Two-Level Voltage Source Converters

Figure 5.34a depicts the simplified circuit design for a three-phase, two-level voltage source converter. The converter consists of six switches, labelled S_1 through S_6, each with an antiparallel free-wheeling diode. Depending on the power and voltage specifications of the converter, the switches may be IGBT or IGCT components (Wu and Narimani 2017).

The converter has shown extensive use in a variety of industrial applications. The converter is frequently referred to as an inverter; Fig. 5.34b illustrates the conversion of a fixed DC voltage to a three-phase AC voltage with variable magnitude and frequency for an AC load. It is typically referred to as an active rectifier or PWM rectifier when the converter converts an adjustable DC voltage for a DC load from a fixed amplitude, fixed frequency AC grid voltage as shown in Fig. 5.34c (Kazimierczuk 2015). The DC side of the converter circuit may provide power to the AC side, and vice versa, regardless of whether it acts as an inverter or a rectifier.

The converter in wind energy conversion systems frequently connects a grid of electricity, and transfers the power produced by the grid and the generator, as depicted in Fig. 5.34d. In this scenario, the converter is referred to as a grid-connected or grid-tied converter. The converter is sometimes referred to as an inverter since it often switches electricity from its DC side to its AC side.

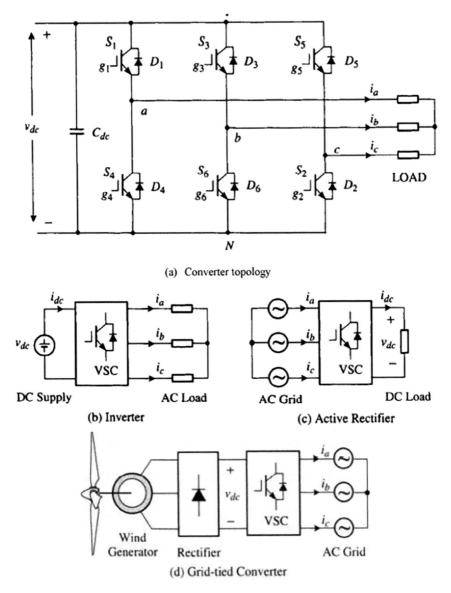

Fig. 5.34 Simplified voltage source converters with two levels (VSC)

The further section mainly discussed the pulse-width modulation (PWM) methods used in two-level voltage source converters. The modulation techniques are applicable to the converter, thus the inverter used may function as an inverter or a rectifier. A thorough examination of space vector modulation (SVM) algorithms follows carrier-based sinusoidal PWM (SPWM) introduction techniques in this section.

I. Sinusoidal PWM

The three-phase sinusoidal modulating waveforms v_{ma}, v_{mb}, and v_{mc} as well as the triangular carrier signal v_{cr}, are used in Fig. 5.35 to illustrate the use of sinusoidal PWM basic operation to the two-level converter. The amplitude-modulation index has control over the inverter output voltage's fundamental frequency component.

$$m_a = \frac{\hat{V}_m}{\hat{V}_{cr}}, \tag{5.43}$$

where $\hat{V}_{cr}$ and $\hat{V}_m$ represent the carrier wave and the modulating wave's respective peaks. The normal method for adjusting the amplitude-modulation index m_a is to change $\hat{V}_m$ while maintaining $\hat{V}_{cr}$ constant. The frequency-modulation index includes

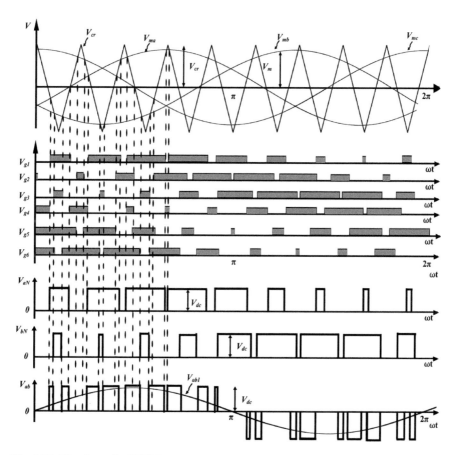

Fig. 5.35 Waveforms for SPWM

the following:

$$m_f = \frac{f_{cr}}{f_m} \qquad (5.44)$$

where the corresponding frequencies of the carrier and modulating waves, are f_{cr} and f_m.

When the carrier wave is compared with the modulating waves, it helps to understand how switches S_1 through S_6 operate. Inverter leg a's upper switch S_1 is switched on when $v_{ma} > v_{cr}$. S_4 the bottom switch is turned off since it functions in a complementary manner. The DC voltage V_{dc} is equal to the resulting inverter terminal voltage v_{an}, which is the phase-a terminal voltage in relation to the negative DC bus N. As illustrated in Fig. 5.35, when $v_{ma} < v_{cr}$, S_4 is on and S_1 is off, leading to $v_{aN} = 0$. The inverter is frequently referred to as a two-level inverter since the waveform v_{aN} only has two levels, V_{dc} and 0, in it. It is stated that a blanking period (or dead time), in which both switches are off, should be included in order to prevent potential short circuiting when an inverter leg's upper and lower components are switching transiently.

The line-to-line voltage of an inverter can be found by using the formula $v_{ab} = v_{aN} - v_{bN}$. The figure also includes the fundamental frequency component's waveform v_{ab1}, m_a and f_m both have autonomous control over the magnitude and frequency of the v_{ab1}.

The switching frequency of the active switches in the two-level inverter can be found from $f_{sw} = f_{cr} = f_m \times m_f$. Figure 5.35 shows that v_{aN} has the fundamental frequency cycles with nine pulses each. S_1 is toggled on and off once for each pulse. The switching frequency S_1, which is also the carrier frequency, is $f_{sw} = 60 \times 9 = 540$ Hz having a 60 Hz fundamental frequency. It is important to keep in mind that in multilevel inverters, it's possible that the carrier frequency and the device switching frequency aren't always the same.

When the carrier wave is synchronized with the modulating wave (m_f is an integer), the modulation scheme is known as synchronous PWM, in contrast to asynchronous PWM, whose carrier frequency f_{cr} is usually fixed and independent of f_m. The asynchronous PWM has an easy implementation with analogue circuitry and a fixed switching frequency. It may, however, produce harmonics that are not typical and whose frequency is not in the form of multiples of the fundamental frequency. The synchronous PWM approach is well applicable to use with a digital processor.

II. Space Vector Modulation

One of the real-time modulation techniques that is widely used for digital control of voltage source inverters is Space Vector Modulation (SVM) (Durgasukumar et al. 2012). The two-level inverter's space vector modulation principle and application are presented in this section.

Switching States Switching states can be used to represent the operational two-level inverter's switch condition in Fig. 5.34a. Switching state P, as illustrated in Table 5.2, denotes that an inverter leg's upper switch is engaged and that the inverter

5.4 Power Electronics in Wind Energy

terminal voltage (v_{aN}, v_{bN}, or v_{cN}) is positive ($+V_{dc}$) while switching state O denotes that voltage at the inverter's terminals is zero as a result of the lower switch's conduction.

According to Table 5.2, with the two-level inverter, there are eight potential switching state combinations. For instance, the S_1, S_6, and S_2 conductivity in the corresponding inverter legs (a, b, and c), correspond to the switching state [POO]. Among the eight switching states, [PPP] and [OOO] are zero states and the others are active states.

Space Vectors: Active and zero space vectors, respectively, can be used to represent the active and zero switching states. The regular hexagon with six equal sections is created by the six active vectors $\overrightarrow{V}_1$ to $\overrightarrow{V}_6$ as shown in Fig. 5.36, a common two-level inverter space vector diagram (I–VI). The hexagon's centre contains the zero vectors $\overrightarrow{V}_0$. The link between space vectors and switching states may be ascertained using the two-level inverter shown in Fig. 5.34a. Considering this inverter operates in a three-phase balance, there are as follows:

$$v_a(t) + v_b(t) + v_c(t) = 0, \tag{5.45}$$

where the instantaneous load phase voltages are v_a, v_b, and v_c. From a mathematical perspective, one of the phase voltages is redundant because the third phase voltage can be easily calculated from any two-phase voltage. As a result, using the $abc/\alpha\beta$ transformation, three-phase variables can become two-phase variables.

$$\begin{bmatrix} v_\alpha(t) \\ v_\beta(t) \end{bmatrix} = \frac{2}{3} \begin{bmatrix} 1 & -\frac{1}{2} & -\frac{1}{2} \\ 0 & \frac{\sqrt{3}}{2} & -\frac{\sqrt{3}}{2} \end{bmatrix} \begin{bmatrix} v_a \\ v_b \\ v_c \end{bmatrix} \tag{5.46}$$

The two-phase voltages in the $\alpha - \beta$ frame can be used to generally express a space vector as follows:

$$\overline{v}(t) = v_\alpha(t) + j v_\beta(t) \tag{5.47}$$

When we convert (5.46) into (5.47) we get

$$\overrightarrow{v}(t) = \frac{2}{3} \left[v_a(t) e^{j0} + v_b(t) e^{j2\pi/3} + v_c(t) e^{j4\pi/3} \right], \tag{5.48}$$

Table 5.2 Definition of switching states

Switching state	Leg a			Leg b			Leg c		
	S_1	S_4	v_{aN}	S_3	S_6	v_{bN}	S_5	S_2	v_{cN}
P	On	Off	V_{dc}	On	Off	V_{dc}	On	Off	V_{dc}
O	Off	On	0	Off	On	0	Off	On	0

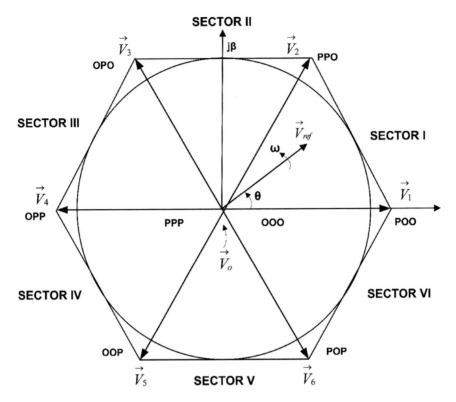

Fig. 5.36 Space vector diagram for the two-level inverter

where $e^{jx} = \cos x + j \sin x$ and $x = 0, 2\pi/3$ or $4\pi/3$. For the active switching state [POO], the generated load phase voltages are

$$v_a(t) = \frac{2}{3}V_{dc}, \quad v_b(t) = -\frac{1}{3}V_{dc} \text{ and } v_c(t) = -\frac{1}{3}V_{dc} \quad (5.49)$$

By putting (5.49) into the Eq. (5.48), the associated space vector, designated as $\vec{V}_1$ may be produced:

$$\vec{V}_1 = \frac{2}{3}V_{dc}e^{j0} \quad (5.50)$$

The six active vectors may be obtained using the same method:

$$\vec{V}_k = \frac{2}{3}V_{dc}e^{j(k-1)\frac{\pi}{3}} \quad k = 1, 2 \ldots 6 \quad (5.51)$$

5.4 Power Electronics in Wind Energy 243

Two switching states, [PPP] and [OOO], are present in the zero vectors $\vec{V}_o$, and one of them is redundant. As will be shown later, the possible switching state may be used to increase other valuable functions or reduce the inverter's switching frequency. Table 5.3 lists the relationship between the space vectors and the relevant switching states.

Keep in mind that the zero and active vectors are referred to as stationary vectors since they are not in motion. The reference vector $\vec{V}_{ref}$ in Fig. 5.36, on the other hand, spins in space with an angular velocity.

$$\omega = 2\pi f, \qquad (5.52)$$

where f is the output voltage of the inverter's fundamental frequency. You may calculate the angle between $\vec{V}_{ref}$ and the α-axis of the $\alpha - \beta$ frame by

$$\theta(t) = \int_0^t \omega(t)dt + \theta_o \qquad (5.53)$$

Three adjacent stationary vectors may be used to provide $\vec{V}_{ref}$ for a specific magnitude (length) and position, using which the switching states of the inverter can be chosen and the gate signals for the active switches can be produced. A variety of switch sets will be switched on or off when $\vec{V}_{ref}$ moving through each sector one at a time. As a consequence, the inverter output voltage changes one cycle over time for every rotation that $\vec{V}_{ref}$ is made in space. Although the magnitude of $\vec{V}_{ref}$ may change the output voltage of an inverter, its output frequency is related to the spinning speed of $\vec{V}_{ref}$.

Table 5.3 Space vectors, switching states, and on-state switches

Space vector		Switching state (three phases)	On-state switch	Vector definition
Zero vector	$\vec{V}_o$	[PPP]	S_1, S_3, S_5	$\vec{V}_o = 0$
		[OOO]	S_4, S_6, S_5	
Active vector	$\vec{V}_1$	[POO]	S_1, S_6, S_2	$\vec{V}_1 = \frac{2}{3}V_{dc}e^{j0}$
	$\vec{V}_2$	[PPO]	S_1, S_3, S_2	$\vec{V}_2 = \frac{2}{3}V_{dc}e^{j\frac{\pi}{3}}$
	$\vec{V}_3$	[OPO]	S_4, S_3, S_2	$\vec{V}_3 = \frac{2}{3}V_{dc}e^{j\frac{2\pi}{3}}$
	$\vec{V}_4$	[OPP]	S_4, S_3, S_5	$\vec{V}_4 = \frac{2}{3}V_{dc}e^{j\frac{3\pi}{3}}$
	$\vec{V}_5$	[OOP]	S_4, S_6, S_5	$\vec{V}_5 = \frac{2}{3}V_{dc}e^{j\frac{4\pi}{3}}$
	$\vec{V}_6$	[POP]	S_1, S_6, S_5	$\vec{V}_6 = \frac{2}{3}V_{dc}e^{j\frac{5\pi}{3}}$

Dwell Time Calculation. Three stationary vectors may be used to create the reference $\vec{V}_{\text{ref}}$, as was previously indicated. The duty-cycle time (on-state or off-state time) of the selected switches during a sampling period T_s is effectively represented by the dwell time for the stationary vectors. The volt-second balancing principle, which states that the product of the reference voltage $\vec{V}_{\text{ref}}$ and sampling period T_s equals the sum of the voltage multiplied by the time interval of selected space vectors, is the foundation for the computation of dwell time.

As long as the sampling interval T_s is short enough, the reference vector $\vec{V}_{\text{ref}}$ may be thought of as constant during T_s. This presumption allows for the approximate representation of $\vec{V}_{\text{ref}}$ as two adjacent active vectors and one zero vector. For example, $\vec{V}_1, \vec{V}_2,$ and $\vec{V}_o$ may synthesize $\vec{V}_{\text{ref}}$ while it is in sector I, as illustrated in Fig. 5.37. The balancing equation for volts per second is

$$\begin{cases} \vec{V}_{\text{ref}} T_s = \vec{V}_1 T_a + \vec{V}_2 T_b + \vec{V}_o T_o \\ T_s = T_a + T_b + T_o, \end{cases} \quad (5.54)$$

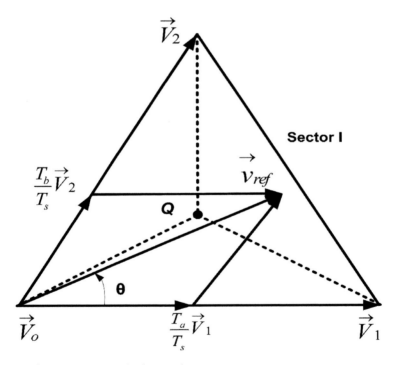

Fig. 5.37 $\vec{V}_{\text{ref}}$ synthesized by $\vec{V}_1, \vec{V}_2,$ and $\vec{V}_o$

5.4 Power Electronics in Wind Energy

where T_a, T_b, and T_o, respectively, are the dwell durations for the vectors $\vec{V}_1, \vec{V}_2$, and $\vec{V}_o$. The space vectors in Eq. (5.51) may be represented as:

$$\vec{V}_{ref} = v_{ref}e^{j\theta}, \quad \vec{V}_1 = \frac{2}{3}V_{dc}, \quad \vec{V}_2 = \frac{2}{3}V_{dc}e^{j\frac{\pi}{3}}, \text{ and } \vec{V}_o = 0 \qquad (5.55)$$

Hence, as illustrated in Fig. 5.37, v_{ref} denotes the magnitude of the reference vector and θ denotes the angle between $\vec{V}_{ref}$ and the α-axis of the $\alpha - \beta$ frame.

In the $\alpha - \beta$ frame, by substituting (5.55) for (5.54) and dividing the resulting equation into its real (α-axis) and imaginary (β-axis) parts, we obtain

$$\begin{cases} \text{Re: } v_{ref}(\cos\theta)T_s = \frac{2}{3}V_{dc}T_a + \frac{1}{3}V_{dc}T_b \\ \\ \text{Im: } v_{ref}(\sin\theta)T_s = \frac{1}{\sqrt{3}}V_{dc}T_b \end{cases} \qquad (5.56)$$

Solving (5.56) together with $T_s = T_a + T_b + T_o$ yields

$$\begin{cases} T_a = \frac{\sqrt{3}T_s v_{ref}}{V_{dc}} \sin\left(\frac{\pi}{3} - \theta\right) \\ \\ T_b = \frac{\sqrt{3}T_s v_{ref}}{V_{dc}} \sin(\theta) \text{ for } 0 \leq \theta < \pi/3 \\ \\ T_o = T_s - T_a - T_b \end{cases} \qquad (5.57)$$

Let's look at a few special examples to see how the dwell times relate to where $\vec{V}_{ref}$ is located. The dwell time T_a of $\vec{V}_1$ will be equal to T_b of $\vec{V}_2$ if $\vec{V}_{ref}$ is precisely halfway between $\vec{V}_1$ and $\vec{V}_2$ (i.e. $\theta = \pi/6$). T_b will be higher than T_a when $\vec{V}_{ref}$ is nearer $\vec{V}_2$. T_a will be 0 if $\vec{V}_{ref}$ and $\vec{V}_2$ are coincident. With the head of $\vec{V}_{ref}$ located right on the central point Q, $T_a = T_b = T_o$. Table 5.4 provides a summary of the association between the $\vec{V}_{ref}$ location and dwell times.

While Eq. (5.57) is obtained when $\vec{V}_{ref}$ is in sector I, it may also be utilized when $\vec{V}_{ref}$ is in other sectors as long as a multiple of $\pi/3$ is deducted from the actual angular displacement θ to make the modified angle θ' fit within the range between zero and $\pi/3$ for use in the equation.

Table 5.4 $\vec{V}_{ref}$ location and dwell times

$\vec{V}_{ref}$ location	$\theta = 0$	$0 < \theta < \frac{\pi}{6}$	$\theta = \frac{\pi}{6}$	$\frac{\pi}{6} < \theta < \frac{\pi}{3}$	$\theta = \frac{\pi}{3}$
Dwell times	$T_a > 0$	$T_a > T_b$	$T_a = T_b$	$T_a < T_b$	$T_a = 0$
	$T_b = 0$				$T_b > 0$

$$\theta' = \theta - (k-1)\pi/3 \text{ for } 0 \le \theta' < \pi/3, \tag{5.58}$$

where $k = 1, 2, 3 \dots 6$ for sectors I, II, …, VI, respectively. For example, when $\vec{V}_{ref}$ is in sector II, dwell durations T_a, T_b, and T_0 for vectors $\vec{V}_2, \vec{V}_3$ and $\vec{V}_0$ are determined based on Eq. (5.57).

Modulation Index Modulation index m_a can also be used to express Eq. (5.57):

$$\begin{cases} T_a = T_s m_a \sin\left(\dfrac{\pi}{3} - \theta\right) \\ T_b = T_s m_a \sin\theta \\ T_0 = T_s - T_a - T_b, \end{cases} \tag{5.59}$$

where

$$m_a = \frac{\sqrt{3}v_{ref}}{V_{dc}} \tag{5.60}$$

The fundamental frequency component's peak value in the inverter output phase voltage is represented by the length of the reference vector $\vec{V}_{ref}$, which is,

$$v_{ref} = \overset{\wedge}{\underset{a1}{V}} = \sqrt{2}V_{a1}, \tag{5.61}$$

where V_{a1} is the rms value of the primary component in the phase-a (output) voltage of the inverter.

The relationship between m_a and V_{a1} may be discovered by substituting (5.61) into (5.60):

$$m_a = \frac{\sqrt{3}v_{ref}}{V_{dc}} = \frac{\sqrt{6}V_{a1}}{V_{dc}} \tag{5.62}$$

The inverter output voltage V_{a1} is proportional to the modulation index m_a for a given DC voltage V_{dc}.

The largest circle that can be inscribed inside the hexagon has a radius that corresponds to the maximum length of the reference vector, $v_{ref, max}$, as illustrated in Fig. 5.36. As the hexagon is made up of six active vectors with lengths of $2V_{dc}/3$, it is possible to get $v_{ref, max}$ from

$$v_{ref,max} = \frac{2}{3}V_{dc} \times \frac{\sqrt{3}}{2} = \frac{V_{dc}}{\sqrt{3}} \tag{5.63}$$

The maximum modulation index is obtained by substituting (5.63) into (5.60).

$$m_{a,max} = 1, \tag{5.64}$$

5.4 Power Electronics in Wind Energy

where the SVM scheme's modulation index falls between the range of

$$0 \leq m_a < 1 \qquad (5.65)$$

Switching Sequence The switching sequence must be set up once the space vectors have been decided upon and their dwell periods have been determined. The switching sequence design for certain $\vec{V}_{ref}$ is often not unique, but it must meet the following two criteria to keep the device switching frequency to a minimum:

1. Just two switches, one turned on and the other off, in the same inverter leg, are used to go from one switching state to the next.
2. $\vec{V}_{ref}$ transitions from one sector to the next in the space vector diagram require no or a minimum number of switchings.

An example of a seven-segment switching pattern is shown in Fig. 5.38, along with the inverter output voltage waveforms for $\vec{V}_{ref}$ sector I, where $\vec{V}_{ref}$ is produced by $\vec{V}_1$, $\vec{V}_2$, and $\vec{V}_0$. For the chosen vectors, the sample time T_s is split into seven parts. The observation is that

- The sample period is equal to the sum of the dwell durations for the seven segments ($T_s = T_a + T_b + T_o$).

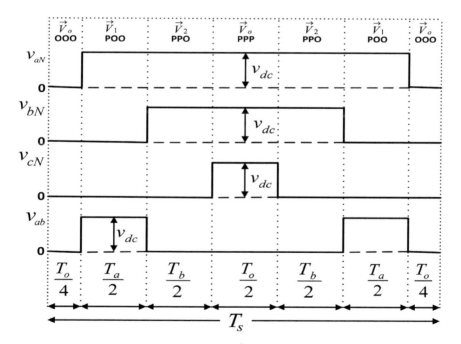

Fig. 5.38 Seven-segment switching sequence for $\vec{V}_{ref}$ in sector I

- Design criterion (1) has been met. For instance, switching S_1 on and S_4 off results in the transition from [OOO] to [POO], using just two switches.
- To lower the number of switchings per sampling period, the redundant switching states for $\vec{V}_o$ are used. The switching state [PPP] is chosen for the $T_o/2$ segment in the middle of the sample period, while [OOO] is utilized for the $T_o/4$ segments on either side.
- Once per sample period, each switch in the inverter is turned on and off. Hence, $f_{sw} = f_{sp} = 1/T_s$, where f_{sw} is the switching frequency of the devices and f_{sp} is the sampling frequency.

(c) Three-Level Neutral Point Clamped Converters

The diode-clamped multilevel inverter creates multiple-level AC voltage waveforms using clamping diodes and cascaded DC capacitors (Cai et al. 2017). The neutral point clamped (NPC) inverter, commonly known as a three-level inverter, has a broad range of practical applications, especially in medium-voltage (MV) variable-speed drives. The inverter can normally be set as a three, four, or five-level topology. The NPC inverter is also a good choice for MV wind energy systems (3–4 kV) (Hoon et al. 2017). In comparison with a two-level inverter previously discussed, the NPC inverter has lower THD in its AC output voltages. The inverter doesn't need switching devices into series to be employed in MV wind energy systems, which is more critical (Lazoueche et al. 2021). For example, the 4 kV WECS can be used with the 6 kV IGBT or IGCT devices used by the NPC inverter without the requirement to connect the switches in series (Rivera et al. 2012).

I. Converter Configuration

The three-level NPC inverter's simplified circuit diagram is shown in Fig. 5.39. Four antiparallel diodes D1–D4 and four active switches S1–S4 are used to make up the inverter leg a. In actual use, a switching device can be either an IGBT or an IGCT. The DC bus capacitor is divided in two on the inverter's DC side, creating a neutral point Z. The clamping diodes are D_{z1} and D_{z2}, which are wired to the neutral point. The inverter output terminal a is linked to the neutral point through one of the clamping diodes when switches S_2 and S_3 are turned on. The voltage E, which is typically equal to half of the total DC voltage V_{dc}, is applied across each of the DC capacitors.

The operating condition of the switches in the NPC inverter may be determined by switching states, as shown in Table 5.5. Switching state P denotes that the upper two switches in the leg a are on and the inverter terminal voltage v_{aZ}, which is the voltage at terminal a with respect to the neutral point Z, is $+E$, whereas N indicates that the lower two switches conduct, leading to $v_{aZ} = -E$.

Switching state O denotes that v_{az} is clamped to zero by the clamping diodes and the inner two switches S_2 and S_3 are both on. One of the two clamping diodes is activated depending on the direction of the load current i_a. For instance, terminal a is connected to the neutral point Z through the conduction of D_{z1} and S_2 when a positive load current ($i_a > 0$) pushes D_{z1} to turn on. Table 5.5 shows that the operation of

5.4 Power Electronics in Wind Energy

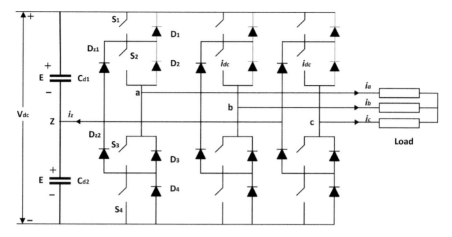

Fig. 5.39 NPC inverter with three-level

Table 5.5 Switching states definition

Switching state	Devices switching states (Phase a)				Inverter terminal voltage
	S_1	S_2	S_3	S_4	V_{az}
P	On	On	Off	Off	E
O	Off	On	On	Off	0
N	Off	Off	On	On	$-E$

switches S_2 and S_3 is complimentary. One must be turned on while the other is off. The complementary pair S_2 and S_4 is also similar.

Switching states and gate signal configurations are shown in Fig. 5.40, and the gate signals for S_1 through S_4 are represented by v_{g1} to v_{g4}, respectively. Selective harmonic elimination techniques, space vector modulation, or carrier-based modulation can all be used to produce the gate signals. The waveform for V_{dc} has three voltage levels: $+E$, 0, and $-E$, which is why the inverter is known as a three-level inverter.

The process for obtaining the line-to-line voltage waveform is shown in Fig. 5.41. The voltages at the inverter terminals v_{az}, v_{bz}, and v_{cz} are three-phase balanced with a $2\pi/3$ phase shift between them. The equation $v_{ab} = v_{az} - v_{bz}$, which contains the five voltage levels ($+2E$, $+E$, 0, $-E$, and $-2E$), may be used to figure out the line-to-line voltage v_{ab}.

II. Space Vector Modulation

As previously mentioned, P, O, and N, the three switching states, may be used to describe how each inverter phase leg functions. As illustrated in Fig. 5.42, the inverter has features a total of 27 different switching state combinations and 19 different stationary space vectors $\left(\vec{V}_0 \text{ to } \vec{V}_{18}\right)$ when all three phases are taken into

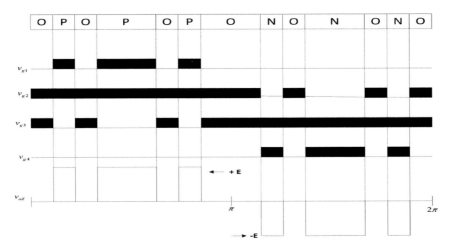

Fig. 5.40 Switching states, gate signals, and inverter terminal voltage V_{az}

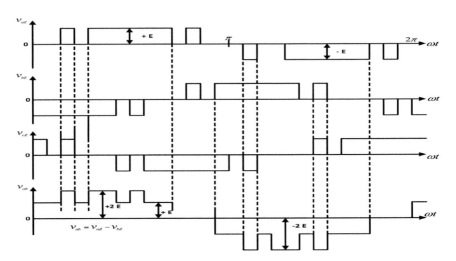

Fig. 5.41 Inverter terminal and line-to-line voltage waveforms

account. The NPC inverter uses the same space vector modulation technique as the two-level inverter, but the implementation is more difficult. The three adjacent stationary vectors that make up the reference vector $\vec{V}_{ref}$ at a specific location in space can be used to compute the dwell periods for every single stationary vector that was selected. The switch turn-on timings during a sample time T_s may be computed based on the calculated dwell times, from which the switching sequence can be designed. Depending on which of the inverter output voltages is obtained, the inverter switches' gate signals may then be generated. At the fundamental frequency, the inverter's

5.4 Power Electronics in Wind Energy

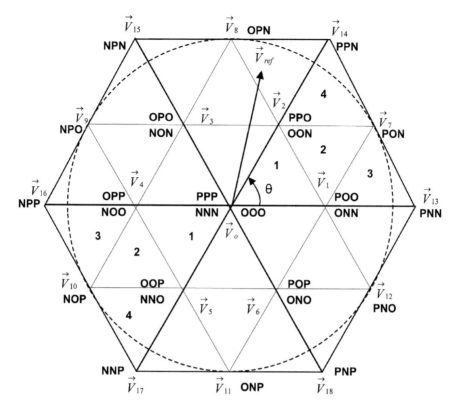

Fig. 5.42 Space vector diagram of the NPC inverter

output voltage fluctuates by one cycle whenever the reference vector $\vec{V}_{ref}$ rotates in space for one cycle.

Figure 5.43 shows the case study [16] where THD profile has been compared between the Three-Level NPC Inverter with Space Vector Modulation and the two-level inverter.

(d) **PWM Current Source Converters**

Current source converters (CSCs) and voltage source converters (VSCs) are two broad categories for solid-state converters. The voltage source converter provides a specified three-phase PWM output voltage waveform, while the current source converter generates a defined PWM current waveform. The PWM current source converter has a basic converter topology, waveforms that are almost sinusoidal, and reliable short-circuit protection. It is especially well suited for high-power applications like wind energy conversion systems and megawatt variable-speed drives.

The current source inverter modulation techniques are the main topic of discussion in this section. Two modulation techniques for the inverter are discussed: selective

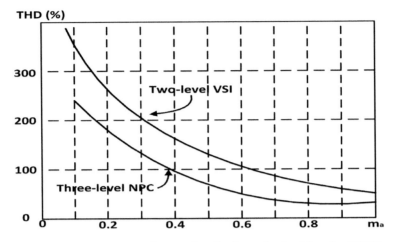

Fig. 5.43 THD profile of the output voltage produced by the two-level and three-level NPC inverters (Wu et al. 2011)

harmonic elimination (SHE) and space vector modulation (SVM). Some modulation approaches were developed to power inverters operating at switching frequencies below 1 kHz to reduce switching losses.

I. Current Source Inverter Topology

Figure 5.44 shows a typical current source inverter (CSI) circuit for three-phase PWM. The inverter is composed of six IGCT devices of symmetrical type or reverse-blocking IGBTs. The inverter generates a predetermined PWM output current i_{aw} and requires a DC source i_{dc} at its dc input.

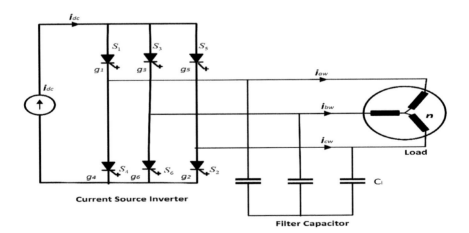

Fig. 5.44 PWM current-source inverter

5.4 Power Electronics in Wind Energy

The current source inverter normally requires a three-phase capacitor C_i, at its output to assist in the commutation of the switching devices. For illustration, the inverter PWM current i_{aw} rapidly decreases to zero when switch S_1 is turned off. The capacitor provides a current path for the energy trapped in the phase-a load inductance. Otherwise, a high-voltage spike would be induced, causing damage to the switching devices. The capacitor also acts as a harmonic filter, improving the load current and voltage waveforms. The value of the capacitor is normally in the range of 0.3–0.6 per unit for an inverter with a switching frequency of 200–400 Hz (Gao et al. 2019).

II. Selective Harmonic Elimination

The switching pattern design for the CSI should generally satisfy two conditions: (1) the DC current i_{dc} should be continuous and (2) the inverter PWM current i_{aw} should be defined. The two conditions can be translated into a switching constraint: at any instant of time (excluding commutation intervals), there are only two switches conducting: one in the top half of the bridge and the other in the bottom half. With only one switch turned on, the continuity of the DC current is lost. A very high voltage will be induced by the constant DC current, causing damage to the switching devices. If more than two devices are on simultaneously, the PWM current i_{aw} is not defined by the switching pattern. For instance, with S_1, S_2, and S_3 conducting at the same time, the currents in S_1, and S_3, which are the PWM currents in the inverter phases a and b, are load-dependent although the sum of the two currents is equal to i_{dc}.

Selective harmonic elimination is an offline modulation scheme, which is able to eliminate a number of low-order unwanted harmonics in the inverter PWM current i_{aw}. The switching angles are precalculated and then imported into a digital controller for implementation. Figure 5.45 shows a typical SHE waveform that satisfies the switching constraint for the CSI. There are five pulses per half cycle ($N_p = 5$) with five switching angles in the first $\pi/2$ period. However, only two out of the five angles, θ_1 and θ_2, are independent. Given these two angles, all other switching angles can be calculated.

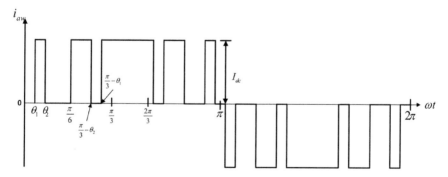

Fig. 5.45 Selective harmonic elimination (SHE) schemes

The two switching angles provide two degrees of freedom, which can be used to either eliminate two harmonics in i_{aw} without modulation index control or eliminate one harmonic and provide an adjustable modulation index m_a. The first option is preferred since the adjustment of i_{aw} is normally done by varying i_{dc}. The number of harmonics to be eliminated is then given by $k = (Np - 1)/2$.

To determine switching angles such as θ_1, and θ_2 in Fig. 5.45 for harmonic elimination, Fourier analysis can be performed, from which a set of nonlinear equations can be formulated. These equations can be solved by numerical methods. Table 5.6 gives a list of switching angles for the elimination of up to four harmonics in i_{aw}. It is noted that the nonlinear equations for harmonic elimination may not always have a valid solution. If this happens, optimization techniques can be used to find optimal switching angles that minimize the magnitude of low-order harmonics.

III. Space Vector Modulation

SVM can be used together with the SHE technique to control the inverter for current sources (Akhtar and Behera 2021). This section presents the principle of the SVM scheme for current source inverters.

Switching States—As stated earlier, the PWM switching pattern for the CSI shown in Fig. 5.44 must satisfy the constraint that only two switches in the inverter conduct at any time instant, one in the top half of the CSI bridge and the other in the bottom half. Under this constraint, the three-phase inverter has a total of nine switching states, as listed in Table 5.7. These switching states can be classified as zero switching states and active switching states.

Table 5.6 SHE switching angles

Harmonics eliminated	Switching angles			
	θ_1	θ_2	θ_3	θ_4
5	18.00	–	–	–
7	21.43	–	–	–
11	24.55	–	–	–
5,7	7.93	13.75	–	–
5,11	12.96	19.14	–	–
5,13	14.48	21.12	–	–
5,7,11	2.24	5.60	21.26	–
5,7,13	4.21	8.04	22.45	–
5,7,17	6.91	11.96	25.57	–
5,7,11,13[a]	0.00	1.60	15.14	20.26
5,7,11,17	0.07	2.63	16.57	21.80
5,7,11,19	1.11	4.01	18.26	23.60

[a] Harmonics are reduced in size but not entirely removed

5.4 Power Electronics in Wind Energy

Table 5.7 Switching states and space vectors

Type	Inverter PWM current					
	Switching states	On-state switch	i_{aw}	i_{bw}	i_{cw}	Space vector
Zero states	(1,4)	S_1,S_4	0	0	0	$\overrightarrow{I}_o$
	(3,6)	S_3,S_6				
	(5,2)	S_5,S_2				
Active states	(6,1)	S_6,S_1	I_{dc}	$-I_{dc}$	0	$\overrightarrow{I}_1$
	(1,2)	S_1,S_2	I_{dc}	0	$-I_{dc}$	$\overrightarrow{I}_2$
	(2,3)	S_2,S_3	0	I_{dc}	$-I_{dc}$	$\overrightarrow{I}_3$
	(3,4)	S_3,S_4	$-I_{dc}$	I_{dc}	0	$\overrightarrow{I}_4$
	(4,5)	S_4,S_5	$-I_{dc}$	0	I_{dc}	$\overrightarrow{I}_5$
	(5,6)	S_5,S_6	0	$-I_{dc}$	I_{dc}	$\overrightarrow{I}_6$

There are three zero switching states: $(1, 4)$, $(3, 6)$, and $(5, 2)$. The zero state $(1, 4)$ signifies that switches S_1 and S_4 in inverter-phase leg a conduct simultaneously and the other four switches in the inverter are off. The DC current source i_{dc} is bypassed, leading to $i_{aw} = i_{bw} = i_{cw} = 0$. This operating mode is often referred to as bypass operation.

There exist six active switching states. State $(1, 2)$ indicates that switch S_1 in leg a and S_2 in leg c are on. The DC current flows through S_1, the load, S_2, and then back to the DC source, resulting in $i_{aw} = I_{dc}$ and $i_{cw} = -I_{dc}$. The definition of the other five active states is also given in the table.

Space Vectors—The active and zero switching states can be represented by active and zero space vectors, respectively. A space vector diagram for the CSI is shown in Fig. 5.46, where $\overrightarrow{I}_1$ to $\overrightarrow{I}_6$ are the active vectors and $\overrightarrow{I}_O$ is the zero vector. The active vectors form a regular hexagon with six equal sectors, whereas the zero vector $\overrightarrow{I}_O$ lies at the centre of the hexagon.

Assuming that the operation of the inverter in Fig. 5.44 is three-phase balanced, that is,

$$i_{aw}(t) + i_{bw}(t) + i_{cw}(t) = 0 \tag{5.66}$$

In the $\alpha - \beta$ frame, the three-phase currents may be transformed to two-phase currents, where the instantaneous PWM output currents in phases a, b, and i_c are i_{aw}, i_{bw}, and i_{cw}, respectively:

$$\begin{bmatrix} i_\alpha(t) \\ i_\beta(t) \end{bmatrix} = \frac{2}{3} \begin{bmatrix} 1 & -\frac{1}{2} & -\frac{1}{2} \\ 0 & \frac{\sqrt{3}}{2} & -\frac{\sqrt{3}}{2} \end{bmatrix} \begin{bmatrix} i_{aw}(t) \\ i_{bw}(t) \\ i_{cw}(t) \end{bmatrix} \tag{5.67}$$

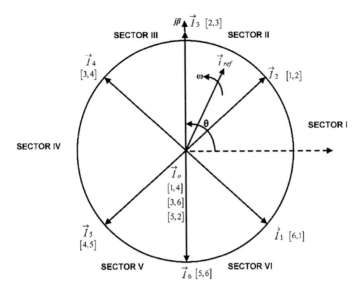

Fig. 5.46 Space vector diagram for the current-source inverter

In general, two-phase currents can be used to represent a current-space vector as follows:

$$\vec{i}(t) = i_\alpha(t) + ji_\beta(t) \tag{5.68}$$

$\vec{i}(t)$ can be shown by the terms i_{aw}, i_{bw}, and i_{cw} when Eq. (5.67) is put in Eq. (5.68)

$$\vec{i}(t) = \frac{2}{3}\left[i_{aw}(t)e^{j0} + i_{bw}(t)e^{j2\pi/3} + i_{cw}(t)e^{j4\pi/3}\right] \tag{5.69}$$

S_1 and S_6 are turned on for the active state (6, 1), and the inverter PWM currents can be expressed as follows:

$$i_{aw}(t) = I_{dc}, \; i_{bw}(t) = -I_{dc}, \; i_{cw}(t) = 0 \tag{5.70}$$

Putting the Eq. (5.70) into (5.69) gives

$$\vec{I}_1 = \frac{2}{\sqrt{3}} I_{dc} e^{j(-\frac{\pi}{6})} \tag{5.71}$$

Similarly, the other five active vectors can be derived. The active vectors can be expressed as follows:

$$\vec{I}_k = \frac{2}{\sqrt{3}} I_{dc} e^{j((k-1)\frac{\pi}{3} - \frac{\pi}{6})} \; \text{ for } k = 1, 2, \ldots, 6. \tag{5.72}$$

5.4 Power Electronics in Wind Energy

It should be noted that since the active and zero vectors are fixed in space, they are referred to as stationary vectors. On the contrary, the current reference vector $\vec{i}_{ref}$ Fig. 5.46 rotates in space at an angular velocity.

$$\omega = 2\pi f, \tag{5.73}$$

where f indicates the inverter output current i_{aw} fundamental frequency. The angular displacement between the $\vec{i}_{ref}$ and the α-axis of the α–β frame can be obtained by the following:

$$\theta(t) = \int_0^t \omega(t)\mathrm{d}t + \theta_o \tag{5.74}$$

For a given length and position, $\vec{i}_{ref}$ can be synthesized by three nearby stationary vectors, based on which the switching states of the inverter can be selected and gate signals for the active switches can be generated. When $\vec{i}_{ref}$ passes through sectors one by one, different sets of switches are turned on or off. As a result, when $\vec{i}_{ref}$ rotates one revolution in space, the inverter output current varies one cycle over time. The frequency and magnitude of the inverter output current correspond to the rotating speed and length of $\vec{i}_{ref}$, respectively.

Dwell Time Calculation—As indicated above, the reference $\vec{i}_{ref}$ can be synthesized by three stationary vectors. The dwell time for the stationary vectors essentially represents the duty-cycle time (on-state or off-state time) of the chosen switches during a sampling period T_s. The dwell time calculation is based on the ampere-second balancing principle, that is, the product of the reference vector $\vec{i}_{ref}$ and sampling period T_s equals the sum of the current vectors multiplied by the time interval of chosen space vectors. Assuming that the sampling period T_s is sufficiently small, the reference vector $\vec{i}_{ref}$ can be considered constant during T_s. Under this assumption, $\vec{i}_{ref}$ can be approximated by two adjacent active vectors and a zero vector. For example, with $\vec{i}_{ref}$ falling into Sector I as shown in Fig. 5.47, it can be synthesized by $\vec{I}_1, \vec{I}_2$, and $\vec{I}_O$. The ampere-second balancing equation is thus given by

$$\begin{cases} \vec{i}_{ref}T_s = \vec{I}_1 T + \vec{I}_2 T_2 + \vec{I}_o T_o \\ T_s = T_1 + T_2 + T_o, \end{cases} \tag{5.75}$$

where T_1, T_2, and T_o represents the dwell times for the vectors $\vec{I}_1, \vec{I}_2$, and $\vec{I}_o$, respectively.

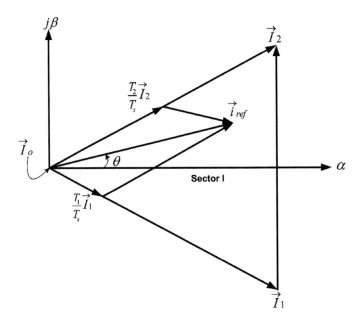

Fig. 5.47 Synthesis of $\vec{i}_{ref}$ by $\vec{I}_1, \vec{I}_2$ and $\vec{I}_o$

$$\vec{i}_{ref} = i_{ref}e^{j\theta}, \quad \vec{I}_1 = \frac{2}{\sqrt{3}}I_{dc}e^{-j\frac{\pi}{6}},$$

$$\vec{I}_2 = \frac{2}{\sqrt{3}}I_{dc}e^{j\frac{\pi}{6}} \text{ and } \vec{I}_o = 0 \qquad (5.76)$$

Putting in Eq. (5.75) and then dividing the resultant equation into two parts i.e. real (α-axis) and imaginary (β-axis) components which provides

$$\begin{cases} \text{Re: } i_{ref}(\cos\theta)T_s = I_{dc}(T_1 + T_2) \\ \text{Im: } i_{ref}(\sin\theta)T_s = \frac{1}{\sqrt{3}}I_{dc}(-T_1 + T_2) \end{cases} \qquad (5.77)$$

By solving Eq. (5.77) $T_s = T_1 + T_2 + T_o$ gives

$$\begin{cases} T_1 = m_a \sin(\pi/6 - \theta)T_s \\ T_2 = m_a \sin(\pi/6 + \theta)T_s \text{ for } -\pi/6 \leq \theta < \pi/6, \\ T_o = T_s - T_1 - T_2 \end{cases} \qquad (5.78)$$

where m_a indicates the modulation index which can be represented as follows:

$$m_{\mathrm{a}} = \frac{i_{\mathrm{ref}}}{I_{\mathrm{dc}}} = \frac{\hat{I}_{aw1}}{I_{\mathrm{dc}}} \tag{5.79}$$

Here, $\hat{I}_{aw1}$ is the maximum value of the fundamental frequency component in i_{aw}.

Note that although Eq. 5.78 is derived when $\overrightarrow{i}_{\mathrm{ref}}$ is in sector I, it can also be used when $\overrightarrow{i}_{\mathrm{ref}}$ is in other sectors provided that a multiple of $\pi/3$ is subtracted from the actual angular displacement θ such that the modified angle θ' falls into the range of $-\pi/6 < \theta' < \pi/6$ for use in the equation, that is,

$$\theta' = \theta - (k - 1)\pi/3 \quad \text{for} -\pi\big/6 \le \theta' < \pi/6, \tag{5.80}$$

where k is in the range of 1–6 for sectors I, II,…, and VI, respectively.

The maximum length of the reference vector, $\overrightarrow{i}_{\mathrm{ref,max}}$, corresponds to the radius of the largest circle that can be inscribed within the hexagon as shown in Fig. 5.46. Since there are six active vectors with lengths of $2I_{\mathrm{dc}}/\sqrt{3}$ and $\overrightarrow{i}_{\mathrm{ref,max}}$ that make up the hexagon,

$$\overrightarrow{i}_{\mathrm{ref,max}} = \frac{2I_{\mathrm{dc}}}{\sqrt{3}} \times \frac{\sqrt{3}}{2} = I_{\mathrm{dc}} \tag{5.81}$$

Putting Eq. (5.81) into Eq. (5.79) provides the peak value of $m_{\mathrm{a}}m_{\mathrm{a}}$

$$m_{\mathrm{a,max}} = 1 \tag{5.82}$$

from which the modulation index is in the range of

$$0 \le m_{\mathrm{a}} \le 1 \tag{5.83}$$

Switching Sequence: Similar to the space vector modulation for the two-level VSI, the switching sequence design for the CSI should satisfy the following two requirements for the minimization of switching frequencies:

1. The transition from one switching state to the next involves only two switches, one being switched on and the other switched off.
2. It takes a minimum amount of switching for $\overrightarrow{i}_{\mathrm{ref}}$ to move from one sector to the next.

Figure 5.48 represents a typical three-segment sequence for the sector I reference vector $\overrightarrow{i}_{\mathrm{ref}}$. The gate signals for switches S_1 through S_6 are shown as v_{g1} to v_{g6}. $\overrightarrow{I}_1$,

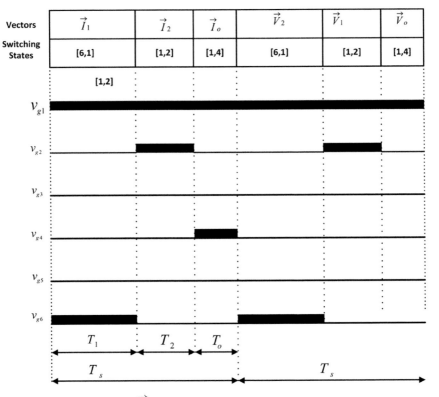

Fig. 5.48 Switching sequence $\vec{i}_{ref}$ for the sector I

$\vec{I}_2$, and $\vec{I}_o$ combine to form the reference vector $\vec{i}_{ref}$. The sampling period T_s is divided into three segments composed of T_1, T_2, and T_0. The switching states for vectors $\vec{I}_1$, and $\vec{I}_2$, are (6,1) and (1,2), and their corresponding on-state switch pairs are (S_6, S_1) and (S_1, S_2). The zero state (1,4) is selected for $\vec{I}_o$ such that design requirement 1 above is satisfied.

IV. **PWM Current Source Rectifier**

Figure 5.49 shows a typical configuration of a PWM current source rectifier (CSR) in a wind energy conversion system. Like the current source inverter (CSI), the PWM rectifier requires a filter capacitor Cr to assist the commutation of switching devices and filter out current harmonics. The capacitor size is dependent on a number of factors such as the rectifier switching frequency, LC resonant mode, required line current THD, and type of generator. It is normally in the range of 0.1–0.3 pu for WECS with a large synchronous generator and 0.3–0.6 pu for induction-generator-based WECS with a switching frequency of a few hundred hertz. Both the SHE and SVM schemes developed for the CSI can be used for the CSR.

5.4 Power Electronics in Wind Energy

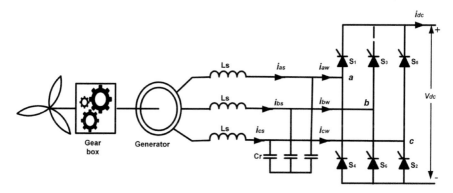

Fig. 5.49 PWM-CSR layouts in a wind energy system (Wu et al. 2011)

The DC output voltage v_{dc} of the rectifier can be adjusted by two methods: modulation index (m_a) control and delay angle (α) control. The operating principle of delay angle control is the same as that of phase-controlled SCR rectifiers.

The input rectifier's active power is generally expressed as follows:

$$P_{ac} = 3V_{a1}I_{aw1}\cos\alpha = \sqrt{3}V_{ab1}I_{aw1}\cos\alpha, \tag{5.84}$$

where I_{aw1}, V_{ab1}, and V_{a1} represent the PWM input current, are the fundamental frequency rms phase input voltage and voltage, line to-line input voltage respectively. The angle between V_{a1} and I_{aw1} is referred to as the delay angle, α. The source of the DC output power is

$$P_{dc} = V_{dc}I_{dc}, \tag{5.85}$$

where V_{dc} and I_{dc} represents the average DC output voltage and current, respectively. The AC input power is equivalent to the DC output power, neglecting the power losses in the rectifier:

$$\sqrt{3}V_{ab1}I_{aw1}\cos\alpha = V_{dc}I_{dc} \tag{5.86}$$

$$\therefore \quad V_{dc} = \sqrt{3/2}V_{ab1}m_a\cos\alpha, \tag{5.87}$$

where

$$m_a = \hat{I}_{aw1}\Big/I_{dc} = \sqrt{2}I_{aw1}\big/I_{dc} \tag{5.88}$$

Equation (5.87) illustrates that for a given line-to-line input voltage V_{ab1}, the average DC voltage of the rectifier can be controlled by both modulation index m_a and delay angle α.

5.4.4 Frequency Converters

A cycloconverter or cycle inverter synthesizes the output waveform from portions of the input waveform without the need for an intermediary direct-current connection, converting one AC waveform, such as the mains supply, to another AC waveform of a lower frequency. Another name for it is a frequency converter. Three-phase applications are where they are most frequently employed. The amplitude and frequency of the input voltage to a cycloconverter typically have constant or fixed values in power systems, but the output voltage of a cycloconverter typically has both variable amplitude and frequency. A three-phase cycloconverter output frequency has to be between one-third and half the input frequency or less. More switching devices are employed, and the output waveform's quality improves as a result (a higher pulse number). With ratings of many megawatts, cycloconverters are employed in very large variable frequency drives. The connections between the thyristors in a cycloconverter have been presented in Fig. 5.50.

The frequency of the input waveform determines a large portion of the noise or harmonics that is produced in the system during the AC waveform switching. These harmonics have the potential to destroy delicate electrical devices. The converter can

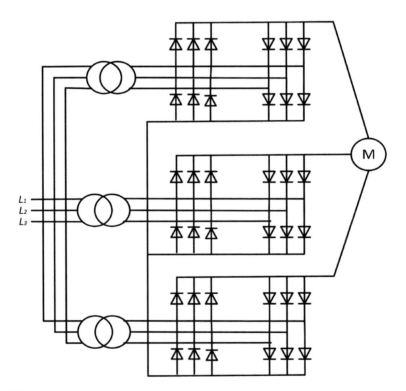

Fig. 5.50 Cycloconverters

generate sub-harmonics if the relative difference between the input and output waveforms is minimal. The sub harmonic noise cannot be removed by the load inductance because its frequency falls under the output frequency. In relation to the input, this restricts the output frequency. For the majority of applications, cycloconverters are frequently outperformed by a DC-link converter system because of these drawbacks.

Power transmission mostly utilizes high-voltage alternating currents (HVAC) and high-voltage direct-current (HVDC) transmission methods. When the transmission distance is smaller than around 70 km, the HVAC transmission (works at 60 or 50 Hz) is appropriate. The HVDC should be the ideal option for sending the power over a long distance, particularly for this case. However, the expensive converter stations at each end of the transmission line cost much higher. In order to reduce the electrical length of the AC transmission line and increase its transmission capacity, a cycloconverter was used in 1994 to produce a lower frequency (50/3 Hz) transmission system.

A cycloconverter's typical applications include regulating synchronous motor starting and an AC traction motor speed (Darba et al. 2010). The silicon-controlled rectifiers (SCRs) in these circuits have tremendous power output, often in the range of several megawatts. On the other hand, low-power, low-cost, cycloconverters for low-power AC motors are also in use; many of these circuits frequently substitute TRIACs for SCRs. A TRIAC has three terminals and can conduct in either direction, unlike an SCR which can only do so in one direction.

The use of cycloconverters for offshore wind farms has been covered in a number of articles during the last few years. According to extensive research, low-frequency transmission systems for offshore wind farms are less expensive to invest in HVDC systems, and their maintenance costs are also significantly lower. Only when the generators are operated at lower frequencies, the system require a stepping-up cycloconverter at the transmission line's end. If the generators run at 50 Hz or 60 Hz, frequency converters are required at both ends (Wu et al. 2008).

5.4.5 Maximum Power Point Tracking Control and Converter Control

Variable-speed wind turbines may operate at their optimal rotational speed as a function of wind speed. The power electronic converter may control the turbine's rotational speed using a maximum power point tracking (MPPT) algorithm to provide the maximum power. By doing this, it is also feasible to prevent exceeding above the nominal power in a situation where the wind speed increases (Pucci and Cirrincione 2011). At the same time, the dc-link capacitor voltage is kept as constant as possible, achieving a decoupling between the turbine-side converter and the grid-side converter. The grid-connected converter will work as an inverter, generating a PWM voltage whose fundamental component has the grid frequency, and also being able to supply the active nominal power to the grid (Raju et al. 2003).

It is possible to modify the wind generator so that it operates as efficiently as possible within the range of wind speeds while still following the maximum power point. There are several approaches for conducting MPPT control on wind turbines.

A. **Torque Speed Ratio (TSR) Control**: The optimal relative speed (or TSR) is constant regardless of the wind speed. It is at this optimum specific speed when the generator speed is optimum. The 'TSR control' method forces the energy conversion system to keep the rotational speed of the turbine at the optimum speed. Figure 5.51 shows this kind of MPPT controller, which needs the wind speed measured by an anemometer. The controller regulates the wind turbine speed to maintain an optimal TSR. However, the accurate wind speed may be difficult to obtain. In addition, the use of an external anemometer increases the complexity and cost of the system.

B. **Power Signal Feedback (PSF) Control**: To run the controller represented by Fig. 5.52, the maximum power curves of the turbine must be understood. This may be done using the modelling and real experimentation. The speed of the wind turbine is used to determine the stored power curve, allowing the system's target power to be tracked. This power curve may commonly be replaced with a predictor or observer of wind speed as a function of power and wind turbine speed.

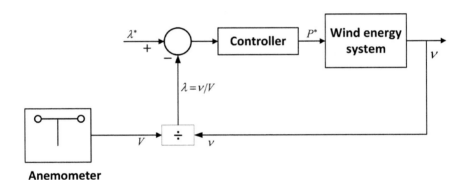

Fig. 5.51 TSR control

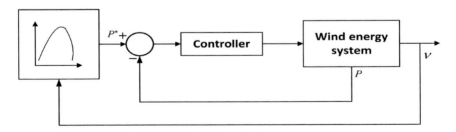

Fig. 5.52 PSF control

5.4 Power Electronics in Wind Energy

C. **Hill Climbing Searching (HCS) Control**: The MPPT techniques employed in solar systems are remarkably similar to this control approach. The output power of a wind turbine should typically grow as the speed does; if not, the speed should be decreased as can be observed from Fig. 5.53. For massive wind turbines, however, this approach might not work since it is challenging to quickly change the speed of massive turbines.

D. **Optimal Torque Control (OTC)**: The fundamental concept behind the OTC technique is to change the generator's speed in accordance with a maximum power reference torque for a specific wind speed. The TSR must be fixed at its ideal value (λ) in order to allow the greatest possible capture of the available wind energy. Figure 5.54 represents the typical torque as a rotor speed function. This technique uses the PMSG torque, which is the highest power torque at a specific speed of wind. The turbine power can be evaluated as a function of tip speed ration (λ) and wind speed (w_m).

$$V_w = \frac{w_m R}{\lambda},$$

where R and V_w are the radius of the turbine and wind velocity, respectively.

$$\text{And also the wind power } P_m = \frac{1}{2}\rho \pi R^5 \frac{w_m^3}{\lambda^3} C_p \quad (5.89)$$

ρ and C_p represents the air density and power coefficient, respectively.

When the rotor runs at λ_{opt}, it will run at $C_{p\,max}$, then the optimum power output

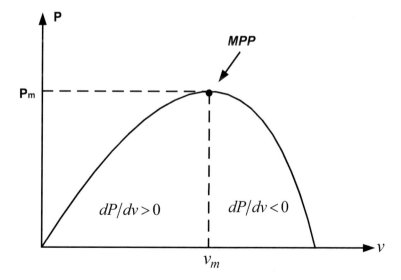

Fig. 5.53 HCS control

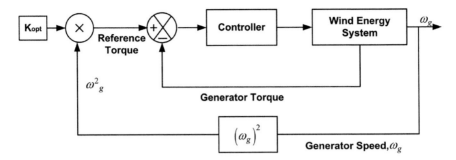

Fig. 5.54 Optimal torque control MPPT method

$$P_{m\_opt} = \frac{1}{2}\rho\pi R^5 \frac{C_{p\,max}}{\lambda_{opt}^3} w_m^3$$

$$P_m = w_m \cdot T_m \tag{5.90}$$

The optimum torque can be expressed as follows:

$$T_{m\_opt} = \frac{1}{2}\rho\pi R^5 \frac{C_{p\,max}}{\lambda_{opt}^3} w_m^2 \tag{5.91}$$

It is a torque-control approach, and Eq. (5.91) may be used to define the ideal torque curve. The reference torque for the controller that is attached to the wind turbine has been represented in Fig. 5.54. This approach is often efficient and effective. Its effectiveness, however, is lower than the TSR control technique's since it does not directly assess wind speed (Abdullah and Yatim 2011).

Review Questions

(1) Explain the different blocks of WECS.
(2) Differentiate between the fixed and variable-speed wind turbines.
(3) Discuss in short the suitability of the generators used in the WECS.
(4) What do you mean by soft starters? Why they are used in the WECS system?
(5) Briefly explain the working of the 3-phase AC voltage controller with the star-connected load.
(6) Discuss the uses of capacitor banks in WECS.
(7) Explain the single-channel boost converter operation in DCM.
(8) What are the basic differences between single-channel and two-channel boost converters?
(9) With the help of a suitable diagram explain the three-level neutral point clamped converters.
(10) Discuss the working of PWM's current source converters.
(11) What do you understand about frequency converters? Discuss in brief.
(12) What are the different methods of MPPT control for wind turbines?

Problems

(1) A 1-phase on–off type AC voltage controller with an input of 230 V, 50 Hz is connected to a resistive load of 20 Ω. The system is on and off for 30 cycles each. Calculate the rms output voltage and current and also the input power factor.

(Ans—162.63 V and 8.13 A, 0.707)

(2) The 1-phase AC voltage controller has an input of 230 V, 50 Hz connected to a resistance of 10 Ω. The thyristor switches on for 25 cycles and switch off for 75 cycles. Then determine (i) the rms output voltage and output current and (ii) the input power factor.

(Ans—60 V and 6 A, 0.5)

(3) The 1-phase full wave controller is connected to the RL load. The input voltage is 120 V at 60 Hz. The inductance and resistance have values of 6.5 mH and 2.5 Ω respectively. The delay angles for both thyristors are 90°. Calculate the (i) Thyristor conduction angle of T1 (ii) rms output voltage (iii) input power factor.

(Ans—130.43°, 68.09 V, 0444)

(4) A 1-phase full wave AC voltage controller has a load of 5 Ω and the input voltage is 230 V with 50 Hz. If the load power is 5 kW. Calculate (i) Thyristor firing angle (ii) Input power factor (iii) Output voltage (rms).

(Ans—92.5°, 0.6871 lag, 158.03 V)

(5) A boost converter has an input voltage of 150 V and the output voltage needed is 450 V. Given, that the thyristors have a conducting time of 150 μs. Calculate the chopping frequency.

(Ans—4.44 kHz)

(6) A boost converter has a resistance of 10 Ω and an input voltage of 220 V. Calculate the average output voltage and current at switching frequency 1 kHz and $D = 0.5$.

Ans—440 V, 44 A

(7) A boost converter has input and output voltage of 200 and 400 V respectively. Calculate the output voltage pulse width if the switch is on for 150 μs. Also calculate the new average value, if the pulse width of output voltage becomes 1/4th for constant frequency operation.

(Ans—150 μs, 1600 V)

(8) A PMSG-based WECS has ratings of 1.2 MW/690 V. The generator runs at a speed of 0.7 pu and supplies 412 kW (0.73 pu) power at a fixed wind speed. The boost converter's input voltage is 680 V. The boost converter's output voltage

268   5 Wind Energy Conversion System

is maintained at 1200 V by the inverter, which is the voltage needed to supply 690 V electrical power to the grid. The output filter capacitor's capacitance is 300 μF, and the two-channel converter's inductance is 270 μH. The change frequency used by the boost converter is 2 kHz. Calculate the (i) average output current of each channel, (ii) total input current of the boost converter, and (iii) total input current fluctuation, considering that all the converters are ideal with no power losses.

(Ans—171.7, 605.8 A, 21.2%)

(9) A 3-phase unidirectional controller connected to a star-connected resistive load has a resistance of 10 Ω in each phase and a line-to-line rms input voltage of 208 V at 60 Hz. The delay angle is 60°. Calculate (i) rms output voltage in phase (ii) input power factor.

(Ans—110.86 V, 0.924)

(10) A 1-phase full bridge inverter controls the power in a resistive load. The nominal value of input DC voltage is 220 V and a uniform PWM with 5 pulses per half cycle is applied. For the required control, each pulse width is 30 degrees. Calculate (i) load rms voltage (ii) If the DC supply rises to 10% evaluate the width of the pulse for maintaining the previous power.

(Ans—200.8 V, 24.75°)

Objective Type Questions

(1) The below-given circuit has $V_s = 230$ V and $R = 20$ Ω. Calculate the average output voltage at the R load if the firing angle is 45°.

    a. 224 V
    b. 15.17 V
    c. − 15.17 V
    d. − 224 V

(2) Pulse gating is needed for

    a. R loads
    b. R and RL loads
    c. RL loads
    d. All load types

(3) High-frequency gating normally uses a

    a. train of pulses
    b. continuous gating block
    c. carrier signal
    d. none

5.4 Power Electronics in Wind Energy 269

(4) In continues gating

    a. the overlap angle is very high
    b. size of the pulse transformer is small
    c. SCR is heated up
    d. commutation cannot be achieved effectively

(5) What is the equation for the Boost converter output voltage?

    a. $6D \times V_{in}$
    b. $4D \times V_{in}$
    c. $5D \times V_{in}$
    d. $D \times V_{in}$

(6) Evaluate the Boost Converter output voltage if the input voltage is 156 V and the duty cycle equals 0.4

    a. 264 V
    b. 260 V
    c. 268 V
    d. 261 V

(7) A dielectric placed in the electric field

    a. There is no charge induced on the dielectric surface.
    b. There is some charge induced on the dielectric surface.
    c. All the charges induced are due to the free electrons in the dielectric
    d. None of the above

(8) PWM switching is used in voltage source inverters for the purpose of

    a. Controlling output current
    b. Controlling input voltage
    c. Controlling input power
    d. Controlling output harmonics and output voltage

(9) A cycloconverter has how many stages for changing the frequency

    a. First stage
    b. Second stage
    c. Third stage
    d. Four stage

(10) Cycloconverer drives are generally employed in

    a. Milling
    b. Traction
    c. Generating low frequencies
    d. Generating pulses

270 5 Wind Energy Conversion System

(11) Frequency converters changes

    a. DC to DC
    b. AC to AC
    c. AC to DC
    d. DC to AC

(12) Which of the following is NOT an example of an energy substitute as a measure for energy conversation?

    a. Wind power to thermal power
    b. Electric heaters to steam heater
    c. Coal to coconut shells, rice husk etc.
    d. Steam-based hot water to solar systems

(13) Which type of generator is used for harnessing the wind energy

    a. Turbine generator
    b. Electron generator
    c. Steam generator
    d. Vapour generator

(14) Which of the following is used for filtering out input frequency and converting it into DC in VAWT?

    a. Passive filter
    b. Rectifier
    c. Bandpass filter
    d. Active filter

(15) Soft starters are used in SCIG based Wind Energy systems to limit

    a. Speed
    b. Inrush current
    c. Frequency
    d. None

(16) Control of frequency and control of voltage in 3-phase inverters can be done

    a. By inverter and converter control circuit
    b. By only the inverter control circuit
    c. By only the converter control circuit
    d. By voltage control of the converter and frequency control of the inverter

Answer

1. c	2. a	3. a	4. c	5. d	6. b	7. a	8. d
9. a	10. c	11. b	12. a	13. a	14. b	15. b	16. d

References

Abdullah MA, Yatim AHM, Tan CW (2011) A study of maximum power point tracking algorithms for wind energy system. In: 2011 IEEE 1st conference on clean energy technol. CET 2011, pp 321–326. https://doi.org/10.1109/CET.2011.6041484

Akhtar MJ, Behera RK (2021) Space vector modulation for distributed inverter-fed induction motor drive for electric vehicle application. IEEE J Emerg Sel Top Power Electron 9:379–389. https://doi.org/10.1109/JESTPE.2020.2968942

Ali MH (2017) Wind energy systems: solutions for power quality and stabilization. Routledge

Bhowmik S, Spec R, Enslin JHR (1999) Performance optimization for doubly fed wind power generation systems. IEEE Trans Ind Appl 35:949–958. https://doi.org/10.1109/28.777205

Boubzizi S, Abid H, El Hajjaji A, Chaabane M (2018) Comparative study of three types of controllers for DFIG in wind energy conversion system. Prot Control Mod Power Syst 3:1–12. https://doi.org/10.1186/S41601-018-0096-Y/FIGURES/21

Breeze P (2015) Offshore wind. In: Wind power generation

Cai G, Liu D, Liu C, Li W, Sun J (2017) A high-frequency isolation (HFI) charging DC port combining a front-end three-level converter with a back-end LLC resonant converter. Energies 10:1462. https://doi.org/10.3390/EN10101462

Capacitor Banks: what is a capacitor bank? Advantages & uses | Arrow.com. https://www.arrow.com/en/research-and-events/articles/capacitor-banks-benefit-an-energy-focused-world

Chen Z (2005) Characteristics of induction generators and power system stability. In: ICEMS 2005: proceedings of the eighth international conference on electrical machines and systems, vol 2, pp 919–924. https://doi.org/10.1109/ICEMS.2005.202679

Darba A, Esmalifalak M, Sarbaz Barazandeh E (2010) Implementing SVPWM technique to axial flux permanent magnet synchronous motor drive with internal model current controller. In: PEOCO 2010, 4th international power engineering and optimization conference, program and abstracts, pp 126–131. https://doi.org/10.1109/PEOCO.2010.5559197

Derbel N, Zhu Q (eds) (2019) Modeling, identification and control methods in renewable energy systems. https://doi.org/10.1007/978-981-13-1945-7

Dewangan S, Vadhera S (2022) Performance evaluation of multilevel inverter in variable speed SEIG-based wind energy system. Arab J Sci Eng 47:3311–3324. https://doi.org/10.1007/S13369-021-06197-Z

Durgasukumar G, Abhiram T, Pathak MK (2012) TYPE-2 fuzzy based SVM for two-level inverter fed induction motor drive. India Int Conf Power Electron IICPE. https://doi.org/10.1109/IICPE.2012.6450468

Gao H, Wang S, Xu D, Wu B, Zargari NR (2019) Leakage current mitigation in current-source inverter based transformerless photovoltaic system using active zero-state space vector modulation. In: 2019 IEEE Energy Convers. Congr Expo ECCE 2019, pp 3775–3779. https://doi.org/10.1109/ECCE.2019.8912849

Hoon Y, Radzi MAM, Hassan MK, Mailah NF (2017) Neutral-point voltage deviation control for three-level inverter-based shunt active power filter with fuzzy-based dwell time allocation. IET Power Electron 10:429–441. https://doi.org/10.1049/IET-PEL.2016.0240

Iov F, Blaabjerg F (2009) Power electronics and control for wind power systems. In: 2009 IEEE power electronics and machines in wind applications. PEMWA 2009, pp 1–16. https://doi.org/10.1109/PEMWA.2009.5208339

Kazimierczuk MK (2015) Pulse-width modulated DC-DC power converters

Kumar R, Das S (2014) A modified approach to both conventional and ANN based SVPWM controllers for voltage fed inverter in sensorless vector control in drive. In: 2014 IEEE international conference on power electronics, drives and energy systems. PEDES 2014. https://doi.org/10.1109/PEDES.2014.7042110

Lazoueche Y, Hamoudi F, Merazka A, Ayachi Amor Y, Metidji B, Kheldoun A (2021) Design and implementation of three-level T-type inverter based on simplified SVPWM using cost-effective STM32F4 board. Int J Digit Signals Smart Syst 5:20. https://doi.org/10.1504/IJDSSS.2021.10034968

Lee K-B, Lee J-S (2017) Reliability improvement technology for power converters. https://doi.org/10.1007/978-981-10-4992-7

Manias SN (2017) AC voltage controllers and thyristor-based static VAR compensators. Power Electron Mot Drive Syst 613–656. https://doi.org/10.1016/B978-0-12-811798-9.00008-1

Morimoto S, Nakayama H, Sanada M, Takeda Y (2005) Sensorless output maximization control for variable-speed wind generation system using IPMSG. IEEE Trans Ind Appl 41:60–67. https://doi.org/10.1109/TIA.2004.841159

Mousavi SM, Fathi SH, Riahy GH (2009) Energy management of wind/PV and battery hybrid system with consideration of memory effect in battery. In: International conference on clean electrical power, pp 630–633. https://doi.org/10.1109/ICCEP.2009.5211989

Nouh A, Mohamed F (2014) Wind energy conversion systems: classifications and trends in application. In: 2014 5th International Renewable Energy Congress (IREC). https://doi.org/10.1109/IREC.2014.6826922

Perera S, Elphick S (2023) Implications of equipment behaviour on power quality. Appl Power Qual 185–258. https://doi.org/10.1016/B978-0-323-85467-2.00002-0

Pucci M, Cirrincione M (2011) Neural MPPT control of wind generators with induction machines without speed sensors. IEEE Trans Ind Electron 58:37–47. https://doi.org/10.1109/TIE.2010.2043043

Qazi SH, Mustafa MW (2016) Review on active filters and its performance with grid connected fixed and variable speed wind turbine generator. Renew Sustain Energy Rev 57:420–438. https://doi.org/10.1016/J.RSER.2015.12.049

Raju AB, Chatterjee K, Fernandes BG (2003) A simple maximum power point tracker for grid connected variable speed wind energy conversion system with reduced switch count power converters. In: PESC Rec.—Annual IEEE conference on power electronics specialists vol 2, pp 748–753. https://doi.org/10.1109/PESC.2003.1218149

Ramos T, Medeiros MF, Pinheiro R, Medeiros A (2019) Slip control of a squirrel cage induction generator driven by an electromagnetic frequency regulator to achieve the maximum power point tracking. Energies 12:2100. https://doi.org/10.3390/EN12112100

Rivera M, Rodriguez J, Yaramasu V, Wu B (2012) Predictive load voltage and capacitor balancing control for a four-leg NPC inverter. In: 15th international power electronics and motion control conference Expo. EPE-PEMC 2012 ECCE Eur. https://doi.org/10.1109/EPEPEMC.2012.6397344

Singh B, Singh SN, Kyriakides E (2010) Intelligent control of power electronic systems for wind turbines, pp 255–295. https://doi.org/10.1007/978-3-642-13250-6_10

Tan K, Islam S (2004) Optimum control strategies in energy conversion of PMSG wind turbine system without mechanical sensors. IEEE Trans Energy Convers 19:392–399. https://doi.org/10.1109/TEC.2004.827038

Vaseee S, Sankerram BV (2013) Voltage balancing control strategy in converter system for three-level inverters. Int J Electr Comput Eng 3. https://doi.org/10.11591/IJECE.V3I1.1471

Wang Q, Chang L (2004) An intelligent maximum power extraction algorithm for inverter-based variable speed wind turbine systems. IEEE Trans Power Electron 19:1242–1249. https://doi.org/10.1109/TPEL.2004.833459

Wu B, Pontt J, Rodríguez J, Bernet S, Kouro S (2008) Current-source converter and cycloconverter topologies for industrial medium-voltage drives. IEEE Trans Ind Electron 55:2786–2797. https://doi.org/10.1109/TIE.2008.924175

Wu B, Lang Y, Zargari N, Kouro S (2011) Power conversion and control of wind energy systems. Power Convers Control Wind Energy Syst. https://doi.org/10.1002/9781118029008

Wu B et al (2011) Power conversion and control of wind energy systems. IEEE eBooks, IEEE Xplore. https://ieeexplore.ieee.org/book/6047595

Wu B, Narimani M (2017) Two-level voltage source inverter. High-power convert. AC Drives 93–117. https://doi.org/10.1002/9781119156079.CH6

Yaramasu VNR, Wu B (2016) Model predictive control of wind energy conversion systems

Ziogas PD (1980) Optimum voltage and harmonic control PWM techniques for three-phase static UPS systems. IEEE Trans Ind Appl IA-16:542–546. https://doi.org/10.1109/TIA.1980.4503826

Chapter 6
Grid Integration Techniques in Solar and Wind-Based Energy Systems

Abstract This chapter deals with the hybrid renewable energy systems, which combine wind and solar energy, their characteristics, implementation strategies, challenges, constraints and financial implications. It provides insights into the difficulties associated with integrating solar and wind energy into the grid-connected system and provides a feasible solution for the production of sustainable power.

Keywords Power quality issues · Grid connected PV systems · Wind energy system · Power quality requirements · Active and reactive power control

6.1 Introduction

Depending on the amount of generation, non-conventional energy may be integrated into the utility grid at the transmission or distribution levels. While small-scale distributed production is often connected to medium or low-voltage distribution networks, wind farms, which are part of large non-conventional energy output, may be coupled straight to the transmission network. Before constructing the system for both kinds of interconnections, a thorough study must be conducted to address the various issues of the system.

6.1.1 Integration of Small-Scale Generation into Grids

Due to the dispersed nature of the resources, power is frequently produced on a modest scale utilizing a variety of renewable energy sources, such as small hydro, solar photovoltaic, biogas, biomass, and tiny wind turbines. The output power lies in the range from a few hundred kW to a thousand MW. Under distributed generation (DG) or dispersed resources (DR), small-scale power businesses are frequently connected to the grid system at the primary or secondary distribution level. Small-scale generation from renewable and non-renewable sources, as well as energy storage, are all examples of distributed resources.

© The Author(s), under exclusive license to Springer Nature Singapore Pte Ltd. 2024
K. Namrata et al., *Wind and Solar Energy Systems*, Energy Systems in Electrical Engineering, https://doi.org/10.1007/978-981-99-9710-7_6

Table 6.1 Interfacing techniques

Category	Interfacing techniques
Wind	Power electronic converter/induction generator/
Fuel cells	Power electronic converter
Small hydro	Power electronic converter, synchronous or induction generator
PV	Power electronic converter

Renewable energy generators having small capacity cannot be linked directly to the grid to produce energy. The utility distribution grid and the generating grid must have some sort of interaction. For example, a DC-to-AC converter based on power electronics is needed to connect the grid system and the photovoltaic (PV) panel generator. Although if a direct link to the AC grid is possible with an induction generator powered by small hydro or wind, problems regarding initial transients, energy conversion effectiveness, and power quality are solved using a power electronic converter. Table 6.1 provides a summary of the typical energy-generating types and their recommended interface technologies (Rajapakse et al. 2009).

Each distribution feeder in a traditional power distribution system has at least one voltage source. However, when clean energy sources having less capacity are linked to the distribution network, this does not happen. Therefore, during interconnection, there are a few particular conditions that must be satisfied for grid safety and reliable operation. Fault clearance, reclosure, and accidental islanding conditions are the main protection-related concerns. The grid-DG interface frequently relies on asynchronous generators or power electronics inverters. The influence of the utilities' operations on power quality, which contains issues like voltage sag and swell, harmonics, and also large fluctuation of voltages, worries the utilities. The technical constraints set forth by utilities to overcome these issues have been cited in several studies as a significant technical challenge to the grid integration of DG. The IEEE P1547 standards are used as a reference for the basic technical rules for small-scale generation integration.

The small-scale generation integration to the grid generally faces the following issues:

1. Protection issues

Protection is the most severe problem with dispersed generation's connectivity. Mostly, a distribution network topology of the radial type is utilized with a time-graded protection strategy. When a DG is attached, it may change how the current protection plan is coordinated and may cause the protection apparatus to malfunction. The protection concerns associated with the interconnectivity of distributed generation have been described in detail which are:

A. Variation in short circuit levels

Reclosure is the primary factor taken into consideration while choosing fuses. The coordination between over-current relays, current transformers, and circuit breakers all depends on the short circuit level. The expected fault current level and the equivalent system fault point impedance are both determined by the short circuit level.

6.1 Introduction

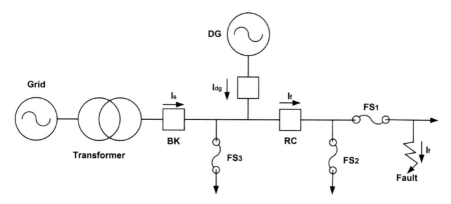

Fig. 6.1 Fault current variations caused due to DG connection

Time variations in the fault current are also determined by rotating machine features in the region. The majority of DG is at first constructed as passive form, however as DGs are connected, the corresponding network impedance decreases, increasing the failure level. Due to the unexpectedly high fault currents that result in CT saturation, the present circuit breakers may not be able to handle a problem. Additionally, the altered fault levels can interfere with the over-current relay's synchronization, resulting in extremely poor protection system performance.

B. Power transfer in reverse direction

Power flows in radial distribution networks are unidirectional. The initial design of protection systems is based on the power flow. Due to a power reversal, the protective relay coordination gets affected as soon as the DG is attached. A schematic diagram of fault current variations caused due to DG connection has been shown in Fig. 6.1.

C. Absence of long-term fault current

If a problem arises, the relays can only recognize and differentiate between the fault currents when the fault current measured by the protective relays is noticeably higher than the typical load current. Effective fault detection using over-current-based protection relays is challenging when the DG fault current contribution is restricted. Small synchronous generators, induction generators, or power electronic converters are typically used to produce energy from renewable sources. The induction generators are not capable of supplying fault currents continuously to 3-ϕ faults, and their ability to contribute fault current to asymmetrical faults is likewise constrained. It is impossible to sustain a fault current that is much greater than the rated current of small synchronous generators. Since power semiconductor devices cannot withstand a considerable over-current for an extended length of time, the internal outputs current is generally restricted by power electronic converters. The absence of prolonged fault conditions affects the relays' capacity to detect defects.

D. Islanding

The term "islanding" refers to the operation of a portion of the utility that is cut off from the main grid and operated as a separate setup powered by either one or more connected generators. Abnormal frequency and voltage changes might result from islanding. An auto-reclosure opening during a failure creates two distinct systems that are run at two different frequencies. Generally, if the auto-reclosure process is done forcefully to reclose, disastrous effects may happen. Additionally, the process of islanding might result in an unground system that is dependent on the transformer connection. The personnel doing the repairs would risk their lives on an unknown island. Because of the aforementioned factors, islanding is seen as a dangerous circumstance, and it is advised that DGs be immediately disconnected from the grid when an island is formed.

2. Voltage control

Interconnection of DG causes voltage swell problems when the connected load is low or distributed production is high, which affects power flows in the feeder and also the voltage profile. In weak networks, the voltage restrictions specifically evaluate the DG capacity, and in addition, the utility has no control over the DG's connection status. When reconnecting a DG under low load conditions or disconnecting a DG under high load conditions, overvoltages or undervoltages could occur.

Due to the poor power quality situation, this may cause over/undervoltage relays to operate. The distribution transformer tap settings selection became challenging due to the increased DG penetration. The problem becomes very challenging if the same transformer does not distribute the DGs uniformly among the feeders. The DG concentration on just one out of two feeders that are fed by the same transformer is shown in Fig. 6.2.

Normally, connecting DGs to nearby loads supplies power, reducing the total current flowing in the transformer. Therefore, the light load settings on the transformer tapping need to be modified. As demonstrated in Fig. 6.2, the sending end voltage fall may cause a voltage violation at the feeder's remote end in the absence of DGs. The feeder with DGs may experience overvoltages if the transformer tap is left on the heavy load setting. Although static VAr compensators and switched capacitors can be incorporated for controlling the voltage of the feeder voltages, the cost is too high. At the distribution level, unbalanced voltage profiles also have an impact on how DGs operate. Both the loads and the distributed generations may be either 3-ϕ or 1-ϕ, as shown in Fig. 6.3.

Normally we do interconnection of single-phase sources, the system becomes more unbalanced. Additionally, three-phase DGs linked to unbalanced distribution networks may experience frequent shutdowns and overheating as a result of imbalanced currents in the DGs.

3. Flicker and harmonics

The supply of high-quality power to consumers is the main issue of service providers. Most DGs are interfaced via power electronic converters; however, this introduces

6.1 Introduction

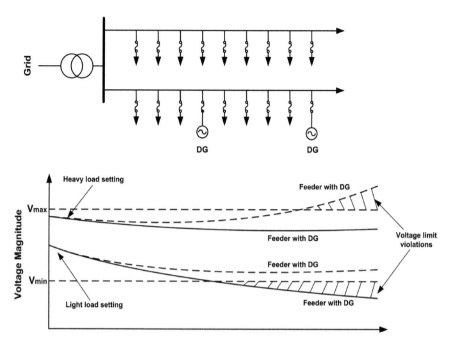

Fig. 6.2 **a** Distribution system with DGs **b** Voltage variation with different load setting (Rajapakse et al. 2009)

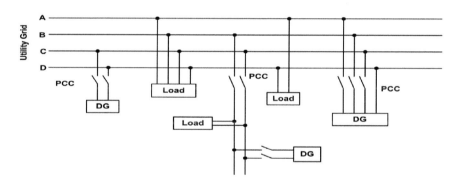

Fig. 6.3 Possible DG interconnection configurations (Rajapakse et al. 2009)

the system harmonics and gives customers a very bad quality of electrical power. This problem can be addressed with harmonic filters. Active, passive, or hybrid filters can all be chosen correctly based on the requirements. When induction generators are directly started while connected, they may produce voltage spikes and flickering. Some soft-starting systems can lessen these voltage swings. The tower shadow effect might cause wind generator voltage to fluctuate on a regular basis.

4. Grid codes examples and standards for interconnection

278   6 Grid Integration Techniques in Solar and Wind-Based Energy Systems

Table 6.2 Response of the interconnected network to an abnormal voltage condition

Voltage range limits (in terms of base voltage percentage)	Time for clearing fault
$V < 50$	0.16
$50 \leq V < 88$	2.00
$110 \leq V < 120$	1.00
$V > 120$	0.16

Table 6.3 Response of the interconnection system to abnormal frequencies

Size of DR	Frequency limits (in Hertz)	Time for clearing fault
≤ 30 kW	> 60.5	0.16
	< 59.3	0.16
> 30 kW	> 60.5	0.16
	$< \{59.8–57.0\}$ (flexible set value)	0.16–300 (flexible)
	< 57.0	0.16

DG interconnection guidelines and norms were developed to solve safety and security problems. IEEE Std. 1547-2003, "IEEE Standards for Interconnecting DR with Electric Power Systems 2003," is the most extensively used standard. For connecting DG to their networks, the majority of utilities have accepted the IEEE Std. 1547-2003 standard. This standard advises disconnecting the DR when the frequency and/or voltage interconnected points vary from their starting values because of faults and other disruptions. The permissible variations are outlined in Tables 6.2 and 6.3.

6.1.2 Large-Scale Generation Integration into Grids

Large-scale renewable energy production using traditional methods, such as hydropower and steam turbines powered by geothermal or biomass energy, depends upon synchronous generators. This traditional technique is well-known to the utilities. However, because of the interest in variable speed generators and the unpredictable nature of wind, large-scale wind farms and solar energy system integration into the electrical grid presents another sort of problem.

6.1.2.1 Grid Integration for Wind Farms

Large-scale wind generation farms with capacities of approximately more than a hundred megawatts have been created as a result of the current increases in wind energy output, and these wind farms are often coupled to the transmission network. In a few nations, including countries like Spain and Germany, the capacity of wind

6.1 Introduction

energy now being installed has surpassed several thousand megawatts and is still increasing. However, wind generators are unable to distribute the generated electricity. Accurate wind speed forecasting is still challenging, despite significant developments for both the short and long-term wind prediction technologies. To balance the variations in wind farm generation, stand-by capacity and a sizable amount of spinning reserve are demanded. Even while the widespread geographic deployment of wind forms has some moderating effects, it is possible that all wind farms might simultaneously produce low amounts of energy. A high-voltage transmission network short circuit results in a short-duration voltage dip, which disconnects a large amount of wind power from the grid system and creates serious stability problems for the overall system. As a result, several system operators impose additional connection standards for sizable wind farms. Wind turbines must support grid operations due to new rules like fault ride-through capabilities. The predicted demand for fault ride-through capabilities in newly installed wind turbines is shown in Fig. 6.4. The figure shows various wind farm voltage drop levels and corresponding connected times. This is an effort to make sure that after the issue has been resolved, the system will remain stable and that the wind farm is set up and ready to start generating power right away.

Depending on the transmission operator, the actual fault ride-through curve may change.

Through the use of "strict" fault ride-through requirements, significant difficulties for the designer of wind turbine systems are given (Singh and Singh 2011).

- During a failure, the wind turbine generator's (WTG) capacity to transmit electricity is decreased. Due to the over speeding, blades, and other spinning components will experience mechanical stress.
- According to the generator technology being used, the WTG may demand more reactive power from the system after a failure, which causes a sluggish fault recovery reaction.

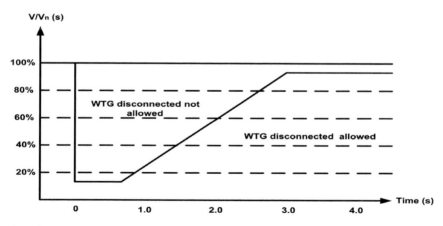

Fig. 6.4 Typical criteria for fault ride-through

- The fault current's characteristics and amplitude are determined using WTG technology. It helps to have a good understanding of this concept when determining ratings of equipment for protection and control settings.
- To handle the mechanical energy, the turbine blades' pitch must be adjusted. The mechanical speed of spinning components during faults can be decreased by the instantaneous blade pitch control.
- Using the proper FACTS-based reactive power compensation equipment will enhance the wind farm's overall performance. One of the main issues with the generation of wind energy is the requirement of reactive power for voltage regulation. Induction generators are generally utilized in the majority of wind power plants. Reactive power cannot be produced by induction machines, unlike synchronous machines.

If the wind generators are unable to satisfy the aforementioned requirements, dynamic reactive power, and static var compensators (SVCs), together with a capacitor bank, are employed sometimes.

- Overvoltage caused by disconnecting the generators of wind turbines at the same time when the capacitor bank is still attached is the main protection issue with big wind turbines. Following the resolution of a system breakdown, the wind farm's performance is dependent on the reactive power compensation strategy and equipment.

Reactive power capacity determines the capacitor bank and SVC's available terminal voltage. Because of this, capacitor banks' reduced reactive power capacity, which is most required during fault recovery from the perspective of reactive power capacity, voltage source converters based on FACTS (such as STATCOM) are often a better choice because they do not have this flaw.

It is important to give proper consideration to the reactive power compensation process and its control strategies. Synchronous condensers, which likewise supply inertia to the system but have a longer reaction time than STATCOM, can also be used as reactive power compensators in large wind farms. This could be useful, especially if there are connections between weak AC systems that have problems with dynamic stability. A thorough examination may be necessary to address the following other potential issues:

- Optimization of the controller and electrical design for turbine power.
- Design of the protection and filter systems.
- Several turbine interactions.
- Problems with wind farms that are linked to series-compensated systems.
- Flickering voltage and other issues with the power supply.
- The startup and grid synchronization of wind farms.

Sub-synchronous resonance issues are caused by the complicated shaft/gear wind turbine system connected to the electric power network.

6.1.2.2 Grid Integration for Solar Energy System

The incorporation of sunlight-powered systems into the power grid is essential for the global shift to a less polluted, more environmentally friendly energy future. Recent years have seen a spectacular increase in solar power, making it one of the sources of clean energy with the fastest rate of development. As a result of its broad adoption, emissions of greenhouse gases are decreased, the security of energy is improved, and economic development is sparked.

- **Advantages of Integrating Solar Energy**

There are several advantages of integrating solar energy systems into the electrical grid, in addition to environmental stability. The first benefit of solar energy is that it diversifies the energy mix, lessening reliance on fossil fuels and boosting energy security. Countries may reduce the dangers brought on by varying fossil fuel costs and geopolitical tensions by utilizing the sun's plentiful and renewable energy.

The second benefit is that widespread solar integration boosts the economy and creates jobs. The solar sector has the ability to offer jobs at several levels, from production and installation to maintenance and control. This industry promotes innovation and technology developments, which helps to create an innovative and stable economy. It also supports economic growth.

In addition, using solar electricity lowers the release of greenhouse gases, preventing global warming and raising air quality. The use of solar power reduces the emission of hazardous gases like carbon dioxide and sulphur dioxide by substituting combustion fuel-based energy production. The transition to cleaner energy sources is essential for achieving global climate goals and preserving the health and welfare of people all over the world.

- **Technical Aspects**

The advantages of integrating solar energy into the electrical grid are clear, but there are a number of technological issues that must be resolved in order to make the process easy and effective. The intermittent nature of solar power generation is one of the main difficulties. Because solar power output is dependent on the weather, its output might change during the day as well as throughout different seasons. Because the electrical grid frequently depends on transportable resources to supply power demand, this intermittency raises questions about stability and dependability. Advanced forecasting methods, energy storage options, and system flexibility measures are needed to tackle such a challenge. Grid operators may anticipate variations and improve grid management with the use of accurate solar power forecasting, and surplus solar energy can be stored and distributed using energy storage technology like batteries.

Additionally, by increasing grid adaptability with demand response initiatives and smart grid technology, solar energy may be successfully integrated into the present electrical grid architecture.

The requirement for grid maintenance and system improvements is another factor. Grid infrastructure that can handle the increased power capacity and control two-way power flows is needed for large-scale solar integration. This can require installing cutting-edge monitoring and control systems, strengthening voltage regulation mechanisms, and updating transmission and distribution lines. In order to preserve grid stability and dependability while tolerating increased levels of solar power penetration, grid operators must also make sure that there are appropriate grid connections and grid code compliance.

- **Market Structure and Regulatory Frameworks**

Solar energy grid integration needs supportive regulatory frameworks and market structures that encourage investment, promote creativity, and facilitate a smooth switch to clean energy sources.

Implementing feed-in tariffs (FITs) or power purchase agreements (PPAs), which give solar energy producers long-term contracts and fixed prices, is a crucial policy instrument. These procedures provide stability and assurance to investors, promoting the wide-scale installation of solar energy installations. Additionally, prosumers—those who produce and use their solar energy—can inject excess power into the grid and get payment or credits for their contribution thanks to net metering and feed-in tariff arrangements.

Policymakers should give the construction of energy storage systems a priority in order to overcome the intermittent nature of solar electricity and enhance system stability. The installation of energy storage devices can be encouraged by incentives like investment tax credits and subsidies, enabling the effective usage of extra solar energy and improving grid dependability.

The policy structures should also give priority to grid modernization initiatives, such as the implementation of smart grid technology and demand management initiatives. Smart grids make it possible to monitor, regulate, and optimize electricity flows in real-time, making it easier to include intermittent renewable energy sources like solar power. Consumers are encouraged through demand response programmes to modify their power use in response to supply-demand mismatches, which lessens the need for new generation capacity and increases system flexibility.

Stakeholder cooperation is essential to creating a favourable policy climate. The technological, regulatory, and financial impediments to the integration of solar energy should be addressed via conversation between governments, regulatory agencies, utilities, and industry participants. The policy framework will be comprehensive, adaptable, and in line with the long-term objectives of the energy transition thanks to this collaborative approach.

In order to achieve an ecological and carbon-free energy future, solar energy sources must be integrated into the grid on a broad scale. Solar power integration has several advantages, including broadening the energy mix, lowering the release of greenhouse gases, and promoting economic development. However, because solar energy is sporadic, there are technological difficulties that must be resolved by sophisticated forecasting, energy storage options, and system flexibility measures.

A key factor in assisting the integration process is the existence of supportive legislative frameworks and market structures. Solar energy adoption is encouraged through feed-in tariffs, power purchase agreements, and net metering procedures, while grid modernization and energy storage infrastructure expenditures improve system stability and dependability. To create comprehensive policies that address the technological, governmental, and financial obstacles to the integration of solar energy, parties must work together.

A feasible path to a cleaner, safer, and wealthier future is provided by the widespread integration of solar power plants into the power grid as we work to create a robust and sustainable energy system. We can create a sustainable energy infrastructure using the power of the sun to ensure a livable and sustainable earth for future generations.

6.2 Integration Issues Related to Wind and Solar Power

6.2.1 Consumer Requirements

In addition to the main small renewable energy system elements, customers may be required to purchase certain additional equipment (referred to as "balance-of-system") to transfer electrical power to their loads and satisfy the rules and regulations of the power provider's grid integration. Users might require the respective tools or devices:

- Safety equipment.
- Power conditioning equipment.
- Instrumentation and metres.

Because there are different grid connection requirements, the customer's system supplier or installer should speak with their power company to find out about those needs before buying any component of their renewable energy system.

The requirements for connecting distributed generation systems, such as household wind or non-conventional power systems, to the electrical grid, now range widely. Yet, every energy distributor faces a similar set of difficulties when integrating smaller renewable energy systems into the grid. For this reason, laws frequently include matters like security and power quality, contracts (which may need liability insurance), metering, and costs.

Consumers must speak with their electrical supplier directly if they want further information about the requirements. If the power supplier does not have someone appointed to manage queries for grid connections, try contacting the state utility commission, the state utility consumer advocate group (which reflects consumers' preferences before state and federal regulators and in courts), the state consumer representation office, or the state-owned power office.

- Handling power quality and safety issues for electrical grids

Electric utilities want to be certain that the consumer system contains components that ensure electricity is of high quality and security. These include power conditioning equipment that assures consumers' power precisely meets the voltage and frequency of the electricity flowing through the grid and switches that disconnect their system from the grid in the case of a power surge or power failure (preventing electricians from being electrocuted). In an effort to address safety and power quality issues, multiple organizations are developing national standards for equipment manufacture, operation, and setup (the vendor, a regional clean energy entity, or the power distributor will recognize which of the specifications may be applied to their conditions and how to integrate them):

a. All distributed generation connected to the grid network, including non-conventional systems, is covered under a standard that was created by the Institute of Electrical and Electronics Engineers (IEEE). Technical specifications and testing for grid-connected operation are provided by IEEE 1547–2003.
b. Underwriters Laboratories (UL) developed UL 1741 to validate converters, inverters, charge controllers, and output controllers for power-producing standalone and grid-connected renewable energy systems. For grid-connected applications, UL 1741 has confirmed that inverters adhere to IEEE 1547.
c. Electrical equipment and wiring safety are covered under the National Electrical Code (NEC), a creation of the National Fire Protection Association.

Some utility commissioners and legislatures now insist that rules regulating distributed generation systems be based on the IEEE, UL, and NEC standards, even though states and electrical suppliers are not obligated by the federal government to implement these standards or codes. Many countries are also "pre-certifying" particular equipment models as appropriate to connect to their electrical systems at the moment.

- Contractual issues for grid-connected systems

Customers' local renewable energy systems will likely require an interconnection agreement with their power supplier before they can be connected to the grid. Power companies may require in the agreement that consumers must:

a. Carry liability insurance—When a customer's system fails, the power supplier is covered by liability insurance in case of accidents. Nowadays, many users have at least a hundred thousand dollars in liability coverage via their homeowner insurance policies, which is frequently enough (although consumers should make sure their policy will cover the system). But be careful that the power supplier can demand that the user carry extra. Customers of some power companies could also be required to pay excessively high indemnification fees for any possible harm, loss, or damage brought on by their system.
b. Pay fees and other charges—Permitting costs, engineering/inspection fees, metering charges (if a second metre is placed), and stand-by charges (to cover the expense of the power provider maintaining the customer system as a backup power supply) may be levied against customers. Don't be hesitant to query any of these expenditures that appear out of place so that clients can account for them in the price of their system.

6.2 Integration Issues Related to Wind and Solar Power

Users may find that their power distributor requires extensive documentation, insurance, and expenses before they can continue with their system. So far, electricity providers in many jurisdictions are already working to accelerate the procurement process by simplifying the contracts putting deadlines on the processing of paperwork, and designating agents to answer questions about grid connections.

- Metering and pricing agreements for systems linked to the grid

When a client has a grid-connected system, excess electricity produced by their renewable energy system flows onto the electric grid so that their utility may use it elsewhere. Power companies are required under the Public Utility Regulatory Policy Act of 1978 (PURPA) to buy excess electricity from grid-connected small renewable energy systems at a price equivalent to what it costs the company to generate the electricity itself. This need is often implemented through different metering techniques used by power companies. The metering arrangements that clients are most likely to get into are as follows:

a. Net sale and purchase—Both the power taken from the grid and the surplus electricity generated and sent back into the grid are tracked by the system's two unidirectional metres. Companies charge a retail value for the electricity customer's use, while the utility company purchases any excess output at a lower price (wholesale rate). Customers' retail rates and the power provider's reduced costs could be very varied.

b. Net metering—The customer benefits the most from net metering. In accordance with this design, one bi-directional metre is used to measure both the quantity of electricity that consumers have from the grid and the surplus that their system feeds back into the grid. As customers use power, the metre rotates forward; as the extra is sent into the grid, it spins backwards. The customer must pay the retail price for any excess power if, at the end of the month, they utilize more energy than their allotted limit. Customers often receive payment from the power company at the saved cost of the excess electricity if they generate more than they consume. When a client feeds electricity back into the grid, the power company pays them the retail price for it. This is the actual benefit of net metering.

Users may benefit from being able to transfer any net excess power their system provides on a monthly basis if the energy they utilize to generate their electricity is weather-dependent. This is the case with several power suppliers. Customers must give the surplus generation to the power supplier if, at the end of the year, they generate more energy than they consume.

6.2.2 Requirement for Wind Farm and Solar Farm Operators

Electricity is a forceful thing. The amount of power in the grid must match real consumption requirements in order to make use of its advantages. The steady increase in demand for and popularity of technologies like electric vehicles has presented a number of challenges for power grid operators (Stompf 2020).

1. Increasing Number of Non-Conventional Sources

It is predicted that 62% of the total electricity generated will come from renewable sources by 2050. In comparison, Slovakia's whole generation of clean energy potential in 2019 was just a 13% of the total amount generated worldwide (29%). Connecting sources of clean energy to the grid is not as simple as it may seem since the efficiency of these sources is completely dependent on the climate. This point of view sees renewable energy sources (RES) as an unpredictable power source, and their operation without a robust management system may have negative outcomes.

Unused energy may be stored and saved using batteries or other energy storage devices. Artificial intelligence may enhance prediction systems, enabling more precise forecasts of the weather or energy usage. With this strategy, utility companies may more effectively plan for their customers' electricity requirements, and smart energy management systems can make renewable energy a trustworthy replacement for fossil fuels.

2. Losses in Transmission Lines

Power lines get hotter due to long-distance electrical transmission, which results in significant energy waste in the form of thermal energy. Regular users eventually have to pay for such losses. Families in Slovakia paid 4.57% of the total cost of electricity in 2019 for power transmission losses, while business owners paid 4%. Even while the amount of energy lost in Europe is just about 4–5%, other nations are losing far more energy than Europe. In India, for example, the figure is 19%, while Haiti has a staggering 50%.

A transition from a few major power plants supplying all of the power to a network of local power sources that guarantee power is utilized as close to its source as possible, even at the individual level households, such as prosumers.

3. Frequent Blackouts

The two primary causes of blackouts are severe weather and obsolete electrical infrastructure. Severe power outages have already caused billions of dollars worth of losses in Australia and the USA and put millions of people's lives in danger. Substantial loss of data and damage to technical gadgets are likely consequences of a countrywide outage, in addition to life paralysis in the affected areas.

Energy sources used for backup, like batteries, offer durable protection and ensure the continued working of delicate equipment in the case of a power outage. When used in conjunction with a renewable energy source, a distributing point may retain green energy for later consumption.

4. Electromobility

Although they are still rare, electric vehicles (EVs) are becoming very famous because of rapid advancements in technology. To avoid spending half a day at the gas station due to the time-consuming nature of EV charging, we must utilize a supercharger, which consumes a lot of energy. By comparison, a full charge from a supercharger would be necessary for the simultaneous launch of 70 air-conditioners. Such an abrupt change in power use presents significant issues to the grid.

Online connections are possible for many sources, including solar panels, batteries, EV chargers, and other machinery. Real-time "big data" analysis enables quicker responses to power system changes and ensures a reliable, high-quality supply of electricity. Alternatively, to put it another way, some machinery may utilize energy that might be used to operate other machinery.

5. Grid Modernization

Though transmission lines have a limited operating lifespan and upgrading or building new ones is costly, there will be always demand the increase their power-carrying capacity.

The amount of electricity transmitted across the power system is reduced locally through energy production and consumption, as previously mentioned. As a result, transmission losses are decreased and power lines have a longer lifespan.

6. Cyber-Security Threats

The digitalization of the power sector has additional negative consequences. There have been incidents in which hackers entered the networks of energy providers and exposed thousands of homes to a controlled blackout.

The potential of distributed databases to remove cyber-threats has proven so effective that worldwide financial organizations, such as J.P. Morgan and Nasdaq, are also validating this. Distributed databases guarantee that a cyber-attack on a single grid point, such as a power plant, cannot stop the operation of the entire system, much like decentralized energy generation, where the operation of the grid is not the responsibility of a single supplier.

7. Extended Transmission Lines

Funding from the infrastructure programme to improve and extend transmission lines would necessitate a simultaneous change of power line inspection in order to monitor and manage a bigger, more interconnected network. Dependence on unreliable energy sources will necessitate rapid responses from energy suppliers in order to deliver electricity over large distances. What occurs in one region of the grid will gradually impact other regions. In the case of a solar or wind power shortfall, a power outage in one region might now damage not only that region but also its potential to play a role in the interconnectivity of other regions. Cumulatively, electricity lines must be inspected across a longer distance and with more precision and regularity, as each grid is a component of a larger ecosystem.

Despite significant technological developments, current power line inspection procedures are insufficient to address the imminent issue.

Even while new technologies are now developing that might provide an enormous push to the scale, efficiency, and accuracy of assessments, their adoption remains a barrier for a typically conservative business deeply involved in outdated traditions and a slow rate of change.

8. Ground-Up Issues

Manual inspections have inherent blind spots, and depending on human teams on the ground might generate blind spots throughout the network. A scheduled repair on a portion of the power line system, for instance, may have far more significant effects on a future ecosystem of connected grids. With improving technology and automated aerial inspections that capture a richer and broader scope of visual data than teams limited to a ground-based perspective, it is easier to assess the effects of these minor impacts on the larger network and to respond with greater speed.

9. Going Along Distance/Inspection Issues

Traditional manual network inspection is also sequential and isolated, with findings assembled after completion, inspecting massive networks difficult and time-consuming. Instead, live-feeding data from all inspection assets to a centrally controlled live digital twin would allow different features and aspects to be assessed in parallel, enabling inspections of wide networks and real-time monitoring of the network risk model. AI inference, for example, requires only three minutes per mile of line corridor for light detection and ranging (LiDAR) technology.

AI also makes it possible to take into account a number of variables simultaneously, including temperature, wind speed, and the type position, and behaviour of plants next to electricity lines. By identifying regions and places of concern, data on strong winds and warm temperatures, for example, may be combined with data on plants to forecast and prevent fires. Drone sensors and weather satellite data might also be included in 4D live digital twins to provide complex, multi-layered maps of networks that model the effects of potential future events like floods or droughts.

10. Data Silos Integration

Utilities must aggregate data from many sources in order to manage a continuously rising grid while protecting it from a wider and more variable range of climatic risks. Similar to how the power grid must become more interconnected to prevent outages; utility data must similarly become more integrated.

Network risk evaluations and AI may then be utilized to better prioritize tasks like power line inspections using this data-rich methodology. New technologies, such as digital twins, can provide a "real document" that records dangers, repairs, or upgrades in real-time, but many network maps are out-of-date and do not reflect current upgrades or maintenance.

11. Advanced Techniques

People and places will need inspections to cover a wider range of infrastructure and risks at a faster rate than ever before due to the continuous construction of vital infrastructure. AI is capable of inspecting networks at a larger scale and speed than humans, removing human bias and mistakes from vital network data and enabling proactive and ultimately predictive maintenance and resilience.

6.2.3 The Integration Issues

6.2.3.1 Introduction

Sources of non-conventional energy, like the wind, the sun, geothermal heat, biomass, etc., are abundant in nature and cannot be exhausted. In order to guarantee that total energy output from all sources, including uncertain clean energy generation, balances power requirements in real-time, it is necessary to integrate these resources cost-effectively as the amount of power generated from variable non-conventional resources increases. The causes include the fast-dropping prices of renewable energy power, regulatory and legislative requirements and incentives, and efforts to minimize pollution from fossil fuel-based power generation, particularly emissions of greenhouse gases. Despite the fact that not all clean energy sources are unpredictable, sun and wind presently account for the majority of the rise in renewable power output. The goal of wind and photovoltaic generation is to harness the unlimited but unpredictable amount of solar irradiance and wind.

Supply conditions on the grid need to be standardized to assure the reliable and efficient operation of end-use equipment and infrastructure. These standards are often known as power quality requirements. They mostly deal with power factor correction, harmonics, and voltage and frequency regulation. The technological characteristics and terminal operations of electrical loads pose challenges to achieving these power quality requirements in all distribution networks. Some loads require more power which causes current flow to rise and line voltage down (like air-conditioners and water heaters of large capacity). Some have relatively large power needs at startup that cause voltage fluctuations (like typical induction motors). Some loads (again, including motors) require a substantial amount of reactive power or produce a significant amount of harmonics (such as computer power supplies and fluorescent lighting). The aggregate effects of loads and network equipment have very complicated effects on power quality at many locations in the distribution network at any moment.

6.2.3.2 Power Quality Issues

Power quality is the concept that sensitive electronic equipment should be powered and grounded in a way that is appropriate for the equipment. All electrical equipment is susceptible to malfunction or failure when subjected to one or more power quality issues. The electrical appliance might be a home appliance, a computer, a printer, a transformer, a motor that runs on electricity, or a generator. Depending on how severe the problems are, all of these gadgets and others respond adversely to poor power quality. A shorter and more succinct definition may be: "Power quality is a collection of electrical parameters that allow a piece of equipment to operate as intended without substantial performance loss or life expectancy." This definition includes both performance and expected lifespan, which are qualities we expect from electrical devices. This section introduces the more frequently used vocabulary

290 6 Grid Integration Techniques in Solar and Wind-Based Energy Systems

Table 6.4 Power quality problems

Issues	Description	Causes	Consequences
Voltage sag (or dip)	A decrease of the normal voltage level between 10 and 90% of the nominal root mean square (RMS) voltage at the power frequency, for durations of 0, 5 cycles to 1 min	((a) Faults on the transmission or distribution network (mostly on parallel feeders) (b) Faults in consumer's installation. Connection of heavy loads and startup of large motors	(a) Malfunction of information technology equipment, namely microprocessor-based control systems (PCs, PLCs, ASDs, etc.) that may lead to a process stoppage (b) Tripping of contactors and electromechanical relays. Disconnection and loss of efficiency in electric rotating machines
Very short interruptions	Complete power outage lasting from a few milliseconds to one or two seconds	(a) Opening and automatic reclosure of protection devices to decommission a faulty section of the network (b) Failure of dielectric failure, flashover of insulator, and lightning	(a) Tripping of protection devices, loss of information, and malfunction of data processing equipment (b) Stoppage of sensitive equipment, such as ASDs, PCs, and PLCs, if they're not prepared to deal with this situation
Long interruptions	Total outage of electrical power for a period longer than one to two seconds	Storms, items (trees, automobiles, etc.) impacting lines or poles, fire, user mistakes, poor coordination, or malfunctioning safety equipment are all examples of equipment malfunction in the electrical system	(a) All machinery is put on hold

(continued)

6.2 Integration Issues Related to Wind and Solar Power

Table 6.4 (continued)

Issues	Description	Causes	Consequences
Voltage spike	Voltage value changes quite abruptly over periods ranging from a few microseconds to a few milliseconds. Even at lower voltage, these differences can amount to thousands of volts	Lightning, line or capacitor switching, severe load disconnections, and power factor correction capacitors	(a) Components damage (particularly electronic components) and insulation materials (b) Data loss or data processing errors (c) Interference in electromagnetic fields
Voltage swell	Temporary rise in voltage at the power frequency that is outside of the normal limits, lasting more than one cycle but usually lasting no longer than a few seconds	(a) Badly dimensioned power sources (b) Start/stop of heavy loads (c) Transformers that are poorly controlled (mostly during off-peak hours), if the voltage levels are too high	(a) Data loss (b) Flickering of lighting and screens (c) Stoppage or sensitivity issue
Harmonic distortion	Waveforms of voltage or current of a non-sinusoidal shape. The waveform is the result of adding several sine waves, each with a distinct amplitude-phase, and frequency that is multiple of the frequency of the power source	(a) Traditional sources include welding equipment, arc furnaces, DC brush motors, and rectifiers. These electric devices operate well above magnetic saturation (b) Adjustable Speed Drives (ASDs), switching mode power supplies, data processing devices, and high-efficiency lighting are only a few examples of modern sources for nonlinear loads	(a) Overloading of neutral in three-phase systems, (b) Increased rate of resonance occurring (c) Overheating of all cables and equipment (d) Drop in electric machine efficiency (e) Communication systems being affected by electromagnetic interference (f) Using average reading metres might lead to measurement errors (g) Thermal safeguards are being tripped inadvertently

(continued)

Table 6.4 (continued)

Issues	Description	Causes	Consequences
Voltage oscillation	Voltage value fluctuation, amplitude modulated by a 0–30 Hz signal	(a) Oscillatory loads and frequent start-and-stop operations of electric motors (such as elevators) (b) Arc furnaces	Flickering of Advanced Science Letters E-ISSN: 1936–7317 lighting and screens, giving the impression of unsteadiness of visual perception
Unbalance voltage	Voltage fluctuation in a three-phase system where the three voltage magnitudes or the phase-angle variations between them are not identical	(a) High capacity 1-phase loads (induction furnaces, traction loads) (b) Inaccurate three-phase system phase distribution of all single-phase loads (this may be also due to a fault)	A negative sequence that is damaging to all three-phase loads is implied by unbalanced systems. Three-phase induction machines are the loads that are most impacted

for describing power quality, along with its definitions, causes, and consequences, as given in Table 6.4, in the light of the idea of power quality.

6.2.3.3 Grid-Connected PV Systems

The solar photovoltaic power system that is linked to the utility grid is referred to as a grid-connected photovoltaic (PV) power system as shown in Fig. 6.5. Solar panels, one or more inverters, a power conditioning unit, and grid connection equipment make up a grid-connected photovoltaic system. They may be anything from a single solar panel on a home or business rooftop to a larger solar power plant for the entire community. Due to the high expense of batteries, grid-connected power systems seldom have an integrated battery option. When the conditions are favourable, the PV system connected to the grid sends extra electricity to the utility grid that exceeds usage by the connected load (Nirosha and Kumar Patra 2020).

Photovoltaic is linked to a grid with a capacity of 407 MW on average, capable of meeting the load of large consumers. They are able to supply extra electricity to the grid, where it is used by other consumers. A metre is used for feedback to keep track of the quantity of electricity being provided. When photovoltaic wattage falls short of normal usage, the customer will continue to purchase grid power, only in a smaller quantity. If photovoltaic wattage is substantially greater than typical demand, the energy produced by the panels will be significantly more than what is needed. In this situation, selling the extra electricity to the grid can generate income.

PV panels must employ an inverter connected to the power grid to condition, or prepare for use, solar energy received by the panels that are meant to be distributed to the power grid. An inverter changes the DC voltage that comes from the PV into

6.2 Integration Issues Related to Wind and Solar Power

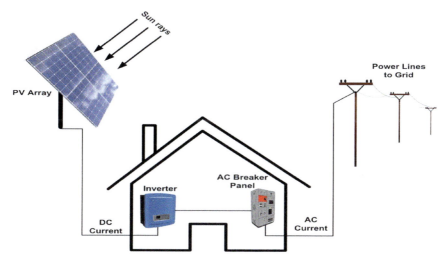

Fig. 6.5 Grid-connected PV system

the AC voltage that is used by the grid. This inverter, which might be either a sizable standalone machine or a group of smaller inverters that are all physically connected to individual solar panels, lies between the solar array and the grid and takes energy from both. Voltage, waveform, and frequency on the grid must be monitored by the inverter. The inverter must not transmit any solar energy if the grid is dead or deviates too much from its normal standards, which is one reason for monitoring. According to safety regulations, an inverter connected to a malfunctioning electrical line will automatically disconnect. In order for electricity to flow out from the solar array smoothly, the converter has to create a voltage slightly higher than the network itself and synchronize with the grid waveform, which is another reason for the inverter to monitor the grid.

Photovoltaic cells (together with other distribution generation units) provide power at the load side of the distribution network, lowering the loading active power of the feeder and enhancing the voltage profile in the process. Grid-connected photovoltaic devices are generally erected to enhance the effectiveness of the electricity system. Hence, photovoltaic systems may increase the lifetime of series voltage regulators and shunt capacitors by delaying their operating period. If PV systems are properly sized and installed, they can reduce distribution feeder losses. Photovoltaic systems may enhance the amount of load that they can handle, or their load-carrying capacity (LCC) while meeting specific requirements of an existing network. Utilities must enhance their generation capacity to meet rising demand and maintain the same reliability standards.

However, PV systems can also have a number of adverse effects on the electrical grid, particularly if they have a significant penetration level. These effects depend on the PV system's size and location. According to their ratings, PV systems may be divided as follows.

1. Small systems rated at 10 kW or less.
2. Intermediate systems rated between 10 kW and 500 Kw.
3. Large systems rated at more than 500 kW.

Normally placed at the transmission and sub-transmission levels as the last category, as opposed to the last two categories, which are placed at the distribution level.

Large PV systems (greater than 500 kilowatts) are likely to have the following effects on networks for transmission and sub-transmission:

1. Severe Variations in Voltage, Frequency, and Power

The output of PV arrays is variable and greatly influenced by external factors like the temperatures and solar irradiance ranges. The production of a PV system will be affected by random temperature and insolation variations, partial shadowing from passing clouds, and other variables, which will cause abrupt changes in the system's produced energy. Reactive power fluctuations produce significant voltage changes, whereas active fluctuations in power cause significant frequency differences in the electrical network. These voltage swings might result in unwanted switching of capacitor banks.

2. Increased Demand for Ancillary Services

In order to manage the photovoltaic energy variations or to move in synchronization with the solar, generating stations' outputs must be frequently modified. This is because the network serves as a power storage system for detecting any energy turbulence and stabilizing the PV sources' output power. The grid must be capable of providing extra power at 1 MW/10 s for instance, if a PV system got shaded by any cloud producing 1 megawatt of energy every 10 s; otherwise, frequency and voltage disruptions will arise in the electrical grid. Hence, energy must implement quickly increasing electricity production to make up for this PV array before power fluctuations and the permitted limits for voltage and frequency deviations are exceeded.

At increasing PV system penetration rates, the prior circumstance also requires a considerable rise in the frequency control requirements. There should be a 10% increase in frequency regulation. The peak permitted penetration of PV in a given area is significantly influenced by the geographical distribution of PV arrays in that area; the nearer the photovoltaic cells are to one another, the greater the energy swings anticipated cloudy conditions, and the most frequent regulatory service is required to mitigate the fluctuations in power.

The observations shown in Table 6.5 suggest that low-powered photovoltaic systems are unlikely to influence frequency control requirements due to their scattered nature and that these specifications can only be chosen depending on the extent to which huge, centralized photovoltaic systems.

6.2 Integration Issues Related to Wind and Solar Power

Table 6.5 Frequency control requirements

Frequency (%)	Area
1.3	Central station
6.3	10 km^2
18.1	100 km^2
35.8	1000 km^2

3. Stability Issues

The output of photovoltaic cells is very dynamic and influenced by the surrounding environment. This unpredictable nature significantly affects the power system's functioning because they are unable to deliver a distributed energy source that is flexible enough to fluctuate requirements. The system must take care of not just unpredictable requirements, but also unpredictable production. Consequently, problems might arise due to increased load stability. The properties of the connecting inverter entirely regulate the dynamic behaviour of the PV arrays because they don't have any rotating parts. The properties of the connecting inverter entirely regulate the dynamic behaviour of the PV arrays because they lack inertia and don't have any spinning masses. The system's damping ratio rises when PV arrays are used in place of more traditional generators as the penetration level of PV rises. Consequently, the system's oscillation is reduced. For PV systems that have not replaced synchronous generators, PV production being present may potentially alter the inter-area mode's mode shape. To ensure adequate damping of the system, some essential synchronous power plants must be kept active (even if they are working outside of their price range for running). When a fault occurs in a network where the PV penetration is high, the rotors of some traditional producers oscillate at greater amplitude.

The study on the effects of highly rated photovoltaics on the stability of subtransmission systems' voltage concluded that static voltage stability is strongly influenced by the locations, sizes, and operating methods of PVs. PV inverters that operate in a manner of continuous power factor degrade voltage stability, while photovoltaic converters that operate in a voltage control step can enhance network voltage constancy.

The following are the expected effects of small or medium Photovoltaic systems (less than 500 kilowatts) on systems distribution:

1. Reverse Flow Excess Power

Power typically flows in a single direction from the low-voltage (LV) system to the medium voltage (MV) network in a normal distribution system. However, at high PV system penetration levels, there are situations when net output surpasses overall demands (particularly at noon time), which causes the power flow's direction to reverse, moving between the LV and MV sides. The feeder overloading at the distribution and severe power losses are caused by this reverse supply of power. The performance of automated installed voltage regulators along the distribution system has also been observed to be affected by reverse power flow since these devices' settings must be adjusted to account for the movement in the load centre. Online tap changers in distribution transformers may be negatively affected by reverse power flow, particularly if they are of the single bridge resistor type.

2. **Overvoltages Along Distribution Feeders**

Overvoltages along distribution, feeders are caused by the flow of reverse power. The voltage may now be pushed further, exceeding the AC acceptable limits, using capacitor banks and voltage regulators that were previously only utilized to slightly enhance the voltage. MV network voltage increase is frequently a barrier to the extensive use of wind turbines. PV system installation may be affected similarly by voltage increase in LV networks. Electrical systems with a significant distributed photovoltaic energy production are more prone to experience this problem.

3. **Voltage Control Challenges**

Voltage control becomes challenging in an energy network with integrated production since there are several supply points. All voltage-regulating components, such as voltage regulators and capacitor banks, are made to work during one-way energy flow systems.

4. **Significant Electrical Energy Losses**

As they shift generation closer to the load, DG systems minimize system losses. Until reverse power flow occurs, this assumption is correct. At a penetration level of around 5%, distribution system losses are at their lowest value; however, as the penetration level rises, the losses also rise.

5. **Unbalance Phase Issue**

Single-phase inverters are frequently used in present residential photovoltaic systems. Phase imbalance may occur if these inverters are not evenly distributed across the various phases, increasing the voltage unbalance and changing the neutral voltage to harmful levels.

6. **Problem with Electromagnetic Interference**

Due to electromagnetic interference caused by the PV inverters with a high switching frequency, peripheral circuits including converters, converter banks, protection devices, and DC linkages may collapse.

7. **Power Quality Issues**

One of the main issues of increased photovoltaic adoption in distribution systems is power quality problems. Harmonic currents are produced by the power inverters that connect solar panel arrays to power grids. As a result, they can increase the voltage's total harmonic distortion (THD), and current at the point of common coupling (PCC). If the network is rigid enough and has lower than average sequence impedance, voltage harmonics are within acceptable limits. On the other side, high pulse power electronic inverters create current harmonics, which often appear at high orders with low magnitudes. A problem with current harmonics of higher order is

6.2 Integration Issues Related to Wind and Solar Power

that they might cause high-frequency resonance in the system. The total magnitude of such current harmonics may be reduced as a result of the diverse impact between various current harmonics. The inter-harmonics that arise in the low harmonic range are another issue with power quality (below the 13th harmonic). Loads around the inverter may interact with these inter-harmonics. Even harmonics, particularly the second harmonic, may increase the undesirable negative sequence currents that impact three-phase loads. Additionally, DC injections may get developed and flow into the distribution transformer, seriously damaging it.

8. Increased Demand for Reactive Power

Photovoltaic inverters are often built for two main reasons: to run with a unity power factor. The first reason is that these types of inverters cannot perform in the voltage control mode according to current specifications (IEEE 929–2000). The second aspect is that owners of smaller residential PV systems are only paid for their kilowatt-hour output, not for their kilovolt-ampere-hour generation, under incentive schemes. So, to increase the active power generation and, consequently, their return people like to use their inverters at a unity power factor. The partial fulfilment of PV systems minimizes the active power consumption of existing loads, reducing the active power supply from the utility. Imaginary power needs, however, remain the same and must be entirely met by the utility. The utilities do not prefer a more reactive power rate supply since distribution transformers would then run at a very low-power factor (in some cases 0.6). The total losses of distribution transformers will rise when the operating power factor of transformers declines, decreasing their efficiency and the efficiency of the total electrical network.

9. Islanding Detection Challenges

Non-detection zones, which are the operational conditions under which an islanding detection method wouldn't function in time and would be prone to failure, are a feature of islanding detection strategies. Costs related to incorporating photovoltaic systems into electricity networks as a whole are further raised by the addition of islanding detecting equipment.

Table 6.6 shows a summary of allowable PV penetration limitations. There is no established peak permitted PV penetration limit power, according to the above results. Based on the limiting factor, PV array size, location, and geographic distribution, it can range from 1.3 to 40%. The peak permitted photovoltaic adoption in this system should be determined by conducting an extensive techno-economic study for each specific network. As an example, Hydro One, the biggest Ontario distribution utility, demands the need for dispersed production connected to a circuit line in a distribution system segment, including the recommended generator, no more than seven per cent of the yearly line segment maximum load."

There are the following advantages of the grid-connected PV system:

A. Some system operators provide programmes like net metering and feed-in tariffs that can reduce a customer's power use expenses.
B. Grid-connected PV systems are easier to install than standalone PV systems since they don't need a battery system.

Table 6.6 Allowable PV penetration limitations summary

Limiting factors	Penetration limit (%)
During cloud transients, generator ramp rates (central station PV)	5
During cloud transients, generator ramp rates (distributed PV)	15
With central station PV, power variations caused by transient clouds	1.3
PV system is spread out across 10 km², and power variations caused by transient clouds will occur	6.3
PV system is spread out across 100 km², and power variations caused by transient clouds will occur	18.1
PV system is spread out across 1000 km², and power variations caused by transient clouds will occur	35.8
Expansion of frequency control vs break-even costs	10
Voltage regulation	40
Lowest losses in the distribution system	5
Overvoltages	33

C. The advantage of photovoltaic (PV) power-generating systems' grid connection is the effective usage of produced electricity since there were no storage issues.

D. A system that uses solar energy is a carbon-negative life cycle because any power produced more than that required to originally construct the panel avoids the requirement to burn fossil fuels.

E. Even if the sun does not always shine, each installation results in a typically predicted reduction in carbon consumption.

There are also some disadvantages of connecting PV systems to the grid which can be stated as:

A. Voltage control problems may arise with grid-connected solar power.

B. The one-way, or radial, flow assumption underlies how the conventional grid functions. However, adding power to the grid raises the voltage and can cause levels to exceed the 5% allowable range.

C. Power quality may be compromised by grid-connected PV.

D. Due to PV's intermittent nature, voltage varies frequently which can cause voltage flicker in addition to wearing out voltage regulators from repetitive adjustment.

E. There are various protection-related issues while connecting to the grid.

6.2.3.4 Grid-Connected Wind Energy System

The grid system is connected to the wind-generating interface system using voltages on every as seen on the side of the impedance Fig. 6.6.

Three-phase electricity that is as symmetrical as feasible is used to transmit power in the electrical network. The total three-phase energy is constant and the voltage

6.2 Integration Issues Related to Wind and Solar Power

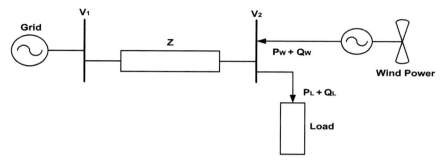

Fig. 6.6 Single-line diagram of grid-connected wind energy system

between lines is three times greater than the phase voltage. The voltage drop through the impedance may be expressed as in (6.1):

$$v_1 - v_2 = \sqrt{3}iz, \tag{6.1}$$

where v_1 and v_2 are the RMS voltage at grid and PCC, z = the transmission line impedance and i = the transmission line current. At the point of common connection (PCC), both the wind farm and the local load are linked. The wind connection's short circuit power (p_{sc}) is represented in (6.2).

$$p_{sc} = v_2/z. \tag{6.2}$$

The current through impedance Z will fluctuate when wind energy output varies. The voltage v_2 varies as a result of these fluctuations in current. Connections with the network that have a short circuit ratio of less than 2.5 are generally avoided because they can cause voltage swings, often known as weak grids.

The impedance $z = r + jx$ at the fundamental frequency. In general, when harmonics are present, the impedance changes can be expressed as (6.3):

$$z(h) = r + jhx_l, \tag{6.3}$$

where h represents the harmonic order and x_l is the inductive reactance that has a direct relation with the frequency. $P + jQ$, where P is the active power and Q is the reactive power, is used to indicate the combination of wind energy generation and load. Reactive power is influenced by the phase difference between voltage and current, as observed in Eq. (6.4).

$$\phi = \tan^{-1}(Q/P). \tag{6.4}$$

The reactive power of the wind influences the voltage v_2. The localized load and grid impedance of the feeding both have an effect.

300   6 Grid Integration Techniques in Solar and Wind-Based Energy Systems

Example 1 Calculate the voltage drop through the impedance (V_Z) when the RMS voltage at the grid (V_grid) is 100 kV, the RMS voltage at the point of common connection (PCC) (V_PCC) is 95 kV, the transmission line impedance (Z) is 0.2 ohms, and the transmission line current (I) is 500 A.

Solution: Given the $V_{grid} = 100$ kV, $V_{PCC} = 95$ kV, Transmission line impedance $Z = 0.2$ ohms and Transmission line current $I = 500$ A.

The voltage drop through impedance

$$
\begin{aligned}
V\_Z &= \left(V_{grid} - V_{PCC}\right) - Z \times I \\
&= (100\,\text{kV} - 95\,\text{kV}) - (0.2\,\Omega \times 500\,\text{A}) \\
&= 5\,\text{kV} - 100\,\text{V} = 4.9\,\text{kV}
\end{aligned}
$$

Example 2 Calculate the wind connection's short circuit power (SSP) when the impedance (Z) is 0.3 ohms and the transmission line current (I) is 600 A.

Solution: Given impedance $Z = 0.3$ ohms, transmission line current $= 600$ A.

$$
\begin{aligned}
\text{SSP} &= Z \times I2 = 0.3\,\Omega \times (600\,\text{A})^2 \\
\text{SSP} &= 0.3\Omega \times (600\text{A})^2 = 0.3\,\Omega \times 360,000\text{A}^2 \\
\text{SSP} &= 0.3\Omega \times 360,000\text{A}^2 = 108,000\,\text{VA} = 108\,\text{kVA}
\end{aligned}
$$

Example 3 Solution: Determine the change in impedance (ΔZ) when the harmonic order (h) is 3, the inductive reactance (X_h) is 0.05 ohms, and the fundamental frequency (f) is 50 Hz.

Solution: Given inductive reactance $X_h = 0.05$ ohms, Fundamental frequency $= 50$ Hz

$$
\begin{aligned}
\Delta Z &= h \times X_h \times f = 3 \times 0.05\,\Omega \times 50\,\text{Hz} \\
\Delta Z &= 3 \times 0.05\,\Omega \times 50\,\text{Hz} = 7.5\,\Omega \\
\Delta Z &= 7.5\,\Omega
\end{aligned}
$$

The power issues and their effects on grid-connected wind energy systems are as follows:

1. **Voltage Variation**

The generator's torque and wind speed are the major causes of voltage fluctuation. Real and imaginary power variations have a direct effect on voltage fluctuations. Reactive power is consumed by the asynchronous generator used in the wind-generating system, which might further harm the grid. The voltages may be varied by turning on and off the wind turbine generator. The voltage difference is often divided into voltage with short and long-duration variations. Table 6.7 shows the voltage variation related to power quality issues.

6.2 Integration Issues Related to Wind and Solar Power

Table 6.7 Voltage variation power quality issues

Quality of power problems	Definition	Causes	Consequences
Voltage sags/voltage dips	It is the reduction of the normal operating voltage within 10 and 90% of the nominal RMS voltage at the line frequency for 0.5 cycles to 1 min	Wind turbine starting issue, transmission/distribution network fault, consumer installation error, high load connection, and startup of huge motors	Equipment malfunctions, such as those in a microprocessor-based control system, programmable logic controller, and adjustable speed drives, may cause a process to halt, contractors to trip, relays to trip for voltage-sensitive loads, and an electric machine to become less efficient
Voltage swell	It is a spike in voltage at power frequency that lasts for more than one cycle but normally lasts for a few seconds	Heavy loads starting and stopping, a system error, and an improperly controlled converter during off-peak hours	Light and screen flickering, harm delicate equipment
Short interruptions	The electrical supply is completely cut off for a period of a few milliseconds to one or two seconds	Mainly as a result of protective systems that automatically open and close	Protection systems trip, critical equipment such as personal computers and programmable logic control systems are shut down
Long-duration voltage variation	It involves a complete stoppage of the electrical supply for a period longer than 1–2 s	Power system equipment failure and protection equipment failure	–

2. Voltage Flicker

Voltage flicker is the term that refers to dynamic changes in voltage in the network brought on by windmills or shifting loads. Consequently, the variation of the power output of wind turbines happens while they are operating continuously. The strength of the grid, the circuit impedance, the phase angle, and the energy efficiency of windmills all influence the voltage fluctuation's magnitude. It is described as a voltage fluctuation with a frequency range of 10–35 Hz. A flicker metre that may be used to directly measure flicker is specified in IEC 61,400-4-15. A measure of flicker provides a normalized without-dimension flicker measurement that is unaffected by network conditions and unaffected by the apparent power of grid short circuits. To obtain a long-term flicker level, it provides a ratio of produced rated apparent power to short circuit power as shown by Eq. (6.5).

$$C(y_g, w_a) = E(p_{sc}/p_a), \tag{6.5}$$

where, $C(y_g, w_a)$ is the dependence of the flicker coefficient on grid impedance angles y_g and the mean wind speed w_a. p_a and p_{sc} denotes both the apparent power of the wind turbine at rated power and the short circuit power of the grid at the PCC. The transient voltage changes are often described using the flicker standards. Long-term flicker (E) is measured over two hours, whereas short-term flicker is measured over a 10-min duration.

Voltage flickering generally causes variations in a wind turbine's active and reactive power, such as yaw error, wind shear, wind turbulence, or fluctuations in the control system and the wind turbine's switching operations. The power output of a wind turbine with a fixed speed decreases each time a rotor blade swings through the tower. In contrast to variable speed turbines, this effect results in periodic power fluctuation at a frequency of roughly 1 Hz. Arc furnaces, arc lights, and capacitor switching all generate flickering.

The consequences result in power quality degradation and cause severe damage to sensitive equipment.

3. **Harmonics**

It occurs as a result of how electronic power devices operate. At the point where the wind turbine is linked to the network, the harmonic voltage and current should be restricted to a suitable level. Harmonic current emissions during the uninterrupted operation of a wind turbine equipped with energy devices must be specified.

4. **Wind Turbine Location in the Power System**

The method used to link the wind power system to the electrical grid has a significant impact on the quality of the power. The influence on power quality is often greater at the consumer terminal connected near to the load than at the terminal connected far from the load. The wind generation system is more cost-effective than other places when it is connected to a medium voltage transmission line since there is less space between the wind generation station and the point of common connection. As a result, the operation and its impact on the power system are dependent on the structure of the adjacent power system network.

5. **Self-Excitation of Wind Turbine Generating System**

After being disconnected from the local demand, a wind turbine generating system (WTGS) with an asynchronous generator starts to self-excite. When WTGS is equipped with a compensating capacitor, the risk of self-excitation increases. Reactive power compensation is provided by the capacitor attached to the induction generator. The balance of the system, however, determines the voltage and frequency. The safety concern and the equilibrium between actual and reactive power are drawbacks of self-excitation.

If the equipment is linked to the generator during the self-excitation, it may be subject to undervoltage, overload, and over-frequency operation.

The advantages of the grid-connected wind energy system are as follows:

A. Due to the fact that this interconnection uses wind energy instead of conventional grid utilities, it can provide power at relatively low costs. It can also assist in lowering electric bills and the amount of energy needed for production.
B. It is highly helpful in lowering the price of electricity and gives customers access to electricity at a very affordable price that everyone can afford.
C. By adding wind power to the system, we can boost energy production to the point where we can sell it to other utilities.
D. The ability to deliver energy practically anywhere is one of the most significant benefits of wind power and grid connection, whether it is used in an area with high-wind power or one with low wind power, such as one with 10 mph per year, for example. In the region with less wind coverage, it can still produce power with the same efficiency.
E. This connection also helps save energy by reducing the amount of fuel used. In contrast to other fuel types like diesel, it may use gasoline as a fuel generator. As it is obtained from the wind, it also aids in lowering environmental pollution by minimizing the emission of radiation resulting from various nuclear processes.

There are also disadvantages associated with the grid-connected wind energy system which may be:

A. One of the major problems with wind power generation is the need for reactive power to sustain voltage.
B. Designing and optimizing controllers for turbine power.
C. Wind farms with series-compensated connections have problems.
D. Issues with power quality, such as voltage flicker.
E. Wind farm startup and grid synchronization.

Issues with sub-synchronous resonance are caused by the complicated shaft/gear system of the wind turbine interacting with the electric network.

6.2.3.5 Non-Technical Issues of Grid-Connected PV and Wind Energy Systems

The non-technical issues that generally occur to the PV and wind energy systems connected to the grid are:

1. A shortage of technically skilled personnel.
2. Insufficient transmission lines are available to deal with RES.
3. The exclusion of RES technologies from the competition by providing them with dispatch priority restricts the building of new power plants for storage purposes.

6.3 Grid Requirements for Solar-Based Energy Systems

6.3.1 Power Quality Requirements

Voltage, flicker, frequency, harmonics, and power factor procedures and rules regulate the quality of the energy generated the utility and the community are both benefited by the solar system AC loads. A solar system may need to be disconnected from the utility if it deviates from certain requirements, which are considered outside of boundaries situations.

A. *DC injection:* Distribution transformers may get saturated by DC injection in the utility, which might cause overheating and tripping. This issue is mitigated for conventional PV systems with galvanic isolation, but with the new transformer and less PV inverter generation, more attention is needed in this area. Therefore, the injection DC limits as seen in Table 6.8 are approved.

With IEEE 1574 and IEC 61727, the harmonic analysis should be used to assess the DC component of the current (fast Fourier transform, or FFT) and no maximum trip time limit exists. The measured DC component during the test must be below the thresholds for the various loading scenarios (33, 23 and 33% of the normal load). For VDE 0126-1-1, this circumstance necessitates the use of a current sensor that has been carefully built to be able to recognize this level and disconnect in the required journey time.

B. Current harmonics: In order to prevent harm from being done to other grid-connected devices, low current distortion should be present in the PV system output. The values listed in Table 6.9 are considered acceptable.

An electronic power source that produces the IEEE 1574/IEC 61,727 test voltage should have a voltage thermos-hydrodynamic (THD) of distinct voltage harmonics and less than 2.5% that are less than half of the harmonic limitations in effect. It

Table 6.8 Limitation of DC injection

IEEE 1574	IEC 61,727	VDE 0126-1-1
$I_{DC} < 0.5\,(\%)$ of the rated RMS current	$I_{DC} < 1\,(\%)$ of the rated RMS current	$I_{DC} < 1$ A Maximum trip time 0.2 s

Table 6.9 Maximum current harmonics

IEEE 1547 and IEC 61,727						
Individual harmonic order odd)* (%)	$h < 11$ 4.0	$11 \leq h \leq 17$ 2.0	$17 \leq h \leq 23$ 1.5	$23 \leq h \leq 35$ 0.6	$35 \leq h$ 0.3	Total harmonic distortion THD (%) 5.0

* Even harmonics cannot exceed 25% of the aforementioned odd harmonic limitations

6.3 Grid Requirements for Solar-Based Energy Systems

Table 6.10 Limits for current harmonics as per IEC 61,000-3-2 (Class A)

Odd harmonics		Even harmonics	
Order (h)	Current (A)	Order (h)	Current (A)
3	2.30	2	1.08
5	1.14	4	0.43
7	0.77	6	0.30
9	0.40	$8 \leq h \leq 40$	$0.23 \times 8/h$
11	0.33		
13	0.21		
$13 \leq h \leq 39$	$0.15 \times 15/h$		

is standard practice to utilize the optimal sinusoidal energy supply to avoid having background distortion affect the results.

Since IEC 61,727 has not yet been authorized in Europe, the standard for class A equipment's harmonic limitations is IEC 61,000-3-2 (as shown in Table 6.10).

The IEC 61000-3-2 current restrictions are stated in amperes and are often greater than those in the IEC 61727. A comparable standard, IEEE 61000-3-12, also applies to equipment with a current of more than 16 A but less than 75 A.

C. *Mean power factor:* It is only mentioned in IEC 61,727 that a photovoltaic DC-to-AC converter must have a mean lagged power factor when the output is less than 0.9 higher than 50%. The majority of photovoltaic inverters made for utility-interconnected operation perform approximately at unity power factor.

The power factor is not needed since IEEE 1574 is an international standard that should allow distributed generation of reactive power. In VDE 0126-1-1, no power factor specifications are stated.

Now, it is generally accepted that the requirement of the power factor for solar converters means that they must run at the semi-unity power factor without controlling the voltage via regulating the grid's use of reactive power. Local grid regulations apply to powerful photovoltaic systems linked straight to the level of distribution since they may take part in grid control. It is also anticipated that utilities will soon let low-power installations trade reactive power, while new rules are still anticipated.

6.3.2 Response to Abnormal Grid Conditions

In the case of abnormal voltage and frequency circumstances on the grid, the PV inverters must disconnect from the system. This action is taken to protect the public's safety, the safety of utility maintenance workers, and associated equipment, such as the solar system.

Table 6.11 Disconnection period for voltage changes*

IEEE 1547		IEC 61,727		VDE 0126–1-1	
Voltage range (%)	Disconnection time (sec.)	Voltage range (%)	Disconnection time (sec.)	Voltage range (%)	Disconnection time (sec.)
$V < 50$	0.16	$V < 50$	0.10	$110 \leq V \leq 85$	0.2
$50 \leq V \leq 88$	2.00	$50 \leq V \leq 85$	2.00		
$110 < V < 120$	1.00	$110 < V < 135$	2.00		
$V \geq 120$	0.16	$V \geq 135$	0.05		

A. Voltage deviations (see Table 6.11)

The local nominal voltage is always used when discussing system voltage. At the utility connection point, the voltages are measured in root mean square (RMS). This break-up time is the period between the occurrences of the abnormal situation to the inverter ceasing to activate the utility line. To enable utility electrical conditions to be sensed for usage by the "reconnect" attribute, the inverter controls must stay connected to the utility. To ride-through temporary disruptions and prevent frequent annoyance trips, the permissible time delay is used.

Fast voltage monitoring is thus necessary because the time of VDE 0126-1-1 disconnection requirement is substantially lower (0.2 sec.).

B. Frequency deviations (Table 6.12)

To minimize excessive annoyance tripping in weak-grid scenarios, the permissible range and time delay are designed to allow for riding through momentary interruptions. Since the frequency limit of the VDE 0126-1-1 is substantially lesser, frequency adaptive synchronization is necessary.

C. Reconnection after trip: Only under the conditions listed in Table 6.13 can the inverter be reconnected following a disconnection brought on by the occurrence of aberrant frequency or voltage circumstances. In order to ensure resynchronization before reconnection and prevent any harm, IEC 61,727 adds a time delay.

6.3.3 Anti-Islanding Requirements

Undoubtedly, the so-called anti-islanding criterion is the most difficult technological requirement. For PV systems linked to the grid, islanding whenever the photovoltaic converter continues to function with the local load for a long period following the grid's route. System disconnect may happen in the usual scenario of a household power system that is with assistance from a rooftop photovoltaic system due to localized device malfunction discovered by the use of ground-fault circuit prevention or a line that has been purposefully disconnected for service.

The following events may happen in any scenario if the PV inverter is not disconnected:

6.3 Grid Requirements for Solar-Based Energy Systems

Table 6.12 Disconnection period for voltage changes

IEEE 1547		IEC 61,727		VDE 0126-1-1	
Frequency range (Hz)	Disconnection time (sec.)	Frequency range (Hz)	Disconnection time (sec.)	Frequency range (Hz)	Disconnection time (sec.)
$59.3 < f < 60.5^{*}$	0.16	$f_n - 1 < f < f_n + 1$	0.2	$47.5 < f < 50.2$	0.2

* For systems with power of < 30 kW, the lower limit may be modified to permit frequency control involvement

Table 6.13 Reconnection requirements after the trip

IEEE 1547	IEC 61,727	VDE 0126-1-1
$88 < f < 110\,(\%)$	$85 < f < 110\,(\%)$ AND	
AND	$f_n - 1 < f < f_n + 1\,(\text{Hz})$ AND	
$59.3 < f < 60.5\,(Hz)$	Minimum delay of 3 min	

- Line retripping or linked devices may cause harm owing to a phase-out shutdown.
- During islanding, there is a safety risk for the people working on transmission lines who think that de-energized transmission lines are safe.

A. AI as per IEEE 1547/UL 1741

According to IEEE 1574, the distributed resources (DR) must identify an islanding condition and stop energizing the region within two seconds after unintended islanding occurs where the DR is still powering the island through the PCC. The test system is outlined in IEEE 1547.1 as seen in Figure 6.7, where EUT stands for either the test equipment or the photovoltaic cell converter.

The PV inverter and the grid must be linked in parallel with an adjustable RLC load in order to conduct the test. In other terms, at the nominal power P and rated grid voltage V, the reactive power generated by C [VAR] should be equal to the reactive power absorbed by L [VAR] and should be equal to the power wasted in R [W]. Adjusting the resonant LC circuit should resonate at the rated grid frequency and quality factor $Q = 1$. Thus, the following equations may be used to compute the local RLC load values:

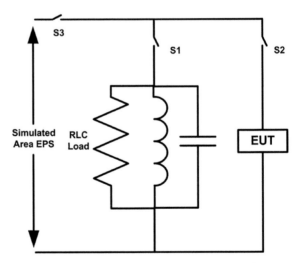

Fig. 6.7 Test configuration for the IEEE 1547.1 anti-islanding requirement

6.3 Grid Requirements for Solar-Based Energy Systems

$$\begin{cases} R = \dfrac{V^2}{P} \\[2ex] L = \dfrac{V^2}{2\pi f P Q^2} \\[2ex] C = \dfrac{P Q_f}{2\pi f V^2} \end{cases} . \tag{6.6}$$

The RLC load's characteristics should be adjusted such that, on a steady-state basis, the grid current via S_3 is less than 2% of the rated value. S_3 must be open in this balanced situation, and the period before disconnection must be evaluated and must be less than 2 s.

Each phase of three-phase solar inverters must be evaluated independently with respect to the neutral. Local RLC loads PV inverters with three wires and three phases should be linked between phases. The anti-islanding criteria described in IEEE 1547 have been synchronized with the UL 1741 standard in the US.

The criterion since the local RLC load's quality factor has been lowered from 2.5 to 1.0 in comparison to the previous IEEE 929–2000 standard, making compliance more technically feasible.

B. AI defined by IEC 62,116

Similar AI criteria to those for IEEE 1547 are suggested in the proposal of IEC 62116-2006. Other DERs that are linked to inverters can also utilize the test. This standard's valid system has a rating of 10 kVA or less in the normative reference IEC 61727-2004; nonetheless, the standard is susceptible to change. Before performing the island detection test, a power balance is necessary on the test circuit, which is the same as that used for the IEEE 1547.1 test (Fig. 6.7). Although there are more test cases in the criteria for passing the test than in the IEEE 1547.1 test, the requirements for confirming island detection are not significantly different.

There are three output power levels: A (100–105% of the inverter's output power), B (50–66%), and C (25–33%) used to evaluate the inverter. Case C is tested at the minimum permissible inverter output power if > 33%, while Case A is evaluated at the highest permissible inverter input power. There are additional requirements that apply to the voltage at the inverter's input (https://www.vde-verlag.de/standards/012 6003/din-v-vde-v-0126-1-1-vde-v-0126-1-1-2006-02.html). All circumstances must be tested with neither a real load nor a reactive load consuming energy deviations greater than condition A's 5% iterated real and reactive power variation from -10 to 10% of the inverter's operational output power. By varying the reactive load by $\pm$ 5% in steps of 1% of the inverter output power, conditions B and C are assessed.

The maximum trip time is 2 s, as per IEEE 1547.1 requirements. There is no detailed explanation of the anti-islanding criteria in IEC 61,727. As an alternative, IEC 62,116 is suggested.

C. Definition of AI by VDE 0126-1-1

One of the following anti-islanding techniques is permissible under VDE 0126-1-1:

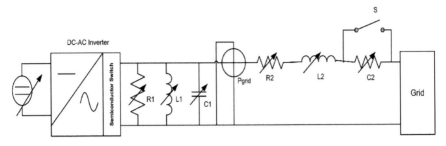

Fig. 6.8 VDE 0126-1-1 test configuration for anti-islanding requirements

1. *Impedance Analysis.* In Fig. 6.8, the test circuit is shown.

The process is built around regional reactive and active electricity balancing utilizing the switch S is opened to, and the variable RLC circuit raises the grid impedance by 1.0. The requisite time, which is 5 seconds, should pass before the inverter disconnects. The test has to be run again with various values of the simulated grid impedance (R2, L2) in the vicinity of 1 ohm (maximum of 0.5-ohm inductive reactance).

2. *RLC Resonant Load Disconnection Detection.* The test requirements require the resonant circuit RLC characteristics to be computed for a component of quality of $Q > 2$ making use of the experiment circuit illustrated in Fig. 6.8 from IEEE 1547.1. 25, 50 and 100% for the aforementioned power levels, the inverter shall shut down with balanced power within a maximum of 5 s of S_2's disconnecting.

A recognized passive anti-islanding technique for three-phase PV inverters involves monitoring the voltage of each phase with regard to the neutral. Having independent management of the three phases' currents is a requirement of this technology.

It has been exceedingly difficult to find a computer-based anti-islanding system solution, which has led to several research projects and publications.

6.4 Grid Requirements for Wind-Based Energy System

Transmission system operators (TSOs) around the world have established new grid connectivity standards known as grid codes (GCs) to mitigate the effects of wind power penetration on power system stability and electricity quality. Although GCs with distinct specifications for distribution system operators (DSOs) have also been established, TSOs are still in charge of managing the power balance in the majority of countries, and the installation of huge wind farms is expected to take place soon.

6.4.1 Voltage and Frequency Variation Under Normal Operation

Wind power plants (WPP) have to stay within a certain range at PCC between the rated voltage and frequency so that they don't get disconnected because of transient problems. This criterion is often defined by different zones:

- Constant functioning within a certain range under and over the assigned rating.
- Time-restricted operation with probable output reduction at wider ranges.
- Immediate termination.

The Danish code (90–106% nominal voltage) and the British code (47.5–52 Hz) include the strictest continuous operation restrictions for frequency and voltage, respectively. Some nations, like the UK, for instance, necessitate keeping the output power constant to provide frequency management for lower-frequency regions outside of the continuous working area.

It is clear that E.ON offshore has the highest frequency limits, at 46.5 and 53.5 Hz. Larger frequency ranges are permitted in nations like Ireland, which have a weakly connected, isolated power infrastructure. The Spanish rule (RD 661–2007) permits the largest constant frequency range of 48–51 Hz, with a 3-s disconnection time for frequencies below 48 Hz. The transmission system operator (TSO) must be consulted on the disconnection time for frequencies more than 51 Hz.

Even though the German GC permits persistent performance at frequencies beyond 50 Hz, power restriction is required starting at 50.2 Hz in order to take part in the primary control. Resynchronization is possible when the voltage reaches about 105 kV in 110 kV networks, 210 kV in 220 kV networks and 370 kV in 380 kV networks.

6.4.2 Active and Reactive Power Control

This is a typical need that aligns with traditional power plants' capacity to change output power in accordance with TSO requirements to enable load balancing with two separate objectives:

- Power curtailments.
- Frequency control.

A. Power curtailments

Each nation has its own set of regulations regarding power interruptions:

- According to Denmark, an accuracy of 5% (5 min. average) and ramp rates of maximum power output per minute are up to 100% with curtailments as low as

312   6 Grid Integration Techniques in Solar and Wind-Based Energy Systems

20% are needed. As indicated in Table 6.14, six distinct curtailment or gradient limiter profiles are defined.

- In Germany, the actual power of the WPP must be altered at a ramp rate of at least ten per cent of the grid connection capacity each minute, but not below 0.1 p.u.
- Although the rate is not specified in Spain, the WPP is required to be able to accept any set-point provided by the TSO and to provide the TSO with the difference between its maximum capacity and the true power for the case in which it uses the derated mode.
- In Ireland, the TSO requires activation in less than 10 s and ramp rates of 1–30 MW per minute, averaged over 1 min and over 10 min, respectively.
- According to Table 6.15, the WPP rating is what determines the ramp rate in China.

Table 6.14 Restrictions on power as outlined by the Danish GC

Regulation type	Purpose	Primary regulation objective
Absolute production limit	Limit the connection point's current power generation from the wind farm to a maximum, clearly indicated MW amount	Optional production limit to MW_{max}
Production-delta constraint	Reduce the wind farm's power production to a level below the available power to establish regulatory reserves	Production limit by MW_{delta}
Balance regulation	Adjust the power production to participate in the downward/upward production regulation power balance that must be practical to the power setup requirements imposed by the TSO	Using the specified gradient, modify current production by MW
Stop regulation	Maintain the existing level of power output (if the wind permits); The function causes the upward control and production constraints to halt if the wind speed rises	Maintain current output
Power gradient limiter	Limit the highest gradient at which power production varies in response to variations in wind speed	Power gradations don't go above the highest settings
System protection	System protection, a protective feature, must be able to automatically lower the wind farm's power output to a level that the power system can tolerate. The power grid may be overburdened and at risk of failure in the event of unanticipated events in the power system, such as forced outages of lines. The system protection regulator needs to be quick to react in order to stop system collapse	Using an external system protection signal as a foundation automatically controls the generation of power

6.4 Grid Requirements for Wind-Based Energy System

Table 6.15 Active power gradients in the Chinese GC

WPP installation capacity (in MW)	10-min maximum ramp (in MW)	1-min maximum ramp (in MW)
< 30	20	6
30–150	Installation capacity/1.5	Installation capacity/5
> 150	100	30

B. Frequency control

The active power output also needs to be adjusted to ensure that the WPP takes part in the main control function continually during frequency variation. In contrast to power curtailments, these ramps are substantially steeper. If the WPP uses another P gradient than the other participants in the balancing act, stability issues could develop. This gradient, which often indicates the needed change in power in response to a frequency fluctuation, is expressed in MW/Hz.

According to the table in Fig. 6.9, various countries participate in frequency regulation in different ways. As seen in Fig. 6.9, all generating units must lower their instantaneous active power when running at a frequency of greater than 50.2 Hz, 40% of the generator's output gradient instantaneous power available per hertz.

According to the German code, wind farms must lower their active power by 0.4 p.u. /Hz (40% of the WPP's available power) when the frequency surpasses the value of 50.2 Hz.

Wind farms with a capacity higher than 50 megawatts are required by the British code to incorporate a frequency control system that can offer both primary and secondary frequency regulation.

The frequency response required by the Irish code can be observed in Fig. 6.9.

The TSO should modify the values of points ABCDE's power and frequency online within the bounds as illustrated in Fig. 6.9. This is due to the fact that in order

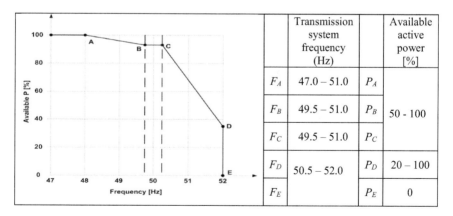

Fig. 6.9 Irish power–frequency curve

to ensure smooth participation of WPP in the TSO frequency management, the power ramp should be imposed by the TSO in accordance with the frequency response of the other members to the balancing act.

The frequency in the electricity system serves as a predictor for the unbalance between production and demand. The frequency must be nearby to its actual value for the power system to operate effectively. When there is an imbalance between supply and demand, primary and secondary controls are used to balance the power. Traditional electrical generators usually have a governor control that serves as the main load frequency control in a power system. The range of this control is from one to thirty seconds. In order to restore the frequency to its nominal value and release used primary reserves, the secondary control is performed with a time window of 10–15 min.

As a result, the secondary control causes generation to grow or drop more gradually. In accordance with several standards, wind farms must be allowed to take part in secondary frequency control which may be done by turning off a few of the wind farm's turbines or lowering the electrical power output by adjusting the pitch control during the over-frequency. Since wind cannot be regulated, it would be intended to keep power production at normal frequencies as low as possible so that the wind farm might offer secondary control at lower frequencies. After that, they will have to give primary, secondary, and high-frequency replies.

Control of Reactive Power at Healthy Condition

By implementing automatic voltage regulation (AVR) the new GC seeks to transform the WPP into an entity that behaves as an old synchronous generator in terms of Q control in reaction to variation of grid voltage. The short circuit power of the grid determines the grid's capacity to change the voltage; therefore, the Q requirements are connected to each grid's features. There are three methods to describe this requirement:

- Q set-point.
- Power factor control.
- Voltage control.

The reactive power control is different as per regulations designed and implemented by the following countries:

I. Germany

While the WPP is operated at rated capacity for variable power ranges within the permitted operating range, areas, as a function of voltage at nominal active power and as a function of active power for the scenarios, show the least need for the production of reactive power. A power factor requirement or a Q requirement may be defined for the requirement. Since the grid features might vary depending on location and strength, the German TSOs have established three variations, as shown in Fig. 6.10a–c.

The complete Q range in voltage or active power (P) planes should be cycled by the WPP in less than 4 min. The positive-sequence system components are referred

6.4 Grid Requirements for Wind-Based Energy System

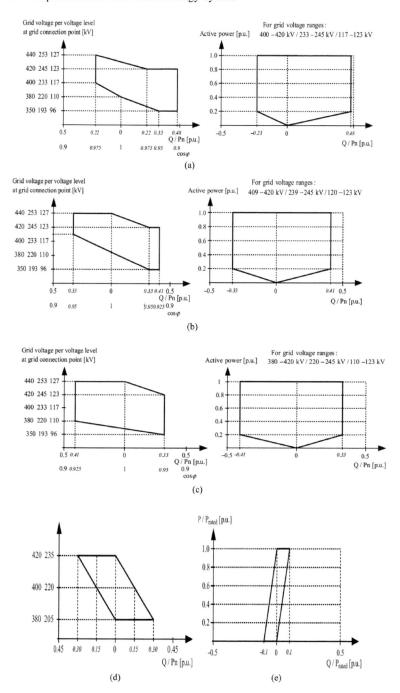

Fig. 6.10 a to c Three variants of *VQ* and *PQ* dependencies defined in Germany, d *VQ* dependence in Spain and e *PQ* dependence in Denmark

to by the Q controller, which should be sluggish with a settling period of several minutes.

II. Spain

All generations on the level, whether conventional or renewable, must meet the Q criteria while operating at normal conditions as outlined by the regulation P.O.7.4 from 2000. According to active power and transmission voltages, the following criteria are defined:

- For all technical P ranges and nominal voltages, the lowest range limit is 0.15i–0.15c.
- As illustrated in Fig. 6.10d, the minimum voltage range is 0.30i–0.30c.
- Denmark

In normal operation, the Danish standard defines the 10 s. average PQ graph represented in Fig. 6.10e, which holds over the whole voltage range. In essence, it establishes a 0.1 p.u. control band. The needed Q is lower than it is for the German and Spanish GCs. This needs to be thought of as the prerequisite. After agreeing with the TSO, the WPP may additionally use the voltage control mode or Q set-point to generate greater amounts of Q.

IV. UK

A power factor having the range of 0.95i to 0.95c at 1 p.u. active power is required by the British code, which was developed only for non-synchronous embedded generation (132/275/400 kV). This criterion, which refers to 0.33 per unit of reactive power, should be retained even if active power is reduced to 0.2 p.u. for the lagging power factor and to 0.5 p.u. for the leading power factor. Following the agreement with the TSO (NGET), Fig. 6.11 displays a lower band of 0.05 per unit of reactive power for low-power leading power factor and an expansion of the Q criteria in the dashed zone for P less than 0.2 per unit.

V. Ireland

Comparatively, the Irish code has a Q of 0.33 p.u. for both the trailing and the leading PF, and its Q must fall linearly to 0 proportional to P for P less than 0.5 p.u., illustrated in Fig. 6.11b.

VI. US

The US FERC 661 rule states that a Q value having a power factor ranging from 0.95i to 0.95c can be needed by the TSO on an individual basis and does not need to be dynamic (STATCOM). This explains why there are still many wind turbines in the USA that have variable rotor resistance and a decreased variable speed range. Only recent turbines with variable speed equipped with full-scale converter (FSC) or doubly fed induction generator (DFIG) technology can meet the demands for dynamic reactive power.

6.4 Grid Requirements for Wind-Based Energy System 317

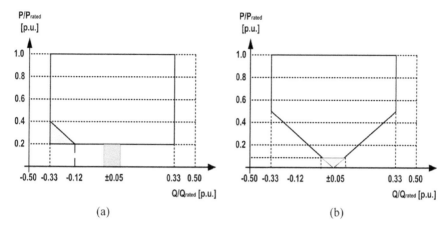

Fig. 6.11 PQ plane's Q criteria for **a** UK and **b** Ireland

6.4.3 Behaviour Under Grid Disturbances

Fault ride-through has started to contribute for improving power system security as a consequence of the increased integration of wind power in the past few years. It is essential that the generators remain connected in the case of a network issue. A significant interruption, such as a malfunction, might result in a voltage drop and an excessive loss of generation if the generators are unable to remain connected. This may disrupt stability and eventually cause other generators to trip in a cascade. The majority of wind turbine generators are built to shut down after a grid disruption before a fault ride-through requirement is introduced. There is always a concern that significant wind power penetrations might result in significant generation loss if wind turbines are disconnected due to grid disturbances, making the system unstable in an otherwise benign disturbance condition. Newly built wind turbines are made to conform to grid connection standards known as grid codes, which stipulate that wind turbines ride-through faults, in order to avoid such circumstances.

In order to connect to the electricity system's network, wind turbine owners have responsibilities that are stated in grid rules and regulations, and their turbines must adhere to certain technical criteria. The regulations further specify that the owner must protect their equipment from loss or damage brought on by internal or external factors, to regulate active and reactive power, to manage frequency, to maintain voltage quality, and to subject it to external control. The testing methods required to verify the wind turbines' ability to ride-through faults and behave in a grid fault are the main emphasis of the grid codes, which vary from one TSO to another.

As per Indian Electricity Grid Code IEGC 2010, during system failures, the wind generators must operate in a limited area. If the operational point drops below the line in the figure, wind farms may be disconnected.

$V_f = 15\%$ of nominal system voltage, V_{pf} = minimum voltages (80% of nominal system voltage). The following Table 6.16 provides the fault clearance time for various system nominal voltage levels.

Wind turbine generators (WTGs) in the wind farm must be able to fulfil the following specifications during fault ride-through:

- Must reduce the grid's reactive power use.
- As soon as the issue is fixed, the wind turbine generators must start producing electricity in accordance with the grid's remaining voltage.

When a voltage dip in the grid develops, wind farms connected to high-voltage transmission systems must remain connected; otherwise, the abrupt removal of a significant volume of wind power may cause a significant voltage dip and have unfavourable effects. When the voltage drop profile is above the line in the figure, wind farms must stay connected. The vertical axis displays the per unit voltage at the site of the grid connection, while the horizontal axis displays the fault's duration in seconds. This code mandates Fault Ride-Through (FRT) capacity during voltage reductions in the Transmission System to 15% of nominal voltage within 300 ms, with the recovery of up to 80% of nominal voltage after three seconds, with the slope depicted in the above-mentioned Fig. 6.12.

Table 6.16 Fault clearing time and voltage limits

Nominal voltage (kV)	Clearing time, T (ms)	V_{pf} (kV)	V_f (kV)
400	100	360	60.0
220	160	200	33.0
132	160	120	19.8
110	160	96.25	16.5
66	300	60	9.9

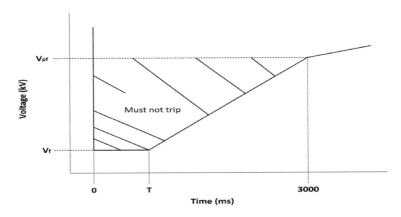

Fig. 6.12 FRT characteristics

6.4 Grid Requirements for Wind-Based Energy System 319

In the case of a zero-voltage grid breakdown, a wind turbine with a non-synchronous generator must stay connected to the grid for 140 ms, under the Scottish grid code (SB/2 2002). Wind turbines linked to transmission networks of 110 kV or above must meet the specifications set out by German transmission utility E.ON Netz. According to this grid rule, in the case of a three-phase short circuit causing a voltage dip of 85% for 150 ms with a voltage recovery to 80% in 3 s, wind turbines shall not be removed from the network. For wind turbines linked to grids with voltages below and above 100 kV, the Danish system operator defines different grid code requirements, and it also specifies the stability of wind farms in the case of asymmetric grid faults and failed reclosure.

6.4.4 Harmonic Requirements for Grid-Connected Wind Power System

The distributed power quality is one of the requirements of all standards pertaining to the systems integrated with the grid. In order to maintain electricity quality, demands are also placed on grid-connected WT systems. Power quality standards have mostly been designed to maintain the amplitude, frequency, and phase quality of the grid voltage waveform. Power fluctuations or system transient activity (such as at startup) are the primary causes of voltage waveform disturbances (due to the stochastic nature of the source) (Preciado et al. 2015). However, the quality of the current is also an issue, and in this regard, the grid converter is the only agent accountable for adhering to both the criteria set out by the transmission system operators and international guidelines and standards for power quality. Asynchronous and synchronous generators that are directly linked to the grid, especially in WT systems, are not constrained by current harmonics (Teodorescu et al. 2010).

The total harmonic distortion (THD) of the injected current in the grid shouldn't be more than 5%. Table 6.17, which applies to all distributed resource technologies with a combined capacity of 10 MVA or less at the point of common connection, provides a comprehensive account of the harmonic distortion with respect to every harmonic when connected to electrical power networks operating at standard primary and/or secondary distribution voltages, respectively (Schwanz 2018).

European guidelines for WT systems suggest applying the requirements for polluting loads, which call for the present total harmonic distortion to be less than

Table 6.17 Distortion limitations for distributed generating systems as a proportion of the fundamental

Odd harmonics	Distortion limit (in %)
3–9	Under 4.0
11–15	Under 2.0
17–21	Under 1.5
23–33	Under 0.6

320   6   Grid Integration Techniques in Solar and Wind-Based Energy Systems

Table 6.18 IEC standard sets distortion limitations for WT systems in terms of the fundamental percentage

Harmonics	Range (in percentage)
5th	5–6
7th	3–4
11th	1.5–3
13th	1–2.5

6–8% based on the kind of network. Table 6.18 provides a thorough description of the harmonic distortion for each harmonic. If several wind turbines are linked to the same PCC, the hth harmonic may be calculated as follows:

$$I_{h\Sigma} = \sqrt[\beta]{\sum_{i=1}^{N} \left(\frac{I_{hi}}{v_i}\right)^{\beta}}, \tag{6.10}$$

where β is 1 for $h < 5$, 1.4 for $5 < \beta < 10$ and 2 for $h > 10$. I_{hi} and v_i are the ith current and voltage at hth harmonic is the number of turbines connected to the PCC.

Example 4 Suppose there are three wind turbines connected to the same point of common connection (PCC). Calculate the total harmonic current at the 5th harmonic.

Ans: Given: Harmonic order: $h = 5$, Number of wind turbines connected to the PCC: $N = 3$,

Individual harmonic currents from each wind turbine at the 5th harmonic: $I_{hi} = 15$ A,

Individual harmonic voltages from each wind turbine at the 5th harmonic: $V_{hi} = 10$ V.

Since β lies in the range of 5–10 so $\beta = 1.4$,

Hence, $I_h = \beta \cdot \sqrt[\beta]{\sum_{i=1}^{N} \left(\frac{I_{hi}}{V_{hi}}\right)^{\beta}}$,

$I_h = 1.4 \cdot \sqrt{\left(\frac{15}{10}\right)^{1.4} + \left(\frac{15}{10}\right)^{1.4} + \left(\frac{15}{10}\right)^{1.4}}, I_h = 1.4 \cdot \sqrt{3 \times 1.761} = 1.4 \cdot \sqrt{5.283}$

$$I_h = 3.22.$$

6.5   Solar and Wind-Based Hybrid Renewable Energy Systems

By 2022, India will attain an important objective of achieving 175 gigawatts of installed green energy capacity, including sixty gigawatts of wind and 100 GW of solar power. To reach this goal, several policy actions have been implemented. The

6.5 Solar and Wind-Based Hybrid Renewable Energy Systems 321

nation's installed renewable power capacity was at almost 86 GW as of December 2019.

Conventional energy sources are no longer a feasible option due to their fast depletion to fulfil the daily rising load demand. Solar power can only generate electricity during the day, mostly between the hours of 8 am and 5 pm, due to its reliance on sunshine. On the other side, the wind is typically more prevalent in the late afternoon and evening hours, and highest at night. With a solar-wind hybrid, electricity output may be evenly distributed throughout the day because of the complementing intermittent characteristics of wind and solar. The grid's dependability is increased with a hybrid by ensuring peak power needs are satisfied (Engin 2013).

The National Wind-Solar Hybrid Policy's launch has been crucial in encouraging the conversion of current wind and solar plants into hybrids and in promoting hybrid plants in India. Additionally, the policy's goal is to maximize and enhance the effectiveness of the use of transmission land and infrastructure, which will reduce inconsistencies related to the production of green energy and aid in achieving improved grid stability.

The National Wind-Solar Hybrid Policy's launch has been crucial in encouraging the conversion of current wind and solar plants into hybrids and in promoting hybrid plants in India. Furthermore, the strategy seeks to maximize the efficiency with which transmission networks and land are used, so reducing the uncertainties associated with green energy power generation and contributing to the enhanced reliability of the grid.

6.5.1 Hybrid Energy Systems

There are several approaches to defining hybrid energy systems. The most general and maybe most beneficial is the following:

In order to overcome any possible limitations provided by either one, "hybrid energy systems" are composed of two or more energy conversion devices (such as electricity generators or batteries) or two or more fuels for the same equipment.

This definition is helpful since it covers a wide variety of scenarios and the crucial aspect of energy conversion diversity. It should be noted that this wide definition allows for transportation energy systems and does not need a device to be based on renewable energy.

For the sake of comparison, it is useful to quickly analyse the characteristics of traditional energy systems that are typically employed in situations where the hybrid system may be utilized instead. In general, there are three categories of conventional systems that are of interest: (1) large utility networks; (2) isolated networks; and (3) small electrical loads with a particular generator.

When used in combination with an isolated or special-purpose application, hybrid systems can be very valuable. As an example, consider the use of solar panels, battery storage, and power electronic converters to provide a small quantity of energy to a load in a remote area. Water pumping and desalination are two more instances.

Electrical loads in these applications can come in a variety of shapes and sizes. They might have variable voltage and frequency, or they could be typical AC or DC.

6.5.2 Hybrid Energy System Characteristics

The features and parts of a hybrid system are mostly determined by the application. The system's isolation from or connection to a centralized power grid is the most crucial factor to take into account (Manwell 2021).

6.5.2.1 Hybrid Systems Connected to the Central Grid

A hybrid system's design is simplified to some extent and its component number may be decreased if it is connected to a central utility grid, as in a DG application. This is so that the hybrid system is not required to regulate the voltage and frequency, which are set by the utility system. Furthermore, the grid usually delivers reactive power. When more energy is needed than the hybrid system can supply, the utility can often fill the gap. In the same way, the utility may take care of any surplus that the hybrid system generates. The grid does not always function as an endless bus, however. Then it is characterized as "weak." It could be necessary to add more controls and components. The grid-connected hybrid system will thus resemble an isolated one more closely.

6.5.2.2 Hybrid Isolated Grid Networks

Hybrid systems with an isolated grid are different from most systems with a central grid in a number of ways. They must first be able to supply the entire amount of energy needed on the grid at any one time or find a smart approach to reduce load when they can't. Both the voltage and the grid's frequency must be under their control. The latter requires that they are able to supply reactive power as demanded.

The two main categories of isolated grid hybrid systems are low penetration and high penetration. The quantity of power produced instantly by the renewable generator divided by the total amount of energy used is referred to as "penetration."

- A renewable generator's impact on the grid is negligible and no special equipment or control is required when the grid penetration is low (20% or less).
- When a renewable energy source has a high penetration, above 50% on average and up to 100% in some cases, it has a significant impact on the power system and will almost certainly require specialized machinery or government oversight. Supervisory control, so-called dump loads, short-term storage, and load management systems may all be included in high-penetration systems.

6.5 Solar and Wind-Based Hybrid Renewable Energy Systems

Table 6.19 Comparative analysis of hybrid versus standalone solar and wind plants

Definition	Integrated solar and wind power plants for having optimum output	Power generation systems that only rely on energy from the sun	Power plants get their energy only from the wind
Energy generated	Because solar and wind energy complement one another in nature, continuously. It decreases seasonal and everyday variations	High when the sky is clear throughout the day. The most productive hours are between 11 am and 4 pm	High in the rainy season, early in the morning, and at night
Transmission capacity utilization	More efficient utilization	Remains underutilized	Remains underutilized
Energy demand met	Even with energy restrictions based on the time of day, it can provide 75–80% of the energy needed	With the time of day restrictions and wheeling modifications, it can only meet 40–50% of the energy needs	Can meet up to 50% of energy needs, depending on how the wind blows and how long it can be stored

6.5.2.3 Isolated or Special Purpose Hybrid Systems

Some hybrid systems operate independently from a true distribution network to serve a specific function. These specific applications might consist of aeration, heating, water pumping, desalination, or powering grinders or other machines. System frequency and voltage management, as well as excess power output, are often not key concerns in the design of these systems. A more traditional generator may be offered in situations when electricity may be needed even when renewable sources are momentarily unavailable. In small isolated systems, renewable generators normally do not operate in conjunction with a fossil fuel generator. A Comparative Analysis of hybrid vs. standalone solar and wind plants has been shown in Table 6.19.

6.5.3 Technology Used in Hybrid Energy

The system's components will be chosen based on its functional and operational needs. A hybrid energy system may use a variety of techniques. Devices typically contain loads, rotating electrical equipment, power converters for renewable energy, control systems, energy storage devices, and load management equipment (Manwell 2021).

6.5.3.1 Energy Consuming Devices or Loads

Similar types of energy-consuming components are frequently used in hybrid energy systems as in conventional energy systems. These consist of electric appliances like lights, heaters, and motors. The total load, often known as load, is the sum of the energy requirements of all the devices. Typically, the load will change significantly during the day and year.

6.5.3.2 Rotating Electrical Machinery

In a hybrid energy system, rotating electrical machinery is present in several parts. Depending on the application, the majority of these devices have a motor or generator function. The generating function is highlighted in this section.

1. **Induction Generators**

The most popular generator design for wind turbines has historically been the induction generator. They are generally combined with other prime movers, for example, internal combustion engines that run on landfill gas or hydroturbines. But there are two crucial factors to take into account when talking about hybrid systems. Induction machines, in the first place, need a lot of reactive power. Although this is not a serious issue, it does have an impact on the system's architecture. The 2nd factor is the starting issue. Compared to when it is running continuously, an induction machine requires a lot more power to start up. It is necessary to make arrangements to guarantee that any associated induction motors or generators can be started by the hybrid system.

2. **Synchronous Generators**

A variety of prime movers can be employed with synchronous generators (SG). The two types of SGs used in hybrid system applications are those employing electromagnetic fields and those having permanent magnets. The first kind of SG utilizes a voltage regulator, which can keep the voltage of the electrical network in stable condition and supply reactive power needed by other system components. In small wind turbines, permanent magnet synchronous generators (PMSGs) are frequently employed. Such generators often work with power electronic converters since they can't maintain voltage. It is possible to activate a synchronous machine so that it can function alongside other generators. When doing this, special care must be given to ensure that all of the generators are in phase with one another.

3. **Renewable Energy Generators**

Generators for renewable energy are machines that convert raw energy from a clean source of energy into electricity. Photovoltaic panels and wind turbines are the renewable energy sources most frequently used in hybrid energy systems. Some hybrid energy systems employ fuel cells, biomass-fuelled generators, or hydropower generators. It must be mentioned that a lot of clean energy generators for non-conventional

6.5 Solar and Wind-Based Hybrid Renewable Energy Systems

energy generation have spinning electrical machinery that functions in the producing mode, also known as a generator.

A. Wind turbines

The energy in the wind is transformed into electricity by wind turbines. The rotor, the drive train, which includes the mainframe, the generator, the tower, the control system, and the foundation are the primary components of a wind turbine. The hub and blades are always mounted on the wind turbine rotor. The purpose of the blades is to transform the wind's force into a torque that ultimately powers the generator.

In a hybrid energy system, the two key components of a wind turbine are the types of generators and rotor controlling techniques. While some wind turbines employ synchronous generators, induction generators are the most common. Depending on the circumstances, a power electronic converter can either be used to link the generator directly to the electrical network or indirectly.

Stall control and pitch control are the two primary rotor control mechanisms used by wind turbines. Rotor control's primary purpose is to provide safety to the wind turbine against very high-speed winds. Rotor control is crucial in hybrid systems because it has an impact on how the hybrid system's energy flows are managed. The rotor can be adjusted under certain circumstances (with pitch control) to speed up a startup or to slow down production when full output is not required. A wind turbine's output fluctuates according to the wind's velocity. This relationship is shown in Fig. 6.13 as an example of a power curve.

B. PV Panels

PV panels are normally utilized to harness solar energy to produce electricity. A variety of solar cells are connected to form PV panels, which produce power at a

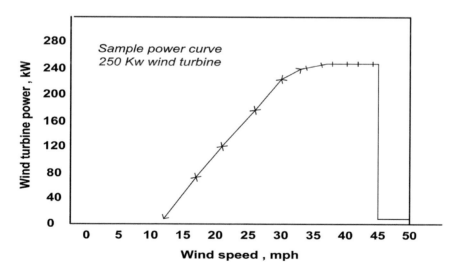

Fig. 6.13 Typical wind turbine power curve

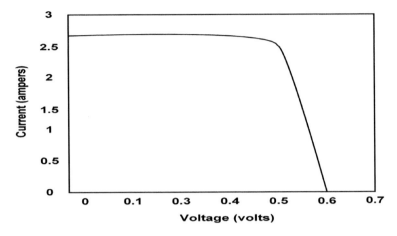

Fig. 6.14 Photovoltaic cell current versus voltage

certain voltage. By nature, solar panels have a DC structure. To produce AC, they need to use an inverter. In PV cells, crystalline silicon plays a key role.

The quantity of solar radiation influences how much current is produced by PV cells (up to a certain voltage). Figure 6.14 depicts the current/voltage relation of the silicon cell for a fixed value of solar radiation. As power is proportionate to the product of current and voltage, a PV cell's output will keep growing until the current starts decreasing.

A PV panel has many cells linked in series since each cell's maximum voltage is less than 1 V. Over a year and a day, the actual radiation level at any given location on the earth's surface changes continuously.

C. Energy storage

In hybrid energy systems, energy storage is often beneficial. There are generally two uses for energy storage. First of all, it can be applied to adjust a discrepancy between the demand and the source of clean energy. Second, it may be utilized to facilitate the entire system's operation and control. Convertible and end-use energy storage are the two main categories. Storage that is easily turned back into power is referred to as convertible storage. End-use storage is a resource that may be put to use for a particular end-use demand, but it may not be transformed back into energy.

Five key battery performance factors are crucial for hybrid energy systems: energy storage capacity, voltage, efficiency, charge/discharge rates, and lifespan of battery. By definition, batteries are DC. A battery is created by connecting many cells in series, with each cell having a nominal voltage of two volts. Typical whole battery voltages range from 2, 6, 12, or 24 V. The true value of the terminal voltage will vary depending on the battery's level of charge, charging or discharging rate, and whether the battery is being charged or drained.

6.5 Solar and Wind-Based Hybrid Renewable Energy Systems

The voltage of the battery and the quantity of charge it can retain before discharging are the main determinants of the energy storage capacity. Current multiplied by time is used to measure charge (Ampere-hours). The phrase "state of charge" is frequently applied to refer to the amount of charge that is held in a battery at any one moment with reference to its state of charge condition (SOC).

A cycle is the process of discharging and then recharging to a specific level (often completely charged). The quantity of materials utilized in a battery's construction essentially determines how much charge it can store overall.

A specific discharge rate is usually used to specify battery capacity. This is due to the fact that batteries' apparent capacities actually depend on the rate of charging and discharging. The outcome of increasing rates is smaller apparent capabilities. Batteries are not entirely effective as energy storage devices. Hence, charging consumes more energy than is available for recovery. Typically, total efficiencies fall between 50 and 80%.

The functional lifespan of batteries is a crucial aspect. According to studies, the habit of employing batteries degrades their storage capacity till they are no longer functional. Battery performance is greatly influenced by both the number of cycles and the depth of discharge (DOD) during each cycle. Depending on the kind of battery, a battery may sustain anywhere between a few thousand and hundreds or even tens of long cycles. A typical battery's cycle lifespan can be observed in Fig. 6.15.

Age, temperature, and maximum depth of discharge (DOD) all have an impact on battery capacity. The SOC of a battery is often represented in the form of a percentage, as follows:

$$SOC(t+1) = SOC(t) \cdot [1 - \sigma(t)] + [I_{bat}(t) \cdot \Delta t \cdot \eta_c(t)/c_{bat}] \quad \text{(During Charging)} \tag{6.11}$$

$$SOC(t+1) = SOC(t) \cdot [1 - \sigma(t)] - [I_{bat}(t) \cdot \Delta t \cdot \eta_{dis}(t)/c_{bat}] \quad \text{(During Discharging)} \tag{6.12}$$

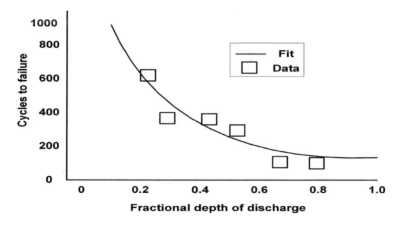

Fig. 6.15 Battery cycle life

with $(1 - DOD) \leq SOC(t) \leq 1$,

where $\sigma(t)$ denotes the hourly self-discharge rate which depends on the state of the battery but is generally considered as a fixed value of 0.02%. c_{bat} represents the battery nominal capacity measured in (Ah). Depending on the SOC and the charging current, the charge efficiency η_c ranges from 0.65 to 0.85, while the discharge efficiency η_{dis} is often maintained at high DOD causing battery problems like stratification, freezing, or saturation, which shorten the battery's lifespan. For this reason, DOD is often set between 50 and 80%. The hybrid system design establishes the battery current, which results from the energy balance between both the input power (wind + solar) and the output power (load). Depending on SOC, the battery voltage can be computed using simple or sophisticated models and parameters. The battery voltage can be considered constant up to a point. It is occasionally taken into consideration how the temperature affects the battery's rated capacity and the floating voltage.

D. AC/DC and DC/AC converters

A photovoltaic and wind hybrid system may employ a number of electrical converters, including:

- DC-to-AC converters or inverters or to supply an alternating current load (between charge regulator and load).
- AC to DC converters or rectifiers, which come after the windmill or engine generator.

The inverters have main three drawbacks:

1. More costly based on output signal characteristics (square, pseudo-sinus, or sinus).
2. Degradation in system stability (inverter performance based on load ratio self-consumption).
3. Failure possibility.

Over a wide range of outputs, advanced electronic inverters operate well. A standalone inverter's efficiency may reach 87% to 95% of two-thirds of its rated capacity, but when the power supply is below this range, efficiency drops quickly and can even drop below 50% at extremely low loads. The efficiency of the inverter will be low while performing at extremely low loads since an inverter needs some electricity to run by itself. There are several instances during the day in a typical home when the electrical load is quite low. One option is to use as many inverters as there are AC loads to serve; this increases the performance of each converter, improving overall system dependability but at a large expense to the system.

Several researchers believe that the mixed system, which splits the hybrid system into two other systems, i.e. a DC one for devices like lights, televisions, and radio, and an AC one for other devices, is the best choice. Any inverters in this system won't run until an AC load requires them to.

As a result, it is clear that choosing the right size for an inverter is essential because the inverter will be shut down if it has low capacity whereas it will not

6.5 Solar and Wind-Based Hybrid Renewable Energy Systems

perform effectively if its size is very large, also the cost will be more. Additionally, some inverters continue to run even when no charge is applied, resulting in a high level of self-consumption.

An inverter can create a square, pseudo-sinus, or sinus electrical signal depending on the kind of connected appliances, but its price can grow by four times for the same operating power as the quality of the signal and how well it performs. In many research articles, inverter performance is assumed to be consistent and between 90 and 95%, which, while occasionally high in comparison to available commercial statistics, varies depending on the load. It is preferable to utilize the curve between the performance of the inverter and load in any modelling when the load is variable.

The battery must be charged using a rectifier if the wind turbine output is alternating. A rectifier must be attached when using an auxiliary engine generator. Similar to an inverter, the efficiency of a rectifier is dependent on the AC power types, the rectifier type, the rectifying components types, and the proportion of the unit's load. The efficiency of a converter is often assumed to be a few per cent lower than the inverter's performance.

The WECS nominal power is used to compute the rectifier peak power of wind turbines. The peak power of an auxiliary generator rectifier is calculated using the highest battery charge current rate, which is at approximately 20% of the battery-rated capacity. When an engine generator is utilized, the inverter is occasionally substituted with an inverter charger that can convert AC from the generator to DC to charge the batteries as well as convert DC from the batteries to AC for the load. Manual or automated transfer switches can be used to go from one mode to another.

E. Energy management and control unit

This essential subsystem, which regulates the system ideally so that the load demand is continually supplied, passes through all energy flows. A wind-PV engine generating system can use a number of typical operating techniques including:

1. Direct alternating current can be sent to the load via the wind turbine, or more commonly, it transforms the AC to DC via a rectifier.
2. Solar and wind power plants directly deliver the load. If the energy produced by clean energy generators is greater than the amount needed, the remaining extra energy is stored in the connected battery; if the storage unit is completely full; the extra energy is wasted or put to other uses.
3. Additional power is drawn from the batteries if the overall amount of electricity produced by generators used for renewable is lower than the requirement.
4. The backup generator is turned on or the electrical load is cut, if batteries are unable to give this extra power. The backup generator may then either charge the battery solely (which is more usual) or charge the battery and power the load concurrently.

The control strategy for batteries manages how well batteries are charged and how effectively energy sources are used. It prevents both excess charging and discharging, which can both reduce a battery's lifespan. The PV array may be completely or partially removed generally when a battery's voltage and the overcharge protection voltage become approximately the same before the wind turbine is unplugged. If the voltage falls below the over-discharge voltage, the load is entirely or partially disconnected, or the electrical energy generated is absorbed by a coupled resistance. When the voltage becomes equal to the connected voltage, the load is once again supplied. When the voltage rises over the initial voltage (higher than the over-discharge voltage), the backup generator turns on, and it's turned off when the voltage drops below a certain point. The charge controller is more "intelligent" and effective in some systems; in addition to voltage, it also considers state of charge (SOC), temperature, and other factors.

Benefits of Solar and Wind Hybrid Systems

1. Optimal land utilization: To effectively utilize the space between wind turbines that is appropriately spaced apart to prevent row impacts.
2. Complementary features of resources: On a daily basis, wind and solar energy resources complement one another, with peak solar times corresponding with periods of lower wind resources and peak wind times occurring before dawn and after sunset, respectively.
3. Cost savings from common infrastructure: For projects with larger potential, O&M service facilities, shared data gathering systems, asset management, and common points of interconnection are advantageous for cost-effectiveness.
4. Comparable technological procedures: Both wind and solar power are generated from renewable resources and may be incorporated into local utility grids using standard AC or DC output.

6.5.4 Strategy for Implementation

MNRE asserts that various configurations and technological applications will affect how a wind-solar hybrid system is implemented (Jethani 2018).

a. Hybrid (wind-solar) AC integration: The alternating output of both the wind and solar systems is connected in this topology, either at the high tension (HT) or low tension (LT) side. In the second scenario, both systems make use of independent step-up transformers, and their HT outputs are linked to a single AC Busbar. To regulate the hybrid system's power output, the appropriate control equipment is used. A schematic diagram of integrated wind, solar, and battery topology diagram using AC-coupling has been shown in Fig. 6.16.

The only benefit of this strategy is that it works well with India's current transmission infrastructure.

b. Hybrid (wind-solar) DC integration: By utilizing a converter-inverter with a wind turbine with a variable speed motor, DC integration is feasible. In this setup, a common DC bus connects the DC output of the wind and solar PV units, and the DC electricity is converted to AC power using a common inverter that can handle the combined output AC capacity. A schematic diagram of wind, solar,

6.5 Solar and Wind-Based Hybrid Renewable Energy Systems

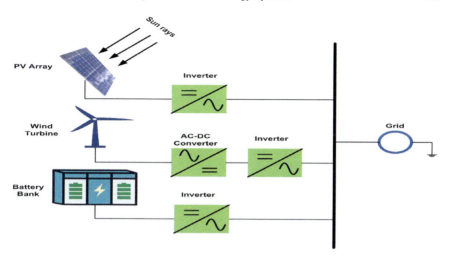

Fig. 6.16 Integrated wind, solar, and battery topology diagram using AC-coupling

and battery integration diagram using DC-coupling topology has been shown in Fig. 6.17.

The benefits of the DC-coupling topology include simpler hardware, reduced costs, and greater energy efficiency.

Selection of Site

Attempt to utilize as much of the land as possible. Nevertheless, wind and solar resources should show complementary output peaks on a daily or annual basis. Wind and solar resources might be studied individually for various locations or in the same location.

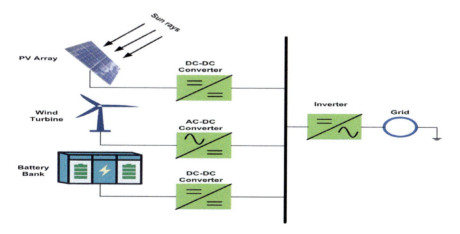

Fig. 6.17 Wind, solar, and battery integration diagram using DC-coupling topology

Hybrid Power Plant Sizing

The fundamental goal of optimization is accomplished by reducing system costs while preserving dependability.

To acquire the optimum results from the combination, sizing is carried out using mathematical models. PV-wind hybrid power system size is heavily influenced by the number of solar panels, wind turbines, batteries, load profiles, and accessible renewable resources. Several academics have investigated the algorithm to determine the capacity of appropriate generator units that may make up a dependable power system at a low cost in order to use renewable energy cost-effectively. It has been stated that a variety of sizing techniques, including iterative and artificial intelligence approaches, may be used to create a hybrid clean energy system that is both technologically and economically optimal (Singh Sandhu and Mahesh 2018).

A limit set by the minimum SOC may be discharged by the battery bank with the complete nominal capacity. Knowing the starting SOC, the charge or discharge time, and the current is essential for determining the true SOC of a battery.

Considerations for Energy Simulations

- Solar PV panel generation is affected by shading. It must be measured in some way. In PV arrays, shading will result in mismatch losses.
- It is necessary to quantify the roughness brought on by solar panels.
- Due to PV arrays, the actual hub height of turbines will vary. Using the displacement height principle, this may be resolved.

Disadvantage:

Losses in Wind Farm

- Availability issues (electrical system).
- Wake effects.
- Electrical (component efficiencies, wiring).
- Turbine efficiency (power curve, operation, inclined flow, high-wind hysteresis).
- Restrictions (environmental, directional, operational, PPA).
- Climatical problems (blade deterioration, temperature shutdown, site constraints, thunderstorms).

Losses in Solar PV Farm

- Solar energy transition (quality degradation, mismatch).
- Efficient radiation (reflection, soiling, and shading).
- Electrical (efficiencies of components and cabling).
- Operating issue (reduction, accessibility).

Uncertainty Hybrid Energy (Mutual Interactions)

Along with individualized uncertainty, the following reciprocal interactions between both sources also need to be taken into account:

Wind Variability (Influenced by Solar Panels)

(a) Uncertain wake loss (due to PV panels).
(b) Uncertain availability loss (because accessing cranes around solar panels is more challenging).

Solar Uncertainty (Influenced by wind turbines)

(a) Mismatch uncertainty induced by shading.
(b) Uncertainty of shading loss in turbines.
(c) Shadow flickering causes uncertainty regarding inverter operation.

6.5.5 Constraints

1. Power instabilities in the grid.
2. Durability limitation.
3. PV panels make turbines physically observable.
4. Battery limitations—The batteries in the system require the greatest attention because they have the shortest lifespan of all the system's components. This approach makes sure that the battery is never totally charged or depleted by maintaining the SOC within the lowest (SOC_{min}) and maximum (SOC_{max}) limits.

A solar-wind hybrid policy was published by the Ministry of New and Renewable Energy (MNRE) in 2018. It offers a framework for promoting grid-connected hybrid power through arrangements that make optimal use of land and transmission infrastructure and, to some extent, control the variability of renewable resources. Almost 50 hybrid projects of megawatt level have either been announced or are being built internationally, with Australia and the USA leading the way. The development of hybrid projects is not limited to India. Potential solutions for the electrical issues, particularly in rural India, might include hybrid energy systems. Nevertheless, in order to make this topic technically practicable and deployable, an additional in-depth study is required (Badwawi et al. 2015).

In the long run, renewable hybrids have the potential to cut energy production costs and hasten India's decarbonization of power generation. Yet, it is necessary to weigh the benefits and drawbacks of developing new coal-fired power plants and hybrid renewable energy sources. Investors must also comprehend the potential and prospective worth of strained coal-fired facilities five to seven years from now when appraising them. Most crucially, for India to fully use the promise of hybrid energy, the governmental and regulatory climate must advance.

6.5.6 Issues and Challenges

Wind-solar hybrids are better than wind and solar by themselves. However, developers in this market are having trouble with things like lower tariffs, uncertainty about policy, and technical problems. Some important issues and problems are:

- Land issues: Adding new wind capacity is hard because most of the useful sites with high-wind potential and grid availability are already completed. India's renewable energy sector has always had trouble getting enough land. India also has a big chance to "repower" old wind farms in good places, especially in the state of Tamil Nadu.
- Integration challenges: On the DC side of the grid, it is hard to connect both wind and solar power. According to MNRE policy, only AC integration is allowed until the DC metering framework is in place. This cuts down on the cost benefits of DC integration in terms of how the balance-of-system is used (BOS).
- Lack of experience: Due to a lack of skilled resources, there isn't enough experience with hybrid plants right now.
- System sizing: The size of the plant needed to make the best use of the generation portfolio varies from site to site based on the amount of wind and sun that can be used. A key question is also what the best size for storage is. A high-capacity installation can lead to almost no use of storage on days when renewable energy is producing a lot of power.

As the sector grows, most of these problems can be solved by putting in place policies and standards that make more sense. The technical problems can be fixed by making sure the hybrid systems are well-designed, have fast-response control systems, and are well-optimized. WSH will get the necessary traction if policy changes are made in conjunction with the elimination of technological inefficiencies.

6.5.7 Applications

Solar-Wind Hybrid Energy Systems (SWHES) are used in practically all fields that use small amounts of electricity. Here are some of the ways SWHES can be used (Chandramouly and Raghuram 2017):

- *Standalone:* Nearly, all SWHES apps operate independently without connecting to the grid.
- *Grid-connected:* The increased power rating of SWHES allows them to be linked to the grid in areas with greater availability of wind and sunlight. If the system fails to produce electricity in these types of operations, the grid will provide the load.
- *Household:* Electricity produced by a hybrid wind and solar plant is used for domestic appliances. Electricity is reliably supplied to various offices or other areas of the building using SWHES.

- *Street lighting:* Solar street lighting is the primary use of SWHES. SWHES illumination is being used in solar streetlights. Using this reduces the demand placed on traditional power plants.
- *Remote Applications:* These SWHES systems are helpful for defence forces where it is hard to supply conventional power.
- *Power Pump:* Almost any building's water supply may be pumped with the use of SWHES. A pump powered by DC may move water around your residence.
- *Ventilation system:* The suggested systems can be utilized to operate bathroom fans, ceiling fans, and floor fans, in buildings for ventilation reasons.
- *Onshore:* SWHES are erected near the water for power generation since the wind blows more often in coastal locations.
- *Village Power:* In villages located in very lower or higher regions, where it is impossible to deliver power, the suggested system is particularly helpful.
- *Commercial:* Hotels and other tourist destinations provide the necessary electric power.

6.5.8 Hybrid System Economics

The system price and the amount of usable power generated are the two main aspects of the energy cost of a hybrid system. Other important factors are the energy's value, the cost of fossil sources, the system's lifespan, maintenance costs and financial costs. The price of the separate parts that make up a hybrid energy system has the most impact on the system's price. The price will also include installing the parts and putting them together into a functional device. Generators for green sources are frequently the most expensive ones.

Energy value in a hybrid energy system depends on the type of energy generated, the cost of the alternatives, and the viewpoint of the system operator. For instance, a diesel power system operator on an island who is thinking about building a wind turbine is generally interested in the turbine cost that would be compensated through the lower cost of diesel fuel (Hassan et al. 2023).

Life cycle costing is a method used to examine the economics of hybrid energy systems. This strategy takes into account that hybrid systems have installation costs that are comparatively high but are robust and have low operating costs over lengthy lifespans. These components are used to estimate present value costs, which are then evaluated using a range of financial measures and compared with costs from non-renewable options (Manwell 2021).

Review Questions

1. What are the major issues with the integration of wind energy into the grid network?
2. What is the integration of solar energy into the grid?
3. What are the challenges for grid integration?
4. What is the impact of a wind power plant on the grid?

336   6 Grid Integration Techniques in Solar and Wind-Based Energy Systems

5. What is the objective of grid integration?
6. What is a renewable microgrid?
7. What is the impact of large-scale integration of renewable energy sources on frequency?
8. Throw some light on control technique methods used for wind turbines.
9. Discuss applications of the hybrid renewable energy system.
10. What are the main factors to be considered for reactive power components (RPC) in a system?

Problems

(1) Determine the voltage drop through impedance when the RMS voltage at the grid is 120 kV, the RMS voltage at the point of common connection is 115 kV, the transmission line impedance is 0.15 ohms, and the transmission line current is 400 A.

Ans: 4.94 kV

(2) Calculate the wind connection's short circuit power when the impedance is 0.25 ohms and the transmission line current is 550 A.

Ans: 75.625kVA

(3) Determine the change in impedance when the harmonic order is 5, the inductive reactance is 0.03 ohms, and the fundamental frequency is 60 Hz.

Ans: 9 ohm

(4) Consider a scenario where four wind turbines are connected to the same point of common connection (PCC). Calculate the total harmonic current at the 8th harmonic.

Ans: 3.394 A

Objective Type Questions

1. What do you mean by wind energy integration to the grid?

    a. Wind power plant physical connection to the grid
    b. Organizing the wind power plant's grid connection
    c. Compilation of all operations associated with wind farm grid connection
    d. Energy sent from the grid to run the wind turbines

2. Which of the following accurately illustrates the steps needed in integrating wind energy into the grid?

    a. Planning $\rightarrow$ system operations $\rightarrow$ physical connection
    b. System operations $\rightarrow$ physical connection $\rightarrow$ planning
    c. Physical connection $\rightarrow$ planning $\rightarrow$ system operations
    d. Planning $\rightarrow$ physical connection $\rightarrow$ system operations

3. What are the two categories of grid integration planning activities?

    a. Both systemic and project-specific

6.5 Solar and Wind-Based Hybrid Renewable Energy Systems

    b. Rotor and shaft
    c. BJT and MOSFET
    d. Designs for low- and high-power systems

4. Plans for network-wide integration of the wind energy grid include

    a. Materials used to manufacture wind turbine blades
    b. System impact analyses carried out for a particular wind project
    c. Solar panel manufacturing plant
    d. Developing grid code

5. What does project-specific planning in the integration of the wind energy grid mean?

    a. Planning actions for a solar panel installation
    b. Planning activities for all future wind power plants
    c. Planning activities related to a unique wind project
    d. Planning actions for maintaining the current thermal power plants

6. Which of the following are the impacts of grid integration of wind power plants?

    a. Economic dispatch and day-ahead unit commitment processes
    b. Unit commitment process day-ahead
    c. Rotor and shaft evaluations
    d. Generator types

7. Which of the following best practices for integrating wind power into the grid?

    a. Unit commitment process day-ahead
    b. Economic dispatch and day-ahead unit commitment processes
    c. Optimum transmission from resource-rich locations to load, flexible generation
    d. Rotor and shaft tests

8. What is a grid code for wind energy integration?

    a. Hexadecimal code
    b. Binary code
    c. Rulebook having generator properties
    d. Conduct code

9. Interactive PV systems operate:

    a. In parallel without grid
    b. In parallel with the grid.
    c. Standalone
    d. None

10. The initial cost of PV systems is:

    a. Medium
    b. High

338   6   Grid Integration Techniques in Solar and Wind-Based Energy Systems

    c. Low
    d. None

11. Energy generation of photovoltaic systems mainly depends on:

    a. Weather
    b. Location
    c. Demand
    d. Both a and b

12. The running cost of photovoltaic systems is:

    a. Medium
    b. High
    c. Low
    d. None

13. To ensure that your PV system will work all day you should use:

    a. Inverter
    b. Battery
    c. Converter
    d. none

14. The photovoltaic system life is:

    a. Short
    b. medium
    c. Long
    d. None

15. The photovoltaic system efficiency in general is:

    a. Low
    b. Medium
    c. High
    d. None

16. Which of the following are the steps involved in designing a standalone photovoltaic system?

    a. Estimation of load
    b. Estimation of solar power
    c. Selection of inverter and size of the battery bank
    d. all

Answers

1. c	2. d	3. a	4. d	5. c	6. a	7. d	8. c
9. b	10. b	11. d	12. c	13. b	14. c	15. a	16. d

References

Badwawi RA, Abusara M, Mallick T (2015) A review of hybrid solar PV and wind energy system. Smart Sci 3:127–138. https://doi.org/10.1080/23080477.2015.11665647

Chandramouly M, Raghuram A (2017) Introduction to solar wind hybrid energy systems. Int J Eng Res Electr Electron Eng 3:2395–2717

DIN V VDE V 0126-1-1 VDE V 0126-1-1:2006-02—Standards—VDE Publishing House. https://www.vde-verlag.de/standards/0126003/din-v-vde-v-0126-1-1-vde-v-0126-1-1-2006-02.html

Engin M (2013) Sizing and simulation of PV-wind hybrid power system. Int J Photoenergy 2013:1–10

Hassan Q, Algburi S, Sameen AZ, Salman HM, Jaszczur M (2023) A review of hybrid renewable energy systems: solar and wind-powered solutions: challenges, opportunities, and policy implications. Results Eng 20. https://doi.org/10.1016/j.rineng.2023.101621

Jethani JK (2018) National wind solar hybrid policy

Manwell JF (2021) Hybrid energy systems

Nirosha C, Kumar Patra D (2020) Power quality issues of wind and solar energy systems integrated into the grid

Preciado V, Madrigal M, Muljadi E, Gevorgian V (2015) Harmonics in a wind power plant. IEEE power energy society general meeting 2015-September. https://doi.org/10.1109/PESGM.2015.7285774

Rajapakse A, Muthumuni D, Perera N, Rajapakse A, Muthumuni D, Perera N (2009) Grid integration of renewable energy systems. Renew Energy. https://doi.org/10.5772/7375

Schwanz D (2018) Harmonics and wind power

Singh Sandhu K, Mahesh A (2018) Optimal sizing of PV/wind/battery hybrid renewable energy system considering demand side management. Int J Electr Eng Inform 10. https://doi.org/10.15676/ijeei.2018.10.1.6

Singh B, Singh SN (2011) Development of grid connection requirements for wind power generators in India. Renew Sustain Energy Rev 15:1669–1674. https://doi.org/10.1016/J.RSER.2010.11.026

Stompf R (2020) 7 major challenges of a power grid and their solutions | FUERGY. https://fuergy.com/blog/7-problems-and-challenges-of-a-power-grid

Teodorescu R, Liserre M, Rodríguez P (2010) Grid converters for photovoltaic and wind power systems. Grid Convert Photovolt Wind Power Syst.https://doi.org/10.1002/9780470667057

Chapter 7
Solar Collectors and Thermal Conversion

Abstract This chapter is useful for comprehending the ideas, layouts, and operational features of different solar collectors and thermal conversion systems, which advance the use of solar energy. It starts with a summary of solar alternatives divided into systems for low, medium and high temperatures followed by systems for thermal collection and storage before diving into solar collectors and their function in thermal conversion. The orientation, tracking modes, performance analysis, and correlations between overall loss coefficients for the cylindrical parabolic collectors are also looked at.

Keywords Thermal temperature system · Solar collectors · Orientation · Tracking modes · Receiver

7.1 The Solar Option

Solar is a very large, inexhaustible source of energy. The power from the sun intercepted by the earth is approximately 1.8×10^{11} MW which is many thousands of times larger than the present consumption rate on the earth of all commercial energy sources. Thus, in principle, solar energy could supply all the present and future energy needs of the world continuously. This makes it one of the most promising of unconventional energy sources.

In addition to its size, solar energy has two other factors in its favour. First unlike fossil fuels and nuclear power, it is an environmentally clean source of energy. Second it is free and available in adequate quantities in almost all parts of the world where people live.

However, there are many problems associated with its use. The main problem is that it is a dilute source of energy. Even in the hottest regions on earth, the solar radiation flux available rarely exceeds 1 kW/m^2 and the total radiation over a day is at best about 7 kWh/m^2. These are low values from the point of view of technological utilization. Consequently, large collecting areas are required in many applications and this results in excessive costs.

© The Author(s), under exclusive license to Springer Nature Singapore Pte Ltd. 2024
K. Namrata et al., *Wind and Solar Energy Systems*, Energy Systems in Electrical Engineering, https://doi.org/10.1007/978-981-99-9710-7_7

A second problem associated with the use of solar energy is that its availability varies widely with time. The variation in availability occurs daily because of the day–night cycle and also seasonally because of the earth's orbit around the sun. In addition, variations occur at a specific location because of local weather conditions. Consequently, the energy collected when the sun is shining must be stored for use during periods when it is not available. The need for storage also adds significantly to the cost of any system. Thus, the real challenge in utilizing solar energy as an energy alternative is economic. One has to strive for the development of cheaper methods of collection and storage so that the large initial investments required at present in most applications are reduced.

A broad classification of the various methods of solar energy utilization is shown in Fig. 7.1. It can be seen that the energy from the sun can be used directly and indirectly. The direct means include thermal and photovoltaic conversion, while the indirect means include the use of water (hydroelectric) power, the winds, biomass, wave energy, and temperature differences in the ocean and marine currents. The last three, viz. wave energy, the energy available from temperature differences in the ocean, and energy in marine currents, along with tidal energy (which is listed later under miscellaneous sources) are often grouped under the broad heading of ocean energy.

Solar thermal power cycles can be classified as low-, medium-, and high-temperature cycles. Low-temperature cycles work at maximum temperatures of about 100 °C, medium-temperature cycles work at maximum temperatures up to 400 °C, while high-temperature cycles work at temperatures above 400 °C.

Low-temperature systems use flat-plate collectors or solar ponds for collecting solar energy. Systems working on the solar chimney concept have also been suggested. Medium-temperature systems use the line-focusing parabolic collector technology. High-temperature systems use either paraboloid dish collectors or central receivers located at the top of towers.

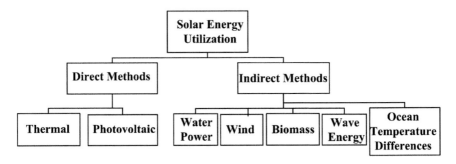

Fig. 7.1 Classification of methods for solar energy utilization

7.1.1 Low-Temperature Systems

Figure 7.2 represents a diagram of a typical low-temperature system with flat-plate collectors and a Rankine cycle. The energy of the sun is collected by water flowing through the array of flat-plate collectors. Booster mirrors that reflect radiation onto the flat-plate collectors are occasionally employed to obtain the highest temperature feasible. In a thermal storage tank that is well-insulated, the hot water is kept at temperatures that are near to 100 °C. From here, it passes through a vapour generator, where the Rankine cycle's working fluid is also passed. The boiling point of the working fluid is low. As a result, the vapour at about 90 °C and a pressure of a few atmospheres leaves the vapour generator. By passing via a prime mover, a condenser, and a liquid pump, this vapour then performs a standard Rankine cycle. Toluene and methyl chloride are examples of organic working fluids, and refrigerants like R-11, R-113, and R-114.

The overall efficiency of this system isn't very high because there isn't much difference in temperature between the vapour coming out of the generator and the condensed liquid coming out of the condenser. The temperature difference for the cycle represented in Fig. 7.2 is just 55 °C. This results in a Rankine cycle efficiency of 7–8%. The collector system's efficiency is in the range of 25%. Consequently, only around 2% of total efficiency is achieved.

In the 1970s, French-designed plants of this kind with a generating capacity of up to 50 kW were constructed in many parts of the world, especially in Africa. By an Indo-German cooperative agreement, a 10-kW plant was also installed at IIT Madras in 1979–1980. However, due to the substantial collector areas required, it has been discovered that such plants are highly expensive. The collectors account for the majority of the installed cost, which is typically around Rs 300,000 per kW for 6–8 h of daily operation.

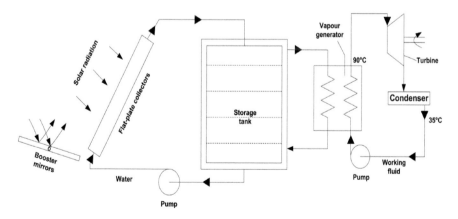

Fig. 7.2 Liquid flat-plate collectors for low-temperature energy generation

Solar ponds have been used in place of flat-plate collectors to cut costs. Israel built the first two solar pond power plants, with a capacity of 6 and 150 kW, roughly 30 years ago. These were followed in 1984 by the biggest plant in the world, which produced 800 kW continuously, at Beit Ha'aravah. The operation of these facilities proved beyond a doubt that solar pond power plants are technically feasible. Despite being less expensive than plants that use flat-plate collector systems, they were also not considered to be economically viable.

In the 1970s, the idea of a solar chimney power station was proposed. A solar updraft tower power plant is another name for it. In such a plant, a circular greenhouse made of a transparent cover held a few metres above the ground by a metal frame surrounds a tall central chimney at its base (Fig. 7.3). The air within the greenhouse warms up by 10–20 °C as sunlight enters through the transparent cover. Thus, a convection system is created in which the heated air is taken up via the central chimney and is continually refilled by new air drawn in at the greenhouse's perimeter. The energy contained in the updraft air is turned into mechanical energy by turbines positioned at the base of the chimney, which is subsequently transferred into electrical energy by standard electrical generators.

In Manzanares, Spain, an experimental pilot plant was the only solar chimney power plant constructed so far. It featured a chimney that was 10.3 m in diameter and 195 m high. With the glass being 1.85 m above the ground, the solar collector area had a radius of 122 m from the chimney. The entire glazing surface area was 46,000 m^2, of which 6000 m^2 was made of glass and 40,000 m^2 of plastic membrane. The turbine, housed at the base of the chimney, had four 5-m-long blades and rotated at 1500 rpm to produce a peak amount of 50 kW. The plant's construction was finished in 1982, and it remained in operation until 1989. The objective was to illustrate the working principle and gather operational information and experience over an extended period. The plant was continuously operated daily basis for 32 months, from mid-1986 to

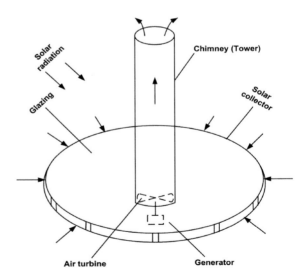

Fig. 7.3 Solar chimney power plant

7.1 The Solar Option

early 1989. It produced 44 MWh per year while operating entirely autonomously for an average of 8.9 h per day throughout this time. The results obtained showed that the plant as a whole was capable of reliable operation over extended periods.

A solar chimney's energy conversion efficiency is inherently low. The expression (Schlaich et al. 2005) may be used to demonstrate that it gives the highest feasible efficiency:

$$\eta_{max} = \frac{gH}{C_p T_a}, \tag{7.1}$$

where $g = 9.81 \, \text{Nm}^2/\text{Kg}^2$,

$H =$ Chimney tower height,

$T_a =$ Ambient air temperature,

$$C_p = 1005 \, \text{J/kg K}.$$

Equation (7.1) shows a basic characteristic of a solar chimney, viz. that the efficiency is directly proportional to the height. The formula may be used to calculate the overall efficiency of a solar chimney power plant.

Example 7.1 It is proposed to set up a solar chimney power plant in Rajasthan with a chimney 300 m high. Calculate the maximum possible conversion efficiency obtainable with the chimney. Also estimate the efficiency of the plant as a whole and the daily electrical output in a typical summer month (in kWh), if the solar collection area of the greenhouse is 50,000 m^2.

Solution We will use Eq. (7.1). Substituting $H = 300 \, \text{m}$, $C_p = 1005 \, \text{J/kg K}$, and $T_a = 305 \, \text{K}$, we get

$$\eta_{max} = \frac{9.81 \times 300}{1005 \times 305} = 0.00960, \quad \text{i.e.} \, 0.96\%.$$

Assume:

1. The turbine generator set converts only 50% out of the maximum available energy into electrical energy, and
2. The collection efficiency of the greenhouse is 25%.

Then

$$\eta_{overall} = 0.25 \times 0.00960 \times 0.50 = 0.0012, \quad \text{i.e.} \, 0.12\%.$$

The solar radiation on a typical summer day in May or June in Rajasthan may be taken as 6.5 kWh/m^2. Therefore, the daily electrical output of the plant

$$= 6.5 \times 50,000 \times 0.0012 = 390 \, \text{kWh}.$$

The overall energy conversion efficiency of a solar chimney power plant is poor, as demonstrated by Example 7.1. Because of this fact, a small-capacity plant (in the kilowatt range) is not economically viable. To achieve economy with scale-up, one has to construct large-capacity plants with tall chimneys and extensive collection areas at the ground level. The construction of a 200 MW plant in Australia has been developed. The design concept shows a chimney that is 1 km tall and a greenhouse collector having 7 km in diameter at a base (EnviroMission 2006).

7.1.2 Medium-Temperature Systems

The most efficient and productive solar thermal-electric power plants to date have operated on medium-temperature cycles and used the line-focusing parabolic collector technology at temperatures about 400 °C. Figure 7.4 displays a schematic layout of a typical plant. In 1984, a 14 MW solar electric generating system known as SEGS I was built as the first plant of its kind to be used commercially. Six plants with a capacity of 30 MW each have since been installed and commissioned (SEGS II–VII) and they were followed by two plants with a capacity of 80 MW each (SEGS VIII and IX). All these plants were built in California by LUZ International, which has a 354 MW installed capacity overall. A 483,960 m^2 collector array is used for SEGS IX.

The axes of the utilized cylindrical parabolic collectors are pointed north–south. The absorber tube in use has a specifically designed selective coating and is made of

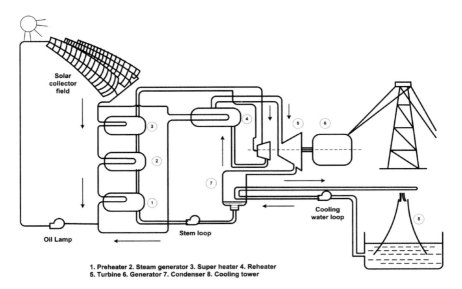

Fig. 7.4 Medium-temperature power generating cycle with cylindrical parabolic concentrators

stainless steel. There is a vacuum and a glass cover around it. With a collector efficiency of around 0.7 for beam radiation, the collectors heat synthetic oil (Dowtherm A) to a temperature of 390 °C. Utilizing a Rankine cycle with a 38% efficiency, superheated high-pressure steam is generated using synthetic oil. The plant generally produces electricity for about 8 h a day and is coupled with natural gas for continuous operation. As installed capacity has increased, the cost of installing this kind of plant has decreased over the years. However, it is very high. According to reports, SEGS VIII, which started operating in 1990, costs $4000 per kW. Since 1991, no new large-size plants have been erected due to high capital costs. Despite this, SEGS I–IX has continued to operate and has accumulated significant operational experience spanning more than 15 years.

Significant advancements in the organic Rankine cycle and parabolic collector technologies have been accomplished since the construction of the SEGS plants. As a result, lower-capacity plants are now more economically viable. Consequently, in 2006, the Arizona Public Service Company built a 1 MWe plant at Red Rock, Arizona, in the United States. The plant has a line-focusing parabolic collector array with an area of 10,346 m^2 which heats a thermic fluid to a temperature of 300 °C. To run an organic Rankine cycle, which uses n-pentane as its working fluid, the heat stored in the thermic fluid is utilized. At the turbine's inlet, the n-pentane vapour is at a pressure of 22.3 bars and temperature of 204 °C, producing a cycle efficiency of 20.7%. The plant produces 2000 MWh yearly as a whole. In addition to the Arizona plant, a 64 MWe solar power plant using parabolic collectors was commissioned in Boulder City, Nevada, USA, in June 2007.

The Indian experience with the line-focusing parabolic collector technology has been restricted so far to a small 50 kW capacity experimental plant installed at the Solar Energy Centre of the Ministry of New and Renewable Energy. Additionally, a project report was prepared for the installation of a 30 MW parabolic collector power plant in Rajasthan. However, the initiative has not been implemented.

Example 7.2 Make suitable assumptions and estimate the collector area required for an 80 MWe line-focusing solar thermal power plant producing electricity for 8 h every day.

Solution We will assume the following assumptions:

1. The collector is operating at a temperature of 400 °C. At this temperature, the collector efficiency may be taken as 0.6.
2. The Rankine cycle has an efficiency of 0.36.
3. The electrical generator efficiency is 0.96.
4. The solar insolation during a typical day is 6 kWh/m^2.

An energy balance over a day yields the following:

Solar radiation incident on the collectors per unit area per day × Collector area × Collector efficiency × Rankine cycle efficiency × Generator efficiency = Electricity production × Hours per day

$$6 \times A_c \times 0.6 \times 0.36 \times 0.96 = 80,000 \times 8$$
$$A_c = 514403 \, \text{m}^2.$$

(Note: This value agrees reasonably well with the area of 483,960 m^2 used in SEGS IX.)

7.1.3 High-Temperature Systems

In the case of high-temperature systems, two concepts have been tested. These are the central receiver and the parabolic dish concept.

In the paraboloid dish concept, the concentrator tracks the sun by rotating about two axes and the sun's rays are brought to a point focus. Heat is generated by a fluid passing through a receiver at the focus, and a prime mover is powered by this heat. Stirling engines have often been preferred as the prime movers. Therefore, these systems are referred to as Dish-Stirling Systems. Paraboloid dish systems are anticipated to produce electricity in kilowatts rather than megawatts due to limitations on the size of the concentrator. Thus, it is believed that they will satisfy the local power requirements of communities, particularly in rural locations.

Table 7.1 provides some information on four systems developed for commercial use. We can observe that the peak outputs range from 8.5 to 25.3 kW and the peak net efficiencies from 19 to 29.4%.

Cost and reliability are two main challenges for commercializing Dish-Stirling systems. Currently, the cost of installing such systems is quite expensive, costing around $10,000 per kW. This is as a result of the lack of systems developed. However, it is predicted that if 500 units are produced annually, the cost of a system might drop below $2500 per kW.

Table 7.1 Paraboloid dish systems (Mancini et al. 2003a)

System name	Sun Dish	Euro Dish	SES	WGA
Location	USA	Spain	USA	USA
Rated output (kW)	22	10	25	9.5
Peak output (kW)	22.9	8.5	25.3	11.0
Peak net efficiency (%)	20	19	29.4	24.5
Area of reflecting glass (m^2)	117.2	60.0	91.0	42.9
Receiver aperture diameter (m)	0.38	0.15	0.20	0.14
Power cycle	Stirling	Stirling	Stirling	Stirling
Working fluid	Hydrogen	Helium	Hydrogen	Hydrogen

Adapted from Mancini et al. (2003a)

In a central receiver system, solar radiation reflected from an array of large mirrors is concentrated on a receiver situated at the top of a supporting tower. These mirrors, known as heliostats, are positioned on the ground surrounding the tower. They each have a regulated orientation, which causes them to reflect beam radiation onto the receiver all day long. The heat is transported to the ground level and utilized to power a thermodynamic cycle like the Rankine or the Brayton cycle by a fluid that is running through the receiver that absorbs the focused radiation. The heat transfer fluids have included molten salts, water (converted into steam), and air. Because of the use of a receiver placed at the top of a tower, a central receiver system is also referred to as a power tower.

Figure 7.5 is a schematic representation of a typical central receiver system that uses molten salt as the heat transfer fluid. The commonly used molten salt is a blend of 60% sodium nitrate and 40% potassium nitrate. Pumped from a tank at ground level to the receiver at the top of a tower, cold salt at 290 °C is heated to 565 °C by concentrated radiation. The salt flows back to another tank at ground level. Hot salt is pushed from the hot tank via a steam generator, which creates superheated steam, to produce power. Following a Rankine cycle, the superheated steam generates mechanical work first, followed by electricity. It is possible to size the heliostat array to capture more energy than the electricity producing system needs. In such situation, the hot tank functions as a thermal storage by accumulating the extra thermal energy in the form of excess salt at 565 °C.

Nine central receiver systems created in the 1980s and 1990s are listed in Table 7.2. They were all constructed as pilot plants with electrical outputs ranging from 0.5 to 10 MW with the intention of proving the feasibility of the concept involved. A few details are given in Table 7.2. These include the size and number of the heliostats, the type of receiver, the fluid used in the receiver, the type of thermal energy storage, and the height of the tower.

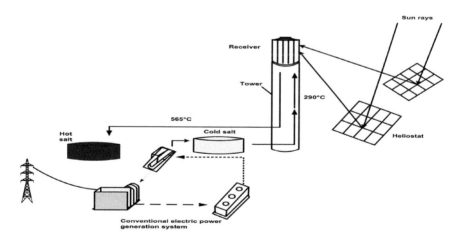

Fig. 7.5 Central receiver system using a molten salt as the heat transfer fluid (Romero et al. 2002)

Table 7.2 Solar central receiver power plants (Romero et al. 2002)

Plant name	SSPS	Eurelios	CESAI	Sunshine	Themis	CES 5	Solar one	TSA	Solar two
Location	Spain	Italy	Spain	Japan	France	USSR	USA	Spain	USA
Output (MWe)	0.5	1	1.2	1	2	5	10	1	10
Number of heliostats	93	112.70	300	807	201	1600	1818	300	1818,108
Area of heliostats (m^2)	39.3	23.52	39.6	16	53.7	25	39.3	39.6	39.395
Total reflecting area (m^2)	3655	6216	11,880	12,912	10,740	40,000	71,447	11,880	81,707
Receiver type	Cavity	Cavity	Cavity	Cavity	Cavity	External	External	Cavity	External
Receiver fluid	Liquid sodium	Stem	Stem	Stem	Molten salt	Stem	Stem	Air	Molten salt
Thermal storage medium	Liquid sodium	Molten salt	Molten salt	Molten salt	Molten salt	Water	Oil/rock	Ceramic	Molten salt
Tower height (m)	43	55	60	69	165	70	80	60	80
Start of operation	1981	1981	1983	1981	1983	1985	1982	1993	1996

The largest of the seven plants constructed in the 1980s was Solar One at Barstow, California. Solar One had an external type, tubular receiver with sub-cooled water entering at the inlet and superheated steam at 102 bar and 510 °C being delivered at the outlet. Most of the time, the steam was directly fed to a turbine which was part of a Rankine cycle. The plant ran for six years, from 1982 to 1988, effectively proving the viability of the power tower principle. Solar One, however, ran into some issues. The receiver had low collection efficiency and the tubes failed frequently because of the thermal stress which they had to encounter in handling single-phase liquid flow, two-phase flow, and single-phase vapour flow. Additionally, the thermal storage system (which was an oil-rock system) had insufficient capacity and only worked between 220 and 305 °C, which was too low.

Due to the issues observed during Solar One's operation, it was decided to modify some components. To ensure that only single-phase liquid flow occurred in the tubes of the receiver, molten salt was employed as the heat transfer fluid rather than water steam. In addition, 108 heliostats (each with an area of 95 m^2) were added, and a new thermal storage system for molten salt with a greater capacity was put in place of the oil-rock storage system. The other hardware of Solar One was essentially reused. The modified plant was known as Solar Two. It was in operation from 1996 for three years. The research effectively proved the potential of molten salt technology and the associated thermal storage system.

The testing of a central receiver system employing air as the heat transfer fluid also took place in the 1990s. This followed the development of cavity-type, volumetric receivers. The CESA-I plant's tower has an open-loop volumetric air receiver that can absorb 2.5 MW of heat. Previously, a cavity-type tubular receiver was used to heat a molten salt. The TSA plant had additional modifications and was used for a shorter time in 1994 and 1999 in addition to its six-month operation in 1993. The application of volumetric receiver air technology was successfully proven when it was demonstrated that a receiver outlet temperature of 700 °C could be easily achieved.

The only plant operating in the present decade is PS 10 in Spain. The plant's construction was finished in 2006, and it started operating for commercial operation in March 2007. Here are some details about the plant:

Output	10 MWe
Number of heliostats	624
Area of heliostat	120 m^2
Receiver fluid	Water (converted into saturated steam at 40 bar)
Thermal storage medium	Water
Tower height	115 m

Despite the fact that all of the central receiver systems given in Table 7.2 were effectively operated, the available data suggests that the construction costs were very high. For example, the price of Solar One was around $14,000 per kW. However, costs have reduced significantly because of improvements in design, and the investment cost of the latest plant PS 10 is reported to be 3500 Euros ($4590) per kW. With increased operational expertise and production scale-up, initial investment costs are projected to decline even further.

7.2 Solar Collectors and Thermal Conversion

Energy is not a good unto itself; it is valued rather as a means of satisfying important needs of a society. In classical thermodynamics, energy is defined as the capacity to do work; but from a more practical point of view, energy is the main stay of any industrial society. In the United States, energy is currently provided by seven primary sources: petroleum, natural gas, coal, hydropower, nuclear fission, geothermal, and wood and waste. The first three of these sources are fossil fuels. They are stored forms of solar energy that received their solar input eons ago, have changed their characteristics over time, and now are in a highly concentrated and convenient form. It is apparent, however, that these stored forms of solar energy are being used so rapidly that they soon will be depleted. To maintain our present social structure, it is desirable; therefore, that we supply an increasing portion of our energy needs from renewable sources. The radiative solar energy reaching the earth during each month is approximately equivalent to the entire world supply of fossil fuels. Thus, from a purely thermodynamic point of view, the global potential of solar energy is many times larger than the current energy use. However, many technical and economic problems must be solved before large-scale use of solar energy can occur. The future of solar power deployment depends on how we deal with these constraints, which include scientific and technological problems, marketing and financial limitations, and political and legislative actions including equitable taxation of renewable energy sources. Approximately 30% of the solar energy impinging on the earth is reflected back into space. The remaining 70%, approximately 120,000 TW (1 TW is equal to 1012 W), is absorbed by the earth and its atmosphere. Solar radiation reaching the earth consists of the beam radiation that casts a shadow and can be concentrated and the diffuse radiation that has been scattered along its path in space from sun to earth. The solar radiation reaching the earth degrades in several ways. Some of the radiation is directly absorbed as heat by the atmosphere, the ocean, and the ground. Another component produces atmospheric and oceanic circulation. A third component evaporates, circulates, and precipitates water in the hydrologic cycle. Finally, a very small fraction is captured by green plants and drives the photosynthetic process.

For solar energy to be used in meeting the demands of a society, it must be converted into heat, mechanical power, or electricity. The conversion methods can be divided into natural and technological conversion systems. In natural conversion, the biosphere, i.e. earth, wind, or water, serves as a solar energy collector and storage. Since no man-made collectors are needed, the cost of energy from natural systems is largely determined by the conversion equipment, such as a wind turbine. In technological conversion systems, solar energy must be absorbed by man-made structures or collectors; the amount of insolation intercepted is determined by the total area and orientation of the collecting surface at a given geographic location (Kreith and Kreider 1978). The source of the sun's energy is a hydrogen-to-helium thermonuclear reaction. The outer layer of the sun, from which the solar radiation emanates, has an equivalent black body temperature of about 5760 K (5487 °C). The solar energy reaching the earth, called insolation, is in the form of photons, or radiation, covering a range of wavelengths corresponding approximately to a 5760 K black body. To convert this radiation into useful energy, one may either use photons in the appropriate wavelength range of the spectrum to generate electricity directly by photovoltaic conversion devices, or one may use the thermal part of the radiation spectrum to heat a working fluid by thermal conversion in a solar collector. The following discussion is concerned only with solar thermal conversion systems. The thermal conversion process of solar energy is based on well-known phenomena of heat transfer (Kreith 1976). In all thermal conversion processes, solar radiation is absorbed at the surface of a receiver, which contains oris in contact with flow passages through which a working fluid passes. As the receiver heats up, heat is transferred to the working fluid which may be air, water, oil, or a molten salt. The upper temperature that can be achieved in solar thermal conversion depends on the insolation, the

7.2 Solar Collectors and Thermal Conversion 353

degree to which the sunlight is concentrated, and the measures taken to reduce heat losses from the working fluid. Since the temperature level of the working fluid can be controlled by the velocity at which it is circulated, it is possible to match solar energy to the load · requirements, not only according to the amount but also according to the temperature level, i.e. the quality of the energy required. In this manner, it is possible to design conversion systems that are optimized according to both the first and the second laws of thermodynamics.

The collection and conversion of the solar radiation to thermal energy depends on the collector design and the relative amounts of direct beam and diffuse radiation absorbed by the collector (Kreider and Kreith 1981). As indicated in the following discussion of solar thermal collectors, the collectors used for higher temperature applications can collect only the direct radiation from the sun. Generally speaking, the southwestern and western regions of the country receive direct normal solar radiation levels sufficiently high for most high-temperature solar thermal conversion applications. High-temperature heat is needed by industry for process heat and by utilities for electricity. In 1980, the last year for which statistics are available, industry and utilities accounted for approximately 73% of the 76.3 quads of energy consumed in the United States (Energy Information Administration 1980). The industrial process heat portion alone was 20.6 quads (17%).

7.2.1 Devices for Thermal Collection and Storage

In any collection device, the principle usually followed is to expose a dark surface to solar radiation so that the radiation is absorbed. A part of the absorbed radiation is then transferred to a fluid like air or water. When no optical concentration is done, the device in which the collection is achieved is called a flat-plate collector (FPC). The flat-plate collector is the most important type of solar collector because it is simple in design, has no moving parts, and requires little maintenance. It can be used for a variety of applications in which temperatures ranging from 40 °C to about 100 °C are required.

A schematic diagram of a liquid flat-plate collector is shown in Fig. 7.6. As stated earlier, it consists of an absorber plate on which the solar radiation falls after coming through a transparent cover (usually made of glass). The absorbed radiation is partly transferred to a liquid flowing through tubes which are fixed to the absorber plate or are integral to it. This energy transfer is a useful gain. The remaining part of the radiation absorbed in the absorber plate is lost by convection and reradiation to the surroundings from the top surface, and by conduction through the back and the edges. The transparent cover helps in reducing the losses by convection and reradiation, while thermal insulation on the back and the edges helps in reducing the conduction heat loss. The liquid most commonly used is water. A liquid flat-plate collector is usually held tilted in a fixed position on a supporting structure, facing south if located in the northern hemisphere.

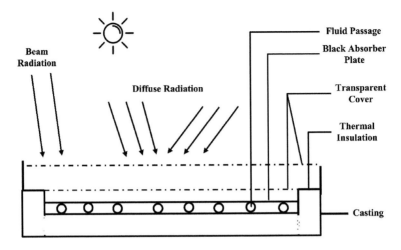

Fig. 7.6 Liquid flat-plate collector

To reduce the heat lost by reradiation from the top of the absorber plate of a flat-plate collector, it is usual to put a selective coating on the plate. The selective coating exhibits the characteristic of a high value of absorptivity for incoming solar radiation and a low value of emissivity for outgoing reradiation. As a result, the collection efficiency of the flat-plate collector is improved. Further improvement in the collection efficiency (or in the operating temperature) is obtained by evacuating the space above the absorber plate and leads to the design of an evacuated tube collector.

An evacuated tube collector (ETC) consists of a number of cylindrical modules mounted side-by-side on a common frame. Figure 7.7 shows a schematic diagram of one type of evacuated tube module. It consists of two concentric glass tubes, with the annular space between them being evacuated. The outer surface of the inner glass tube is selectively coated. The incoming solar radiation is absorbed on this surface and partly conducted inwards through the tube wall. The inner tube is filled with water, and the heat is transferred to the water by thermosyphon circulation. It is to be noted that the heat loss by convection to the surroundings is reduced significantly due to the vacuum in the annular space. This results in an improvement in the collection efficiency.

A schematic cross section of a conventional flat-plate collector for heating air (commonly referred to as a solar air heater) is shown in Fig. 7.8. The construction of such a collector is essentially similar to that of a liquid flat-plate collector except for the passages through which the air flows. These passages have to be made larger in order to keep the pressure drop across the collector within manageable limits. In the diagram shown, the air passage is simply a parallel plate duct.

When higher temperatures are required, it becomes necessary to concentrate the radiation. This is achieved using focusing or concentrating collectors. A schematic diagram of a typical line-focusing concentrating collector is shown in Fig. 7.9. The

7.2 Solar Collectors and Thermal Conversion 355

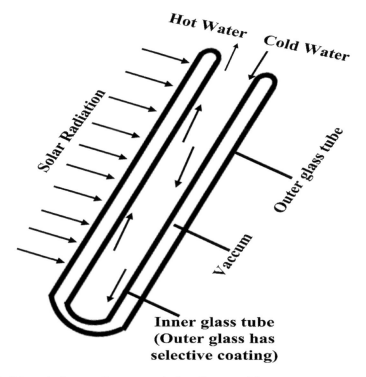

Fig. 7.7 Schematic diagram of an evacuated tube collector module

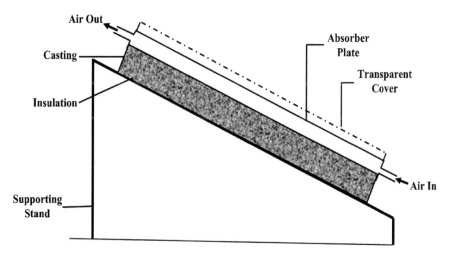

Fig. 7.8 Solar air heater

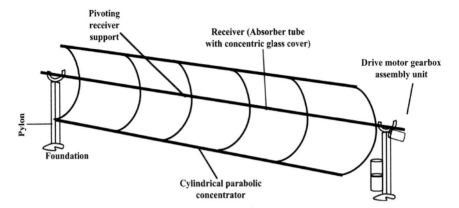

Fig. 7.9 Cylindrical parabolic concentrating collector

collector consists of a concentrator and a receiver. The concentrator shown is a mirror reflector having the shape of a cylindrical parabola. It focuses the sunlight on to its axis, where it is absorbed on the surface of the absorber tube and transferred to the fluid flowing through it. A concentric glass cover around the absorber tube helps in reducing the convective and radiative losses to the surroundings. In order that the sun's rays should always be focused on to the absorber tube, the concentrator has to be rotated. This movement is called tracking.

In the case of cylindrical parabolic concentrators, rotation about a single axis is generally required. Fluid temperatures up to 400 °C can be achieved in cylindrical parabolic focusing collector systems. The generation of still higher working temperatures is possible by using paraboloid reflectors (Fig. 7.10) which have a point focus. These require two-axis tracking so that the sun is in line with the focus and the vertex of the paraboloid.

Typical values of efficiency obtained from flat-plate collectors, evacuated tube collectors, line-focusing collectors, and paraboloid dish collectors are shown in Fig. 7.10. The efficiency is plotted as a function of the operating temperature and as is to be expected, it decreases with increasing temperature. The data presented in Fig. 7.11 is useful for doing approximate calculations of thermal systems using solar collectors. Such calculations help in giving estimates of the collector area required for a given performance. Alternatively, they can be used for obtaining a rough idea of the performance of a solar thermal system using a certain number of collectors of a specified type.

As stated earlier, one of the major problems associated with the utilization of solar energy is its variability. For this reason, most applications require some type of energy storage system. The purpose of such a system is to store energy when it is in excess of the requirement of an application and to make it available for extraction when the supply of solar energy is absent or inadequate. Energy storage can be in various forms—thermal, electrical, mechanical, or chemical. Thermal energy can be stored as sensible heat or as latent heat. Sensible heat storage is usually done in an

7.2 Solar Collectors and Thermal Conversion

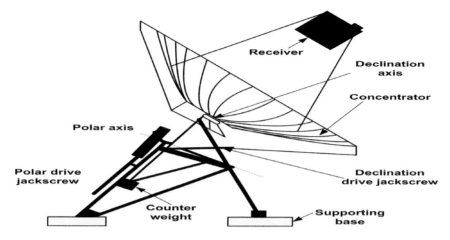

Fig. 7.10 Paraboloid concentrating collector

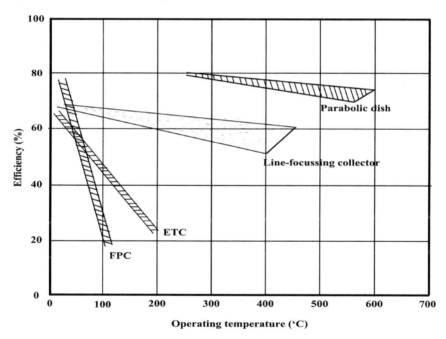

Fig. 7.11 Efficiency of various types of collectors as a function of operating temperature (Adapted from Gehlisch et al. 1982 and Rabl 1985, 1976a)

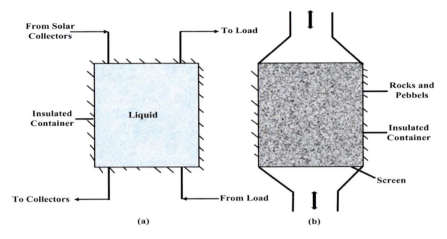

Fig. 7.12 Schematics of two forms of sensible heat storage: **a** liquid and **b** porous solid

insulated container containing a liquid like water or a porous solid in the form of pebbles or rocks (Fig. 7.12). The first type is preferred with liquid collectors, while the second type is compatible with air heaters.

In the case of latent heat storage, heat is stored in a substance when it melts and extracted when the substance freezes. Sensible heat storage systems operate over a range of temperatures, while latent heat storage systems operate essentially at the temperature at which the phase change takes place.

Apart from thermal storage, other forms of storage are possible if the application involves some energy conversion process. For example, if mechanical work is being obtained after conversion, storage could be done in large sized flywheels or in compressed air. It has been suggested that compressed air could be stored in large underground chambers. Similarly, if electrical power is being obtained after conversion, storage could be in the form of electric batteries. Many types of electric batteries are used for the purpose, the most common being lead-acid batteries.

A novel device which combines the functions of both collection and storage is the solar pond. It consists of an expanse of water about a metre or two in depth in which salts like sodium or magnesium chloride are dissolved. The concentration of the salt is more at the bottom and less at the top. Because of this, the bottom layers of water are denser than the surface layers even if they are hotter and natural convection does not occur. Thus, the heat from the sun's rays absorbed at the bottom of the pond is retained in the lower depths, and the upper layers of water act like a thermal insulation. The solar pond is taken up for detailed consideration in subsequent section ahead.

7.3 Solar Concentrating Collectors

7.3.1 Introduction

Concentration of solar radiation is achieved using a reflecting arrangement of mirrors or a refracting arrangement of lenses. The optical system directs the solar radiation on to an absorber of smaller area which is usually surrounded by a transparent cover. Because of the optical system, certain losses (in addition to those which occur while the radiation is transmitted through the cover) are introduced. These include reflection or absorption losses in the mirrors or lenses, and losses due to geometrical imperfections in the optical system. The combined effect of all such losses is indicated through the introduction of a term called optical efficiency. The introduction of more optical losses is compensated for by the fact that the flux incident on the absorber surface is concentrated on a smaller area. As a result, the thermal loss terms do not dominate to the same extent as in a flat-plate collector and the collection efficiency is usually higher.

It has been noted earlier that some of the attractive features of a flat-plate collector are simplicity of design and ease of maintenance. The same cannot be said of a concentrating collector. Because of the presence of an optical system, a concentrating collector usually has to follow or "track" the sun so that the beam radiation is directed on to the absorber surface. The method of tracking adopted and the precision with which it has to be done varies considerably. In collectors giving a low degree of concentration, it is often adequate to make one or two adjustments of the collector orientation every day. These can be made manually. On the other hand, with collectors giving a high degree of concentration, it is necessary to make continuous adjustments of the collector orientation. The need for some form of tracking introduces a certain amount of complexity in the design. Maintenance requirements are also increased. All these factors add to the cost. An added disadvantage is the fact that much of the diffuse radiation is lost because it does not get focused.

In the last few years, significant advances have been made in the development of concentrating collectors and a number of types have been commercialized abroad. Almost all of them are line-focusing cylindrical parabolic collectors and yield temperatures up to 400 °C.

7.3.2 Definitions

In order to be consistent in the use of terms, we will use the phrase "concentrating collector" to denote the whole system. The term "concentrator" will be used only for the optical subsystem which directs the solar radiation on to the absorber, while the term "receiver" will normally be used to denote the subsystem consisting of the absorber, its cover and other accessories.

We will now define four terms: aperture, area concentration ratio, intercept factor, and acceptance angle.

A. Aperture (W): It is the plane opening of the concentrator through which the solar radiation passes. For a cylindrical or linear concentrator, it is characterized by the width, while for a surface of revolution, it is characterized by the diameter of the opening.

B. Area concentration ratio (C): It is the ratio of the effective area of the aperture to the surface area of the absorber. Values of the concentration ratio vary from unity (which is the limiting case for a flat-plate collector) to a few thousand for a paraboloid dish. This quantity is also referred to as the geometric concentration ratio or simply concentration ratio.

C. Intercept factor (γ): It is the fraction of the radiation, which is reflected or refracted from the concentrator and is incident on the absorber. The value of the intercept factor is generally close to unity.

D. Acceptance angle ($2\theta_a$): It is the angle over which beam radiation may deviate from the normal to the aperture plane and yet reach the absorber. Collectors with large acceptance angles require only occasional adjustments, while collects with small acceptance angles have to be adjusted continuously.

7.3.3 Methods of Classification

Concentrating collectors are of various types and can be classified in many ways. They may be of the reflecting type utilizing mirrors or of the refracting type utilizing Fresnel lenses. The reflecting surfaces used may be parabolic, spherical, or flat. They may be continuous or segmented. Classification is also possible from the point of view of the formation of the image, the concentrator being either imaging or non-imaging. Further, the imaging concentrator may focus on a line or at a point.

The concentration ratio is also used as a measure for classifying concentrating collectors. Since this ratio approximately determines the operating temperature, this method of classification is equivalent to classifying the collector by its operating temperature range.

A final possibility is to describe concentrating collectors by the type of tracking adopted. Depending upon the acceptance angle, the tracking may be intermittent (one adjustment daily or every few days) or continuous. Further, the tracking may be required about one axis or two axis.

7.3.4 Types of Concentrating Collectors

A number of concentrating collector geometries are shown in Fig. 7.13.

The first type shown in Fig. 7.13a is a flat-plate collector with adjustable mirrors at the edges to reflect radiation on to the absorber plate. It is simple in design, has

7.3 Solar Concentrating Collectors

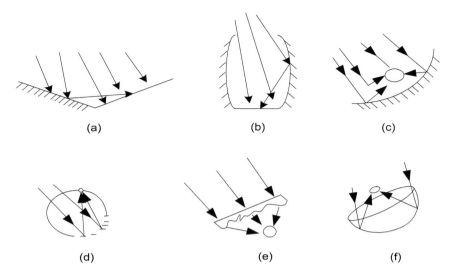

Fig. 7.13 Types of concentrating collector: **a** flat-plate collector with plane reflectors, **b** compound parabolic collector, **c** cylindrical parabolic collector, **d** collector with fixed circular concentrator and moving receiver, **e** Fresnel lens concentrating collector, and (f) paraboloid dish collector

a concentration ratio a little above unity and is useful for giving temperatures about 20 °C or 30 °C higher than those obtained with a flat-plate collector alone.

A compound parabolic concentrating collector (CPC) is shown in Fig. 7.13b. The concentrator consists of curved segments which are parts of two parabolas. Like the first type, this collector is also non-imaging. The concentration ratio is moderate and generally ranges from 3 to 10. The main advantage of the compound parabolic collector is that it has a high acceptance angle and consequently requires only occasional tracking. In addition, its concentration ratio is equal to the maximum value possible for a given acceptance angle.

The next type of collector Fig. 7.13c is a cylindrical parabolic collector in which the image is formed on the focal axis of the parabola. Many commercial versions of this type are now available.

Unlike the cylindrical parabolic collector in which the concentrator has to rotate in order to track the sun, the type shown in Fig. 7.13d has a fixed concentrator and a moving receiver. The concentrator is an array of long, narrow, flat mirror strips fixed along a cylindrical surface. The mirror strips produce a narrow line image which follows a circular path as the sun moves. This path is on the same circle on which the mirror strips are fixed. Thus, the receiver has to be moved along the circular path in order to track the sun.

Concentration is also achieved by using lenses. The most commonly used device is the *Fresnel lens* shown in Fig. 7.13e. The one shown in the figure is a thin sheet, flat on one side and with fine longitudinal grooves on the other. The angles of these grooves are such that radiation is brought to a line focus. The lens is usually made of extruded acrylic plastic sheets. Line focusing collectors like the ones shown in Fig. 7.13c–e usually have concentration ratios between 10 and 80 and yield temperatures between 150 and 400 °C.

In order to achieve higher concentration ratios and temperatures, it becomes necessary to have point focusing rather than line focusing. A sketch is shown in Fig. 7.13f. Such collectors can have concentration ratios ranging from 100 to a few thousand and have yielded temperatures up to 2000 °C. However, from the point of view of the mechanical design, there are limitations to the size of the concentrator and hence the amount of energy which can be collected by one dish. Commercial versions have been built with dish diameters up to 17 m. In order to collect larger amounts of energy at one point, the central receiver concept has been adopted. In this case, beam radiation is reflected from a number of independently controlled mirrors called heliostats to a central receiver located at the top of a tower.

Rabl (1976a) has shown that for a given acceptance angle $(2\theta_a)$, the maximum possible concentration ratio of a two-dimensional (line-focus) concentrator is

$$C_{m,2D} = \frac{1}{\sin \theta_a}. \tag{7.2}$$

For a three-dimensional (point-focus) concentrator, he has shown that

$$C_{m,3D} = \frac{1}{\sin^2 \theta_a}. \tag{7.3}$$

The half-angle subtended by the sun at the earth is 0.267°. Substituting this value in Eqs. (7.2) and (7.3), we see that the maximum value of concentration ratio for a line-focus concentrator is 215 and for a point-focus concentrator, it is 46,000. In actual systems, the values of concentration ratio are much lower since the acceptance angle need to be greater than 0.267° for a number of reasons. These include tracking error, imperfection in the reflecting- or refracting-component of the concentrator, mechanical misalignment between the concentrator and the receiver, etc.

7.3.5 Thermal Analysis of Concentrating Collectors

We will now discuss the thermal analysis of a concentrating collector. Like a flat-plate collector, an energy balance on the absorber yields the following equation under steady-state conditions

$$q_u = A_a S - q_l, \tag{7.4}$$

where

q_u rate of useful heat gain,
A_a effective area of the aperture of the concentrator,

7.4 Flat-Plate Collectors with Plane Reflectors 363

S solar beam radiation per unit effective aperture area absorber in the absorber, and

q_l rate of heat loss from the absorber.

Equation (7.4) is written under the assumption that the contribution of the diffuse component of solar radiation is negligible.

q_l can be written in terms of an overall loss coefficient defined by the equation

$$q_l = U_l A_p (T_{pm} - T_a),\tag{7.5}$$

where

U_l overall loss coefficient,
A_p area of the absorber surface,
T_{pm} average temperature of the absorber surface, and
T_a temperature of the surrounding air.

We combine Eqs. (7.4) and (7.5) to obtain

$$q_u = A_a \left[S - \frac{U_l}{C}(T_{pm} - T_a) \right],\tag{7.6}$$

where $C = \left(\frac{A_a}{A_p} \right)$ is the concentration ratio.

7.4 Flat-Plate Collectors with Plane Reflectors

A flat-plate collector with plane reflectors is a simple non-imaging concentrating collector and represents an effective means of getting slightly higher temperatures than are obtainable with a flat-plane collector alone. With a single collector, it is possible to use four reflectors all around. On the other hand, with an array of flat-plate collectors, it is possible to have only two arrays of reflectors, one of which face north and the other south (Fig. 7.14). The reflectors used may reflect the radiation in a specular or diffuse manner. The concentration ratios obtained are low and normally range from 1 to 4. Operating temperatures up to 130–140 °C can be attained. An advantage associated with this type of concentrating collector is that the diffuse component of the incoming solar radiation is not entirely wasted.

With an array of flat-plate collectors, the usual practice is to use an array of north-facing reflectors only, since these are more convenient to handle and adjust than south-facing reflectors. The inclination of the reflectors is usually adjusted once every few days. For the case of a north-facing specular reflector array whose dimensions are equal to those of the flat-plate collector array, it can be shown that the inclination ψ of the reflectors should be

$$\psi = (\pi - \beta - 2\phi + 2\delta)/3,\tag{7.7}$$

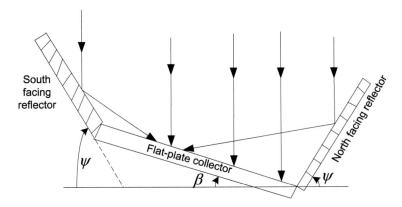

Fig. 7.14 Flat-plate collector with reflectors

where β = slope of the collectors. Equation (7.7) is derived under the assumption that the reflector mirrors are adjusted in such a way that the sun's rays striking the top edge of the mirrors at 1200 h (LAT) are reflected to the top edge of the collectors. If we assume the location to be Mumbai and take the collector slope to be equal to the latitude, the value of ψ is found to vary from 25° to 57° as the declination varies from $-23.45°$ to its maximum value of $+23.45°$.

At times other than noon, only a certain fraction of the radiation falling on the reflectors will be reflected usefully on to the collectors with the remaining falling on the sides of the collector. It has also been shown that specular reflectors are more effective in augmenting the radiation than diffuse reflectors (McDaniels et al. 1975; Grassie and Sheridan 1977).

7.5 Cylindrical Parabolic Collector

7.5.1 Description

The cylindrical parabolic collector is also referred to as a parabolic trough or a linear parabolic collector as shown in Fig. 7.15. The basic elements making up a conventional collector are (i) the absorber tube located at the focal axis through which the liquid to be heated flows, (ii) the concentric transparent cover, (iii) the reflector, and (iv) the support structure. Elements (i) and (ii) together constitute the receiver, while elements (iii) and (iv) constitute the concentrator.

The collectors are available over a wide range of aperture areas from about 1–60 m^2 and with widths ranging from 1 to 6 m. Concentration ratios range from 10 to 80, and rim angles (see Fig. 7.17) from 70 to 120°.

The absorber tube is usually made of stainless steel or copper and has a diameter of 2.5–5 cm. It is coated with a heat resistant black paint and is generally surrounded

7.5 Cylindrical Parabolic Collector

Fig. 7.15 Cylindrical parabolic collector (*Source* Wikipedia)

by a concentric glass cover with an annular gap of 1 or 2 cm. In the case of high-performance collectors, the absorber tube is coated with a selective surface and the space between the tube and the glass cover is evacuated. In some small collectors, the concentric cover is replaced by a glass or plastic sheet covering the whole aperture area of the collector. Such an arrangement helps in protecting the reflecting surface from the weather.

The liquid heated in the collector depends upon the temperature required. Usually organic heat transfer liquids (referred to as thermic fluids) are used. Because of their low thermal conductivities, these liquids yield low heat transfer coefficients. Augmentative devices in the form of twisted tapes or central plugs (which create annular passages) are therefore used to increase the value of the heat transfer coefficient.

The reflecting surface is generally curved back silvered glass. It is fixed on a light-weight structure usually made of aluminium sections. The proper design of this supporting structure and of the system for its movement is important, since it influences the shape and orientation of the reflecting surface. Some of the factors to be considered in designing the structure are that it should not distort significantly due to its own weight and that it should be able to withstand wind loads.

Compared to flat-plate collectors, there are very few manufacturers of concentrating collectors all over the world. The volume of production is also low. In India, many experimental collectors have been built and tested.

366 7 Solar Collectors and Thermal Conversion

7.5.2 Orientation and Tracking Modes

A cylindrical parabolic collector is oriented with its focal axis pointed either in the east-west (E-W) or the north-south (N-S) direction. In the east-west orientation, the focal axis is horizontal, while in the north-south orientation, the focal axis may be horizontal or inclined. The various tracking modes, which can be adopted, are as follows:

Mode I The focal axis is E-W and horizontal. The collector is rotated about a horizontal E-W axis and adjusted once every day so that the solar beam is normal to the collector aperture plane at solar noon on that day.

In this mode, the aperture plane is an imaginary surface with either $\gamma = 0°$ or $\gamma = 180°$. The case of $\gamma = 0°$ occurs when $(\phi - \delta) > 0$, while the case of $\gamma = 180°$ occurs when $(\phi - \delta) < 0$. In order to find the slope β of the aperture plane, we substitute the condition at solar noon, viz. $\omega = 0°$, $\theta = 0°$ in Eq. (2.4) (should be referred from Chap. 2 as per equation number). This yields

$$\beta = (\phi - \delta) \text{ for } \gamma = 0°, \tag{7.8a}$$

$$\text{and } \beta = (\delta - \phi) \text{ for } \gamma = 180°. \tag{7.8b}$$

The angle of incidence of the beam radiation on the aperture plane throughout the day is obtained by putting Eqs. (7.8a) and (7.8b) in Eq. (2.4). For both cases, $\gamma = 0°$ and $\gamma = 180°$, we obtain the same relation

$$\cos\theta = \sin^2\delta + \cos^2\delta \cos\omega. \tag{7.9}$$

Mode II The focal axis is E-W and horizontal. The collector is rotated about a horizontal E-W axis and adjusted continuously so that the solar beam makes the minimum angle of incidence with the aperture plane at all times.

In this mode also, the aperture plane is an imaginary surface with either $\gamma = 0°$ or $\gamma = 180°$. Equation (2.4) is applicable with $\gamma = 0°$ or $180°$. In order to find the condition to be satisfied for θ to be a minimum, we differentiate the right-hand side of the resulting equation with respect to β and equate it to zero. Thus, we get

$$\tan(\phi - \beta) = [\tan \delta / \cos \omega] \text{ for } \gamma = 0°, \tag{7.10a}$$

$$\text{and } \tan(\phi + \beta) = [\tan \delta / \cos \omega] \text{ for } \gamma = 180°. \tag{7.10b}$$

Equations (7.10a) and (7.10b) can be used for finding the slope of the aperture plane. Equation (7.10a) corresponding to $\gamma = 0°$ is used if the magnitude of the solar azimuth angle γ_s is less than $90°$, while Eq. (7.10b) corresponding to $\gamma = 180°$ is used if the magnitude of the solar azimuth angle is greater than $90°$.

7.5 Cylindrical Parabolic Collector

The expression for the corresponding minimum angle of incidence is obtained by substituting Eqs. (7.10a) and (7.10b) in the appropriate version of Eq. (2.4). For both cases, we obtain

$$\cos\theta = (1 - \cos^2\delta\sin^2\omega)^{1/2}. \tag{7.11}$$

Mode III The focal axis is N-S and horizontal. The collector is rotated about a horizontal N-S axis and adjusted continuously so that the solar beam makes the minimum angle of incidence with the aperture plane at all times.

In this mode, the surface azimuth angle $\gamma = +90°$ before noon, and $\gamma = -90°$ after noon. Thus, before noon, Eq. (2.4) becomes

$$\cos\theta = (\sin\phi\sin\delta + \cos\phi\cos\delta\cos\omega)\cos\beta + \cos\delta\sin\omega\sin\beta. \tag{7.12}$$

In order to find the condition to be satisfied for θ to be a minimum, we differentiate the right-hand side of Eq. (7.12) with respect to β and equate it to zero. Thus, we get

$$\beta = \tan^{-1}\left[\frac{\cos\delta\sin\omega}{\sin\phi\sin\delta + \cos\phi\cos\delta\cos\omega}\right]. \tag{7.13}$$

Equation (7.13) is used for finding the slope of the aperture plane at any time before noon. The expression for the corresponding minimum angle of incidence is obtained by substituting Eq. (7.13) in Eq. (7.12), giving

$$\cos\theta = [(\sin\phi\sin\delta + \cos\phi\cos\delta\cos\omega)^2 + \cos^2\delta\sin^2\omega]^{1/2}. \tag{7.14}$$

After noon, i.e. with $\gamma = -90°$, we would obtain

$$\beta = \tan^{-1}\left[\frac{-\cos\delta\sin\omega}{\sin\phi\sin\delta + \cos\phi\cos\delta\cos\omega}\right]. \tag{7.15}$$

The expression for $\cos\theta$ remains the same.

Mode IV The focal axis is N-S and inclined at a fixed angle equal to the latitude. Thus, it is parallel to the earth's axis. This orientation is sometimes referred to as a *polar mount*. The collector is rotated about an axis parallel to the earth's axis at an angular velocity equal and opposite to the earth's rate of rotation (15°/h). It is adjusted such that at solar noon the aperture plane is inclined surface facing due south. Thus, putting $\beta = \phi$ and $\omega = 0$ in Eq. (2.7), we get Equator

$$\theta = \delta. \tag{7.16}$$

This is also seen from Fig. 7.16 in which the circle represents the longitude through the location of the collector. At all other times, since the collector is rotated at speed

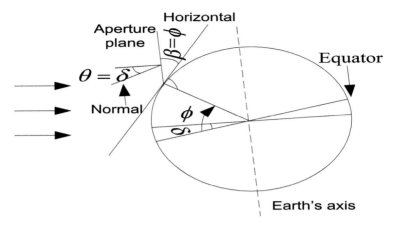

Fig. 7.16 Tracking mode IV for a cylindrical parabolic collector. Angle of incidence θ is equal to the declination angle

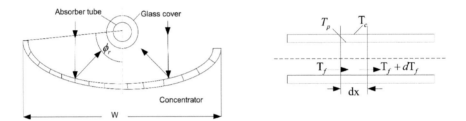

Fig. 7.17 Cylindrical parabolic collector in cross section

equal to the earth's rate of rotation and about an axis parallel to the earth's axis, it follows that Eq. (7.16) is still valid.

Mode V The focal axis is N-S and inclined. The collector is rotated continuously (but not at a constant angular velocity) about an axis parallel to the focal axis, as well as about a horizontal axis perpendicular to this axis, and adjusted so that the solar beam is normally incident on the aperture plane at all times. In this situation, obviously $\cos\theta = 1$. It is easy to show that at solar noon,

$$\beta = |\phi - \delta|. \tag{7.17}$$

It is of interest to compare the amounts of beam radiation which would be incident on a collector's aperture plane over a day if one adopted the various tracking modes. This comparison is made through a numerical example.

Example 7.3 A cylindrical parabolic collector is used in New Delhi (28.58° N, 77.20° E). Compare the beam radiation which would fall on one square metre of the

7.5 Cylindrical Parabolic Collector

aperture plane of this collector from 0600 to 1800 h (LAT) on June 10 for the five tracking modes just described. The following values of I_b are available.

Time (h)	I_b (W/m^2)	Time (h)	I_b (W/m^2)
0630	110	1230	523
0730	240	1330	495
0830	333	1430	445
0930	424	1530	322
1030	495	1630	220
1130	550	1730	118

Solution On June 10, $n = 161$ and $\delta = 23.012°$.
We show a sample calculation for one time, viz. 1030 h.

$$\cos \theta_z = \sin 28.58° \sin 23.012°$$
$$+ \cos 28.58° \cos 23.012° \cos 22.5° = 0.9338.$$

For tracking mode, I, from Eq. (7.9),

$$\cos \theta = \sin^2 23.012° + \cos^2 23.012° \cos 22.5° = 0.9355.$$

Therefore, tilt factor r_b for the aperture plane $= \frac{\cos \theta}{\cos \theta_z} = \frac{0.9355}{0.9338} = 1.002$.
Beam flux incident normally on the aperture plane $= I_b r_b = 495 \times 1.002 = 495.9$ W/m^2.
Similarly for the other tracking modes we obtain the following results:

	Tracking mode			
	II	III	IV	V
$\cos \theta$	0.9359	0.9980	0.9204	1
r_b	1.0023	1.0688	0.9857	1.0709
$I_b r_b$ (W/m^2)	496.1	529.0	487.9	530.1

The values of $I_b r_b$ obtained over the day are given in Table 7.3. The total incident energy from 0600 to 1800 h is also obtained by adding the values of $I_b r_b$. The assumption made is that the instantaneous value is also the average value for a one-hour period. It is seen that the maximum total is obtained with mode V. This is as expected since mode V involves two-axis tracking and gives normal incidence. The other modes involve continuous one-axis tracking or one daily adjustment and yield lesser totals. The results obtained are obviously dependent on the latitude of the location, the day of the year and the input radiation data.

Table 7.3 Comparison of tracking modes

LAT	$I_b r_b$ (W/m^2)				
(h)	Mode I	Mode II	Mode III	Mode IV	Mode V
0630	99.1	153.8	360.4	346.1	376.1
0730	230.7	254.5	476.1	445.1	483.6
0830	327.9	335.0	489.0	451.3	490.4
0930	422.3	424.0	511.9	471.2	511.9
1030	495.9	496.1	529.0	487.9	530.1
1130	552.4	552.4	554.0	512.2	556.4
1230	525.3	525.3	526.9	487.0	529.1
1330	495.9	496.1	529.0	487.9	530.1
1430	443.2	445.0	537.2	494.5	537.2
1530	317.0	323.9	472.8	436.4	474.2
1630	211.4	233.2	436.5	408.0	443.3
1730	106.3	165.0	386.6	371.3	403.4
Total (kWh/m^2)	4.227	4.404	5.809	5.399	5.866

In practice, modes II, III, and IV are the most common. Mode V is not preferred because of the complexity of providing motion about two axes, while mode I is not used because it does not yield an image on the focal axis. From a design standpoint, modes II and III are the simpler of the three modes II, III, and IV. Since the focal axis is horizontal, it is easy to connect collectors in series (the outlet of one collector being the inlet of the next) and to provide common shaft arrangements for tracking. Land space is also used more effectively in these modes.

7.5.3 Performance Analysis

We now consider the performance analysis of a cylindrical parabolic concentrating collector whose concentrator has an aperture W, length L, and rim angle ϕ_r (Fig. 7.17). The absorber tube has an inner diameter D_i and an outer diameter D_o, and it has a concentric glass cover of inner diameter D_{ci} and outer diameter D_{co} around it. The fluid being heated in the collector has a mass flow rate $\dot{m}$, a specific heat C_p, an inlet temperature T_{fi}, and an outlet temperature T_{fo}.

The collector is operated in any one of the modes described in Sect. 7.5.2 and the beam radiation normally incident on its aperture is $I_b r_b$, whose value can be calculated from the equations derived in Sec. 7.5.2. In some of the tracking modes, the sun's rays are incident at an angle and will, therefore, come to a focus a little beyond the length of the concentrator. We assume that the absorber tube is long enough to intercept this image. In practice this would mean that the absorber tube might be a little longer (say about 10%) than the concentrator and that the flux falling on the tube would not

7.5 Cylindrical Parabolic Collector

be uniform along the length. For the purposes of analysis, however, we will not take into account this extra tube length, and we will assume that the radiation flux is the same all along the length. We will also make the assumption that the temperature drops across the absorber tube and the glass cover are negligible.

The concentration ratio of the collector is given by

$$C = \frac{\text{Effective aperture area}}{\text{Absorber tube area}} = \frac{(W - D_o)L}{\pi D_o L} = \frac{(W - D_o)}{\pi D_o}. \tag{7.18}$$

The analysis which follows is in many respects similar to the analysis of a liquid flat-plate collector. An energy balance on an elementary slice dx of the absorber tube, at a distance x from the inlet, yields the following equation for a steady state:

$$dq_u = [I_b r_b (W - D_o)\rho\gamma(\tau\alpha)_b + I_b r_b D_o(\tau\alpha)_b - U_l \pi D_o (T_p - T_a)]dx, \tag{7.19}$$

in which

dq_u useful heat gain rate for a length dx,
ρ specular reflectivity of the concentrator surface,
γ intercept factor, the fraction of the reflected radiation intercepted by the absorber tube,
$(\tau\alpha)_b$ average value of the transmissivity-absorptivity product for beam radiation,
U_l overall loss coefficient,
T_p local temperature of absorber tube, and
T_a ambient temperature.

The first term on the right-hand side in Eq. (7.19) represents the incident beam radiation absorbed in the absorber tube after reflection, while the second term represents the absorbed incident beam radiation which falls directly on the absorber tube. The second term is small in comparison with the first but cannot be ignored when the concentration ratio is small. The third term represents the loss by convection and reradiation.

In a manner similar to that adopted for a flat-plate collector, we define an absorbed flux S as follows,

$$S = I_b r_b \rho\gamma(\tau\alpha)_b + I_b r_b(\tau\alpha)_b \left(\frac{D_o}{W - D_o}\right). \tag{7.20}$$

Equation (7.19) thus becomes

$$dq_u = \left[S - \frac{U_l}{C}(T_p - T_a)\right](W - D_o)dx. \tag{7.21}$$

The useful heat gain rate dq_u can also be written as

$$dq_u = h_f \pi D_i (T_p - T_f)dx, \tag{7.22}$$

$$dq_u = \dot{m}C_p dT_f, \tag{7.23}$$

where

h_f heat transfer coefficient on the inside surface of the tube, and
T_f local fluid temperature.

We combine Eqs. (7.21) and (7.22) in such a manner as to eliminate the absorber tube temperature T_p and obtain

$$dq_u = F'\left[S - \frac{U_1}{C}(T_f - T_a)\right](W - D_o)dx, \tag{7.24}$$

where F' is the collector efficiency factor defined by

$$F' = \frac{1}{U_1\left[\frac{1}{U_1} + \frac{D_o}{D_i h_f}\right]}. \tag{7.25}$$

Again, combining Eqs. (7.23) and (7.24), we obtain the differential equation

$$\frac{dT_f}{dx} = \frac{F'U_1\pi D_o}{\dot{m}C_p}\left[\frac{CS}{U_1} - (T_f - T_a)\right]. \tag{7.26}$$

Integrating and using the inlet condition at $x = 0$, and $T_f = T_{fl}$ we have the temperature distribution

$$\frac{\left(\frac{CS}{U_1} + T_a\right) - T_f}{\left(\frac{CS}{U_1} + T_a\right) - T_{fl}} = \exp\left\{-\frac{F'U_1\pi D_o x}{\dot{m}C_p}\right\}. \tag{7.27}$$

The fluid outlet temperature is obtained by putting $T_f = T_{fo}$ and $x = L$ in Eq. (7.27). Making this substitution and subtracting both side of the resulting equation from unity, we have

$$\frac{(T_{fo} - T_{fl})}{\frac{CS}{U_1} + T_a - T_{fl}} = 1 - \exp\left\{-\frac{F'U_1\pi D_o L}{\dot{m}C_p}\right\}. \tag{7.28}$$

Thus, the useful heat gain rate

$$q_u = \dot{m}C_p(T_{fo} - T_{fl}) = \dot{m}C_p\left[\frac{CS}{U_1} + T_a - T_{fl}\right]\left[1 - \exp\left\{-\frac{F'U_1\pi D_o L}{\dot{m}C_p}\right\}\right]$$
$$= F_R(W - D_o)L\left[S - \frac{U_1}{C}(T_{fl} - T_a)\right], \tag{7.29}$$

7.5 Cylindrical Parabolic Collector 373

where F_R is the heat-removal factor defined by

$$F_R = \frac{\dot{m}C_p}{U_l \pi D_o L} \left[1 - \exp\left\{ -\frac{F' U_l \pi D_o L}{\dot{m}C_p} \right\} \right].$$ (7.30)

The instantaneous collection efficiency η_i is given by

$$\eta_i = \frac{q_u}{(I_b r_b + I_d r_d) \text{WL}}.$$ (7.31)

If ground-reflected radiation is neglected. The instantaneous efficiency can also be calculated on the basis of beam radiation alone, in which case

$$\eta_{ib} = \frac{q_u}{I_b r_b \text{WL}}.$$ (7.32)

7.5.4 Correlations Between the Overall Loss Coefficient and Heat Transfer

In this part, we'll outline the steps for determining the total loss coefficient U_l as well as the correlations needed to figure out heat transfer coefficient individually. The total loss coefficient based only on convection and reradiation losses is calculated. A system of long, concentric tubes is what we refer to as being made up of the absorber tube and the glass cover that surrounds it. We have made the same assumptions, and we have

$$\frac{q_l}{L} = h_{p-c}(T_{pm} - T_c)\pi D_o + \frac{\sigma \Pi D_o \left(T_{pm}^4 - T_c^4\right)}{\left\{ \frac{1}{\varepsilon_p} + \frac{D_o}{D_{ci}} \left(\frac{1}{\varepsilon_c} - 1 \right) \right\}}$$ (7.33)

$$= h_w(T_c - T_a)\pi D_{co} + \sigma \pi D_{co} \varepsilon_c \left(T_c^4 - T_{sky}^4 \right),$$ (7.34)

where

$\frac{q_l}{L}$ rate of heat loss per unit length,

h_{p-c} convective heat transfer coefficient between the absorber tube and the glass cover,

T_{pm} average temperature of the absorber tube, and

T_c cover temperature.

All other symbols were already defined. Equations (7.33) and (7.34) are a pair of nonlinear equations that must be solved for the unknowns $\frac{q_l}{L}$ and T_c, using the values of h_{p-c} and h_w.

Coefficient of Heat Transfer Between the Cover and the Absorber Tube

The enclosed annular area between a horizontal absorber tube and a concentric cover's natural convection heat transfer coefficient h_{p-c} is determined using a correlation developed by Raithby and Hollands (1975).

$$\frac{k_{\text{eff}}}{k} = 0.317(R_a^*)^{1/4},\tag{7.35}$$

where

k_{eff} Effective thermal conductivity defined as the thermal conductivity that the motionless air in the gap must have to transmit the same amount of heat as the moving air, and

R_a^* Modified Rayleigh number related to the usual Rayleigh number by the following equation:

$$(R_a^*)^{1/4} = \frac{\ln(D_{ci}/D_o)}{b^{3/4}\left(\frac{1}{D_o^{3/5}} + \frac{1}{D_{ci}^{3/5}}\right)^{5/4}} R_a^{1/4}.\tag{7.36}$$

The radial gap $b = (D_{ci} - D_o)/2$ is the characteristic dimension that is utilized to calculate the Rayleigh number. Properties are assessed at the mean temperature $(T_{pm} + T_c)/2$. The k_{eff} value cannot be smaller than the value of k. Therefore, (k_{eff}/k) is set to unity if using Eq. (7.35) results in a value that is less than unity.

By equating expressions for the heat exchange rate per unit length, one may determine how the heat transfer coefficient h_{p-c} and the effective thermal conductivity relate to one another. We can show the relationship as follows:

$$\frac{2\pi k_{\text{eff}}}{\ln(D_{ci}/D_o)}(T_{pm} - T_c) = h_{p-c}\pi D_o(T_{pm} - T_c).$$

Hence,

$$h_{p-c} = \frac{2k_{\text{eff}}}{D_o \ln(D_{ci}/D_o)}.\tag{7.37}$$

Ra* must be smaller than 10^7, and b must be less than $0.3D_o$, in order to implement Eq. (7.35).

Coefficient of Heat Transfer on the Cover's Outer Surface

The well-known correlation based on the data of Hilpert's (1933), who performed experiments on air flow at right angles across cylinders of different diameter at low levels of free stream turbulence, can be applied to determine the convective heat transfer coefficient h_w, on the outer surface of the cover (often called the wind heat coefficient). Hilpert's data (1933) can be correlated by the equation

7.5 Cylindrical Parabolic Collector

$$Nu = C_1 Re^n. \tag{7.38}$$

The value of C_1 and n for the above equation are as follows:

$$For 40 < Re < 4000, \ C_1 = 0.615, \ n = 0.466$$
$$For 4000 < Re < 40,000, \ C_1 = 0.174, \ n = 0.618$$
$$For 40,000 < Re < 400,000, \ C_1 = 0.239, \ n = 0.805.$$

Following thorough review of the data available for cross flow across a cylinder by (Hilpert 1933), the correlation shown below was proposed:

$$Nu = 0.3 + \frac{0.62 Re^{1/2} Pr^{1/3}}{\left[1 + (0.4/Pr)^{2/3}\right]^{1/4}} \left[1 + (Re/282,000)^{5/8}\right]^{4/5}. \tag{7.39}$$

All values of Re up to 10^7 are acceptable for Eq. (7.39). Churchill and Bernstein (1977) suggested changing the last term $\left[1 + (Re/282,000)^{5/8}\right]^{4/5}$ to $\left[1 + (Re/282,000)^{1/2}\right]$ for the range $20,000 < Re < 400,000$.

D_{co} is the characteristic dimension that should be used in Eqs. (7.38) and (7.39). Properties must be examined at the average temperature $(T_c + T_a)/2$.

For cross flow and at low turbulence intensity, Eqs. (7.38) and (7.39) have been found. In practice, the flow may not be at right angles and the turbulence intensity in the wind may not be insignificant. As a result, there is an uncertainty in the value of the h_w predicted by Eqs. (7.38) and (7.39). However, this uncertainty has no impact on the total loss coefficient's value.

Heat Transfer Coefficient on the Inside Surface of the Absorber Tube

Under the assumption that the flow is completely developed, the value of h_f on the inside surface of the absorber tube may be determined. The length-to-diameter ratio (L/D_i), which is usually more than 20, supports this assumption. When the Reynolds number is less than 2000, the air flow is laminar and the heat transfer coefficient may be calculated from the equation

$$Nu = 3.66. \tag{7.40}$$

The heat transfer coefficient may be computed using the well-known Dittus-Boelter equation for a Reynolds number greater than 2000, the flow is turbulent.

$$Nu = 0.023 Re^{0.8} Pr^{0.4}. \tag{7.41}$$

D_i is the characteristic dimension that was utilized to determine Nu and Re in Eqs. (7.40) and (7.41). At the mean temperature $(T_{fi} + T_{fo})/2$, properties are examined. It should be noted that the values of h_f for a liquid flat-plate collector may also be determined using Eqs. (7.40) and (7.41).

The mass flow rate $\dot{m}$ is typically low, and the flow is laminar in most situations. Therefore, Eq. (7.40) is applied. As a result, the value of h_f is sometimes so small as to adversely affect the value of F_R. This is particularly true when the liquid used is heat transfer oil. Despite having poor thermal conductivities and high Prandtl numbers, these fluids have high boiling temperatures. In these circumstances, it is preferable to utilize an augmentation method to increase the heat transfer coefficient. One of the most basic methods is to place twisted tape of width D_i all along the inside of the absorber tube. For this case (Hong and Bergles 1976), proposed the correlation shown below:

$$\mathrm{Nu} = 5.172\left[1 + 0.005484\{\mathrm{Pr}(\mathrm{Re}/X)^{1.78}\}^{0.7}\right]^{0.5}, \tag{7.42}$$

where X = tape twist ratio = H/D_i and H = length over which the tape is twisted through $180°$.

D_i is the characteristic dimension used to compute Nu and Re. Calculations using Eq. (7.42) demonstrate that the use of a twisted tape causes a very significant increase in the heat transfer coefficient when the Prandtl number is high. The pressure decrease does not rise in precisely the same ratio at the same time. Based on the research of, the following relationships may be used to compute the pressure drop (Date 2000).

$$f\mathrm{Re} = 38.4(\mathrm{Re}/X)^{0.05} \text{ for } 6.7 \le (\mathrm{Re}/X) \le 100$$
$$= C_2(\mathrm{Re}/X)^{0.3} \text{ for}(\mathrm{Re}/X) > 100, \tag{7.43}$$

where f = friction factor, and

$$C_2 = 8.8201X - 2.1193X^2 + 0.2108X^3 - 0.0069X^4.$$

Example 7.4 Calculate the overall loss coefficient U_1 for the receiver of a cylindrical parabolic concentrating collector system. The receiver consists of a selectively coated absorber tube with one glass cover around it. The following data is given:

Absorber tube, inner diameter D_i	7.5 cm
Outer diameter D_o	8.1 cm
Glass cover, inner diameter D_{ci}	14.4 cm
Outer diameter D_{co}	15.0 cm
Emissivity of absorber tube surface ε_p	0.15
Emissivity of glass ε_c	0.88
Mean temperature of absorber tube, T_{pm}	170 °C
Ambient temperature, T_a	25 °C
Wind speed, V_∞	4 m/s

7.5 Cylindrical Parabolic Collector

Solutions Substituting the given data into Eqs. (7.33) and (7.34), we get

$$\frac{q_l}{L} = h_{p-c}(443.2 - T_c)\pi \times 0.081 + \frac{5.67 \times 10^{-8} \times \pi \times 0.081 (443.2^4 - T_c^4)}{\left\{\frac{1}{0.15} + \frac{0.081}{0.144}\left(\frac{1}{0.88} - 1\right)\right\}}$$
$$= 0.2545 h_{p-c}(443.2 - T_c) + 0.2140 \times 10^{-8} (385.8 \times 10^8 - T_c^4), \quad (7.44)$$

$$\frac{q_l}{L} = h_w(T_c - 298.2)\pi \times 0.15 + 5.67 \times 10^{-8} \times \pi \times 0.15 \times 0.88 (T_c^4 - 292.2^4)$$
$$= 0.4712 h_w(T_c - 298.2) + 2.3513 \times 10^{-8} (T_c^4 - 72.90 \times 10^8). \quad (7.45)$$

The values of the two unknown quantity $\frac{q_l}{L}$ and T_c can be found by solving the Eqs. (7.44) and (7.45), which requires the values of h_{p-c} and h_w. A trial-and-error concept have to be applied since these quantities depends upon T_c. Assume the value of T_c equal to 310 K.

1. Calculation of h_{p-c}

Mean air temperature between tube and cover

$$= \frac{443.2 + 310}{2} = 376.6 \,\mathrm{K} = 103.4\,^\circ\mathrm{C}.$$

The corresponding values of k, v, Pr, and Ra at this air temperature are

$$k = 0.0323 \,\mathrm{W/m\,K}, \, v = 23.52 \times 10^{-6} \,\mathrm{m^2/s}, \, \mathrm{Pr} = 0.688, \text{ and}$$
$$\mathrm{Ra} = 9.81 \times \frac{1}{376.6} \times \frac{(443.2 - 310) \times 0.0315^3}{23.52^2 \times 10^{-12}} \times 0.688 = 134{,}877.$$

From Eq. (7.35)

$$\frac{k_{\mathrm{eff}}}{k} = 0.317 \times \frac{\ln(0.144/0.081)}{0.0315^{3/4}\left(\frac{1}{0.081^{3/5}} + \frac{1}{0.144^{3/5}}\right)^{5/4}} \times (134877)^{1/4} = 3.6349,$$

and from Eq. (7.37)

$$h_{p-c} = \frac{2 \times 3.6349 \times 0.0323}{0.081 \ln(0.144/0.081)} = 5.036 \,\mathrm{W/m^2\,K}.$$

2. Calculation of h_w

Mean temperature of air between the cover and ambient

$$= \frac{310 + 298.2}{2} = 304.1 \,\mathrm{K} = 30.9\,^\circ\mathrm{C}.$$

The corresponding values of k and v at this temperature are

$$k = 0.0268 \text{ W/m K and } v = 16.09 \times 10^{-6} \text{ m}^2/\text{s.}$$

Taking assumption that the wind velocity is at right angle to the collector axis and using Eq. (7.38) we have

$$\text{Re} = \frac{4 \times 0.15}{16.09 \times 10^{-6}} = 37,300$$

$$\text{Nu} = 0.174(37,300)^{0.618} = 116.4,$$

$$h_w = 116.4 \times \frac{0.0268}{0.15} = 20.79 \text{ W/m}^2 \text{ K.}$$

Putting the h_w and h_{p-c} values in Eqs. (7.44) and (7.45) we found the values of T_c and $\left(\frac{q_l}{L}\right)$ using the trial-and-error method.

T_c (K)	$(q1/L)$ from	
	Equation (5.37)	Equation (5.38)
310	233.5	161.3
315	225.9	224.7
315.1	225.6	226.0

The $\frac{q_l}{L}$ values at the bottom of the table are approximately equal, so we can take the mean of the value of 225.8 W/m. Since the values of the h_{p-c} and h_w will not change much if the initial guess of $T_c = 310$ K is changed to 315.1 K, it will not be necessary to repeat the calculation.

Therefore, $U_1 = \frac{225.8}{\pi \times 0.081 \times (170-25)} = 6.12 \text{ W/m}^2 \text{ K.}$

Empirical Equation for the Overall Loss Coefficient:

Mullick and Nanda (1989) have established a semi-empirical equation for directly determining the total loss coefficient using calculations for a significant number of instances encompassing a wide range of scenarios encountered with cylindrical parabolic collectors. The requirement for an iterative computation is removed by this equation.

$$\frac{1}{U_1} = \frac{1}{C_3 (T_{pm} - T_c)^{0.25} + \left[\sigma \left(T_{pm}^2 + T_c^2 \right) (T_{pm} + T_c) / \left\{ \frac{1}{\varepsilon_p} + \frac{D_o}{D_{ci}} \left(\frac{1}{\varepsilon_c} - 1 \right) \right\} \right]}$$
$$+ \left(\frac{D_o}{D_{co}} \right) \left(\frac{1}{h_w + \sigma \varepsilon_c \left(T_c^2 + T_a^2 \right) (T_c + T_a)} \right). \tag{7.46}$$

From the Raithby and Hollands the constant C_3 value can be obtained using the below equation:

7.5 Cylindrical Parabolic Collector

$$C_3 = \frac{17.74}{\left(T_{pm} + T_c\right)^{0.4} D_o\left(D_o^{-0.75} + D_{ci}^{-0.75}\right)}. \tag{7.47}$$

T_c (cover temperature) can be obtained as

$$\left(\frac{T_c - T_a}{T_{pm} + T_a}\right) = 0.04075 \left(\frac{D_o}{D_{co}}\right)^{0.4} h_w^{-0.67} \left[2 - 3\varepsilon_p + \frac{\left(6 + 9\varepsilon_p\right)T_{pm}}{100}\right]. \tag{7.48}$$

If $333 < T_{pm} < 513$ K, and by

$$\left(\frac{T_c - T_a}{T_{pm} - T_a}\right) = 0.163 \left(\frac{D_o}{D_{co}}\right)^{0.4} h_w^{-0.67} \left[2 - 3\varepsilon_p + \frac{\left(1 + 3\varepsilon_p\right)T_{pm}}{100}\right]. \tag{7.49}$$

If $513 < T_{pm} < 623$ K.$513 < T_{pm} < 623$ K.

While using equations form (7.46)–(7.49) the values of T_{pm}, T_c, and T_a have been expressed in K, D_o, D_{ci}, and D_{co} in m, σ in W/m^2 K^4, and h_w in W/m^2 K. U_l value has been expressed in W/m^2 K. The range for Eq. (7.46) has been taken as follows:

$$0.1 \le \varepsilon_p \le 0.95,$$
$$0.0125 \le D_o \le 0.15 \,\text{m},$$
$$15 \le h_w \le 60 \,\frac{\text{W}}{\text{m}^2} - K,$$
$$273 \le T_a \le 313\text{K}.$$

The cover temperature of the glass has been estimated in the range of $-10\,°C$ to $+10\,°C$ using Eqs. (7.48) and (7.49), which is a good estimation for obtaining the U_l value using Eq. (7.46) having accuracy of $\pm1\%$ for $333 < T_{pm} < 513$ K and $\pm2\%$ for $513 < T_{pm} < 623$ K.

7.5.5 A Numerical Example

We will now illustrate the procedure for calculating the performance of a cylindrical parabolic collector through a detailed numerical example.

Example 7.5 A cylindrical parabolic collector located in Mumbai, operating in tracking mode II, is used for heating a thermic fluid. The concentrator has an aperture of 1.25 m and a length of 3.657 m, while the absorber tube (3.81 cm inner and 4.135 cm outer diameter) has a concentric glass cover (5.60 cm inner and 6.30 cm outer diameter) around it. A twisted tape with a tape twist ratio of 4 is used inside the absorber tube.

Values of other design parameters of the collector are as follows:

380   7 Solar Collectors and Thermal Conversion

Specular reflectivity of concentrator surface	0.85
Glass cover transmissivity for solar radiation	0.85
Glass cover emissivity/absorptivity	0.88
Absorber tube emissivity/absorptivity	0.95
Intercept factor	0.95.

Values of the operational and meteorological parameters are as follows:

Date	April 15
Time	1230 h (LAT)
I_b	705 W/m^2
I_g	949 W/m^2
Ambient temperature	31.9 °C
Wind speed	5.3 m/s
Mass flow rate of thermic fluid	0.0986 kg/s
Inlet temperature	150 °C.

Calculate

1. The slope of the aperture plane and the angle of incidence on the aperture plane,
2. The absorbed flux S,
3. The convective heat transfer coefficient on the inside surface of the absorber tube,
4. The collector heat-removal factor and overall loss coefficient,
5. S. The exit temperature of the thermic fluid,
6. The instantaneous efficiency, and
7. The pressure drop.

Solution

1. Slope of the Aperture Plane and Angle of Incidence

In tracking mode II, the slope of the aperture plane and angle of incidence are given by Eqs. (7.10a, 7.10b) and (7.11). On April 15, $n = 105$,

$$\delta = 23.45 \sin\left[\frac{360}{365}(284 + 105)\right] = 9.415°.$$

Therefore, substituting $\delta = 9.415°$, $\omega = -7.5°$ and $\phi = 19.12°$ in Eqs. (7.10a, 7.10b) we have

$$\beta = 9.625° \text{ and } \cos\theta = \left[1 - \cos^2 9.415° \sin^2(-7.5°)\right]^{1/2} = 0.9917,$$
$$\theta = 7.398°.$$

7.5 Cylindrical Parabolic Collector

2. Absorbed Flux S

From Eq. (2.36),

$$r_b = \frac{0.9917}{\sin 19.12° * \sin 9.415° + \cos 19.12° * \cos 9.415° * \cos(-7.5°)} = 1.0143.$$

Therefore, substituting in Eq. (7.20) and taking $(\tau\alpha)_b = \tau\alpha$, we have

$$S = 705 * 1.0143 \left[0.85 * 0.95 * 0.85 * 0.95 + \frac{0.85 * 0.95 * 0.04135}{(1.25 - 0.04135)} \right]$$
$$= 486.03 \text{ W/m}^2.$$

3. Convective Heat Transfer Coefficient h_f

We will use Eq. (7.42) for calculating h_f. The properties of the fluid are given in Fig. 7.18. Keeping in mind that the rise of temperature of the fluid will only be a few degrees in this case, properties will be taken at a mean fluid temperature of 152 °C. Thus,

$$\rho = 750.3 \text{ kg/m}^3, C_p = 2.449 \text{ kJ/kg K},$$
$$v = 2.42 \times 10^{-6} \text{ m}^2/\text{s}, k = 0.119 \text{ W/m K},$$

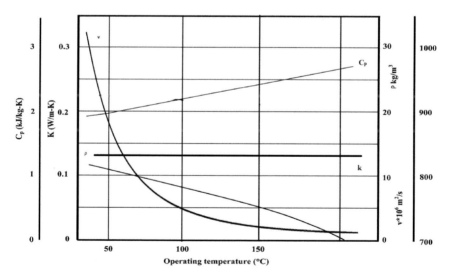

Fig. 7.18 Example 7.5—properties of thermic fluid

Average velocity $V = \frac{\dot{m}}{\frac{\pi}{4}D_i^2\rho} = \frac{0.0986}{\frac{\pi}{4}*0.0381^2*750.3} = 0.1153 \frac{m}{s}$, $V = \frac{\dot{m}}{\frac{\pi}{4}D_i^2\rho} = \frac{0.0986}{\frac{\pi}{4}*0.0381^2*750.3} = 0.1153 \frac{m}{s}$,

Reynolds number $Re = \frac{VD_i}{v} = \frac{0.1153*0.0381}{2.42*10^{-6}} = 1815$,

Prandtl number $Pr = \frac{C_p v\rho}{k} = \frac{2.499*2.42)*10^{-6}*750.3*1000}{0.119} = 37.37$,

Nusselt number $Nu = 5.172\left[1 + 0.005484\left\{37.37\left(\frac{1815}{4}\right)^{1.78}\right\}^{0.7}\right]^{0.5} = 61.70$.

Therefore, $h_f = 61.70 * \frac{0.119}{0.0381} = 192.7 \frac{W}{m^2}$ K.

4. Collector Heat-Removal Factor and Overall Loss Coefficient

As in the case of the flat-plate collector, an iterative procedure will be required since the values of F_R and U_l cannot be directly determined and the value of one depends on the other.

Assume $U_1 = 13.28$ W/m^2 K. From Eq. (7.25), the collector efficiency factor

$$F' = \frac{1}{13.28\left[\frac{1}{13.28} + \frac{0.04135}{0.0381*192.7}\right]} = 0.9304,$$

$$\frac{\dot{m}Cp}{\pi D_o U_1 L} = \frac{0.0986 * 2.449 * 10^3}{\pi * 0.04135 * 13.28 * 3.657} = 38.275.$$

Therefore, from Eq. (7.30), heat-removal factor.

$$F_R = 38.275\left[1 - \exp\left(-0.9304/38.275\right)\right] = 0.9192.$$

Concentration ratio $C = \frac{(1.25-0.04135)}{\pi*0.04135} = 9.304.\pi * 0.04135 = 9.304.$
Thus, from Eq. (7.29),

useful heat gain rate

$$q_u = 0.9192(1.25 - 0.04135) * 3.657 * \left[486.03 - \frac{13.28}{9.304}(150 - 31.9)\right]$$
$$= 1289.8 \, W.$$

Therefore, rate of heat loss

$$= (W - D_o)LS - q_u$$
$$= (1.25 - 0.04135) * 3.657 * 486.03 - 1289.8$$
$$= 858.46 \, W = \pi D_o L U_l(T_{pm} - T_a).$$

Hence $(T_{pm} - T_a) = \frac{858.46}{\pi*0.04135*3.657*13.28} = 136.07 \, °C$,

7.5 Cylindrical Parabolic Collector

$$T_{pm} = 167.97\,^{\circ}C = 441.13\,K.$$

We will now calculate the value of U_1, corresponding to this value of T_{pm} and show that it is equal to the assumed value. The procedure of Example 7.4 will be followed.

Assume $T_c = 60.23\,^{\circ}C = 333.39\,K$.

From Eqs. (7.35) and (7.37), we get

$$h_{p-c} = 5.113\,W/m^2\,K.$$

From Eq. (7.38), $h_w = 34.119\,W/m^2\,K$.

Substituting these values of T_c, h_{p-c} and h_w in Eqs. (7.33) and (7.34), we have

$$\frac{q_l}{L} = 5.113(441.13 - 333.39)\pi * 0.04135$$

$$+ \frac{5.67 * 10^{-8} * \pi * 0.04135\left(441.13^4 - 333.39^4\right)}{\left\{\frac{1}{0.95} + \frac{0.04135}{0.0560}\left(\frac{1}{0.88} - 1\right)\right\}} = 234.5\,\frac{W}{m},$$

and

$$\frac{q_l}{L} = 34.119(333.39 - 305.06)\pi * 0.063 + 5.67 * 10^{-8}$$

$$* 0.063 * 0.88\left(333.39^4 - 299.06^4\right) = 234.5\,\frac{W}{m}.$$

The two values of (q_l/L) match with each other. The corresponding value of U_1 is given by

$$U_1 = \frac{234.5}{\pi * 0.04135 * (441.13 - 305.06)} = 13.27\,\frac{W}{m^2}\,K,$$

which also matches the original guess.

5. Exit Temperature

Equating the heat gained by the fluid to the useful heat gain rate, we get

$$0.0986 * 2.449 * (T_{fo} - 150) = \frac{1289.8}{1000}.$$

Hence $T_{fo} = 155.34\,^{\circ}C$.

6. Instantaneous Efficiency

Using Eq. (7.31).

$$\eta_i = \frac{1289.8}{(705 * 1.0143 + 244 * 0.9930) * 1.25 * 3.657} = 0.295.$$

Using Eq. (7.32)

$$\eta_{ib} = \frac{1289.8}{705 * 1.0143 * 1.25 * 3.657} = 0.395.$$

7. **Pressure Drop**

We use Eq. (7.43) to get

$$\frac{\mathrm{Re}}{X\frac{1815}{4}},$$

$$C_2 = 13.0964.$$

Hence $f\mathrm{Re} = 13.0964 * (453.75)^{0.3} = 82.069^{0.3} = 82.069$

$$f = 0.0452$$

$$\Delta p = \frac{4f\rho L V^2}{2D_i} = \frac{4 * 0.0452 * 750.3 * 3.657 * 0.1153^2}{2 * 0.0381}$$

$$= 86.6\,\mathrm{N/m^2} = 0.88\,\mathrm{cm}\ \mathrm{of\ water.}$$

This is a reasonable value.

The values of η_i and η_{ib} obtained in this example are low and the difference between them is significant. The low values are partly due to the fact that the inlet fluid temperature is rather high for the given concentration ratio, while the difference between the two values is due to the high diffuse component in the global radiation.

In order to obtain a break-up of the losses occurring in the collector, we distinguish between optical losses and thermal losses. Optical losses are those which occur in the path of the incident solar radiation before it is absorbed at the surface of the absorber tube, while thermal losses are due to convection and reradiation from the absorber tube and conduction through the ends. On this basis, we define an optical efficiency η_o, as the fraction of the solar radiation incident on the aperture of the collector which is absorbed at the surface of the absorber tube. Thus,

$$\eta_o = \frac{I_b r_b \rho \gamma (\tau\alpha)_b (W - D_o) L + I_b r_b (\tau\alpha)_b D_o L}{I_b r_b W L}$$

$$= \rho\gamma(\tau\alpha)_b \frac{(W - D_o)}{W} + (\tau\alpha)_b \frac{D_o}{W} = \frac{S}{I_b r_b} \frac{(W - D_o)}{W}. \tag{7.50}$$

Substituting the numerical values of this example, we get

$$\eta_o = \frac{486.03}{705 * 1.0143} * \frac{(1.25 - 0.04135)}{1.25} = 0.657.$$

7.5 Cylindrical Parabolic Collector

7.5.6 Parametric Study of Collector Performance (Kelkar 1982)

We now investigate the impact of a few key factors on the performance of a cylindrical parabolic collector, just as we did in the case of the flat-plate collector. This was accomplished by creating a MATLAB computer programme that executes the process of Example 7.5.

Performance Over a Day with Different Tracking Modes

The performance of the collector in Example 7.5 is first examined over two days, i.e. April 15 and December 15, using incoming data for solar radiation, air temperature, and wind speed. The computations are performed for each of the tracking modes mentioned in Sect. 7.5.2. Tables 7.4 and 7.5 provide the input data and the outcomes. The last column represents the average values of the efficiency over a day.

Table 7.4 Cylindrical parabolic collector performance for the whole day in different tracking modes (15 April, Mumbai)

LAT (h)		0730	0830	0930	1030	1130	1230	1330	1430	1530	1630	Average value
I_b (W/m^2)		145	292	465	593	684	705	665	562	385	205	
I_d (W/m^2)		145	201	235	252	252	244	225	200	175	130	
T_a (°C)		27.8	29.6	31.0	31.8	32.0	31.9	31.8	31.5	31.0	30.2	
V_∞ (m/s)		1.7	2.3	3.1	4.1	4.9	5.3	5.4	5.4	5.3	4.9	
Mode I	q_u (W)	–	195.9	661.4	1000	1239	1289	1177	888.8	398.4	–	
	η_i (%)	–	8.7	20.6	25.7	28.7	29.5	28.7	25.4	15.6	–	22.1
	η_{ib} (%)	–	14.7	30.9	36.5	39.1	39.5	38.2	34.3	22.7	–	31.7
Mode II	q_u (W)	–	200.0	662.5	1001	1240	1289	1183	890.1	403.8	–	
	η_i (%)	–	8.9	20.6	25.7	28.7	29.5	28.7	25.4	15.8	–	22.1
	η_{ib} (%)	–	15.0	30.9	36.5	39.1	39.5	38.2	34.4	22.9	–	31.7
Mode III	q_u (W)	378.5	680.8	971.5	1111	1229	1280	1300	1260	1030	710.3	
	η_i (%)	18.1	23.5	26.7	27.5	28.6	29.3	30.4	31.1	29.5	26.2	27.7
	η_{ib} (%)	23.4	31.7	36.5	37.9	39.0	39.4	39.5	39.1	36.4	31.1	36.4
Mode IV	q_u (W)	366.0	664.8	958.4	1106	1229	1280	1294	1245	1010	692.8	
	η_i (%)	16.3	22.0			28.8	29.5	30.2	30.5	28.8	24.5	27.1
	(%)	23.0	31.4	36.2	37.8	39.0	39.4	39.5	38.9	36.1	30.7	36.2
Mode V	q_u (W)	379.0	682.0	979.7	1129	1255	1304	1321	1270	1032	710.9	
	η_i (%)	18.1	23.5	26.8	27.8	28.9	29.7	30.6	31.2	29.6	26.2	27.9
	η_{ib} (%)	23.5	31.7	36.6	38.1	39.3	39.6	39.7	39.2	36.4	31.1	36.5

Table 7.5 Cylindrical parabolic collector performance for the whole day in different tracking modes (15 December, Mumbai)

LAT (h)		0730	0830	0930	1030	1130	1230	1330	1430	1530	1630	Average value
I_b (W/m^2)		81	250	404	518	580	580	519	405	255	94	
I_d (W/m^2)		61	92	113	129	137	139	133	119	98	61	
T_a (°C)		23.6	25.3	27.4	29.0	30.0	30.4	30.3	29.8	28.8	27.4	
(m/s)		2.3	2.8	2.9	2.9	3.0	3.1	3.4	3.6	3.4	3.1	
Mode I	q_u (W)	–	514.5	998.8	1341	1522	1521	1337	997.0	539.1	–	
	η_i (%)	–	22.4	31.4	35.1	36.6	36.5	34.8	31.0	22.9	–	29.8
	η_{ib} (%)	–	26.8	36.6	40.7	42.2	42.2	40.5	36.4	27.5	–	35.0
Mode II	q_u (W)	–	539.8	1005	1342	1522	1521	1338	1004	564.9	44.5	
	η_i (%)	–	23.4	31.6	35.2	36.5	36.5	34.9	31.3	23.9	3.4	30.1
	η_{ib} (%)	–	27.5	36.7	40.7	42.2	42.2	40.5	36.5	28.2	4.0	35.1
Mode III	q_u (W)	328.6	786.2	928.8	971.2	981.5	980.8	967.9	927.3	815.4	475.4	
	η_i (%)	18.5	29.1	30.3	30.0	29.7	29.6	29.7	30.0	29.4	23.4	28.6
	η_{ib} (%)	20.6	33.0	35.5	36.3	36.5	36.5	36.1	35.4	33.6	25.7	33.7
Mode IV	q_u (W)	378.7	935.3	1178	1304	1362	1362	1301	1177	967.2	533.6	
	η_i (%)	19.4	30.7	33.3	34.2	34.5	34.4	33.9	32.9	31.0	24.0	31.7
	η_{ib} (%)	22.5	35.5	38.8	40.3	40.9	40.9	40.1	38.7	36.0	27.4	37.0
Mode V	q_u (W)	467.6	1074	1339	1474	1537	1537	1471	1336	1109	636.4	
	η_i (%)	23.3	33.9	36.0	36.6	36.8	36.7	36.3	35.6	34.1	27.7	34.6
	η_{ib} (%)	25.5	37.4	40.5	41.8	42.3	42.3	41.6	40.3	37.9	30.0	39.0

7.5 Cylindrical Parabolic Collector

Figures 7.19 and 7.20 provide plots representing how the instantaneous efficiency and angle of incidence changed over time on one specific day, namely December 15. The basic trend of efficiency fluctuation throughout a day is the same for all modes, as can be seen in Fig. 7.19. Efficiency initially rises, achieves a peak value near noon, and then falls. This is because of the efficiency, which follows the change of the incoming beam radiation, is significantly affected by it.

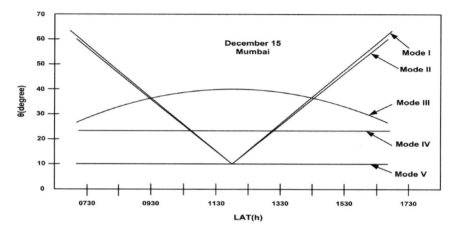

Fig. 7.19 Variation of angle of incidence in different tracking modes—data of Example 7.5

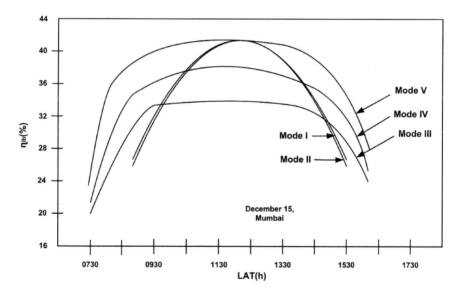

Fig. 7.20 Variation of efficiency of a cylindrical parabolic collector in different tracking modes—data of Example 7.5

The performance of the collector in mode V needs to be the best at all times, regardless of the day or the hour. Consequently, an evaluation of the collector's performance in the other modes may be done by comparing it to the best possible performance.

Modes I and II The collector is oriented about a horizontal E-W axis in mode II. As a result, tracking accounts for changes in the sun's height but not its azimuthal direction. Hence, the incidence angle may sometimes be quite high away from solar noon, as seen in Fig. 7.19. Additionally, since there is less insolation during these hours of the day, there is low absorption efficiency. However, the same is achieved with mode V at noon, when the sun's beams fall normally at the collector surface.

Mode I and mode II are nearly equivalent. Calculations indicate that during all times of the year, performance nearly resembles that attained with mode II. However, because there is only one correction each day, the incident angles are a little bit higher than in mode II, and as a result, the efficiency is slightly poorer. However, it should be emphasized that in the computations, the intercept factor was considered to be constant. Since the image is not developed on the focal axis, this may not applicable for mode I.

Mode III This mode involves rotating the collector along a horizontal N-S axis. As a result, tracking handles the sun's azimuthal swing but not its altitude swing. On both the days under consideration, the incident angle is low in the morning and evening and reaches its peak at noon. However, the change in the incident angle over a day is not as large as in modes I and II. Because of this, this mode performs substantially superior at times other than solar noon than modes I and II during the same hours. On the other hand, since the incident angle is not zero at noon time, the efficiency under this mode is not equivalent to that achieved with mode V. As a result, the efficiency variation curve for mode III intersects the efficiency variation curves obtained with modes I and II and tends to remain flat for 4–5 h each day, as shown in Fig. 7.20. This results in a fairly uniform useful heat gain.

The collector's performance is now quite sensitive to both the operating day and the latitude. Since Mumbai is located at a latitude of 19.12°N and the declination on April 15 is only + 9.4°, the incidence angle for the entire day is low in the example considered. It varies from a minimum of 1.8° at 0730 h to a maximum of 9.7° at noon. Thus, on April 15, the collector's performance in mode III at all hours of the day is extremely close to that attained in mode V. On the other hand, the performance on December 15 is notably different from that achieved in mode V since the declination is -23.3°. The CPC performance for the entire day on 15 April and 15 December for different tracking modes is given in Table 7.4 and 7.5, respectively.

Mode IV In this mode, the collector is rotated along an N-S axis parallel to the earth's axis of rotation and the incidence angle always seems to be equivalent to the declination angle. Consequently, the collector's performance is unaffected by latitude. Since the declination fluctuates only within the ranges of − 23.45° and + 23.45°, the performance in this mode is generally near to that achieved in mode V.

7.5 Cylindrical Parabolic Collector

When the declination angle is zero on two equinox days of March 21 and September 21, the performance is identical to mode V.

Effect of Inlet Temperature As the fluid inlet temperature rises, so does the surface temperature of the absorber tube. Consequently, reradiation and convection losses to the environment rise, leading in a reduction of efficiency. This is evident from Eq. (7.29), which yields the collector's usable heat gain rate. In order to show this effect, calculations are done for the case of the parabolic collector of Example 7.5 operating under the same conditions with only inlet temperature varying from 120 to 180 °C. The results are shown in Fig. 7.21. The value of η_{ib} is seen to drop significantly with T_{fi}, the decrease being slightly nonlinear. The nonlinearity is caused by the fact that as T_{fi} increases the value of the overall loss coefficient increases significantly. Figure 7.21 also represents the optical efficiency value, which does not vary with inlet temperature. Losses due to reradiation and convection are represented by the difference between η_o and η_{ib}.

Effect of Mas Flow Rate Effect: The value of the internal heat transfer coefficient h_f increases as the mass flow rate of the thermic fluid increases. As a result, the collector efficiency factor and collector heat-removal factor increase and the efficiency increase. In Fig. 7.22, the mass flow rate is changed from 0.0329 to 0.1315 kg/s to highlight this impact. With rising values of $\dot{m}$, the slope of the efficiency curve continues to decrease, and the value of η_{ib}, tends towards an asymptotic value. In the same moment, the pressure drop rises, resulting in an increase in pumping power requirement. Fortunately this increase is not so rapid because of the high Prandtl number of the fluid and the presence of a twisted tape. Consequently, the optimal value of $\dot{m}$ would be one for which the asymptotic value of η_{ib} has been almost obtained without a too large pressure decrease. In the current case, this optimal value seems to be approximately 0.12 kg/s.

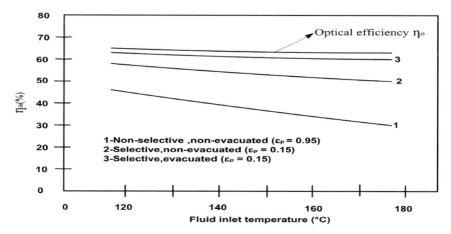

Fig. 7.21 Variation of efficiency with fluid inlet temperature for three receiver designs—data of Example 7.5

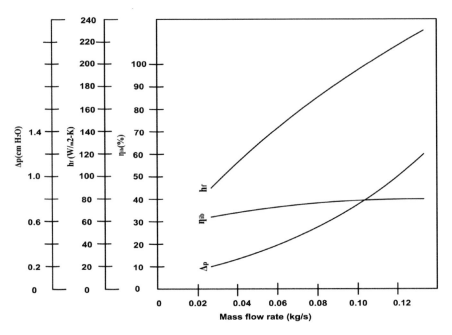

Fig. 7.22 Variation of performance with mass flow rate—data of Example 7.5

Effect of Selectivity of Absorber Tube Surface and Evacuation of Annulus: It is clear from Fig. 7.21 that losses from reradiation and convection are quite considerable. Any method to reduce or suppress these losses would obviously contribute to a large increase in collection efficiency. A selective surface is used to limit losses from reradiation, and a vacuum is created in the annular region between both the absorber tube and the glass cover to minimize the losses from convection. The effect of introducing these measures is also shown in Fig. 7.21 in which the variation of η_{ib} with T_{fi} is shown for (i) a selective surface with $\varepsilon_p = 0.15$ and a non-evacuated annulus, and (ii) a selective surface with $\varepsilon_p = 0.15$ and an evacuated annulus. As it can be observed, efficiency has dramatically increased, and the rate of decrease of η_{ib} with T_{fi} is also reduced. Of course, the decline in U_l's value is due to rise in η_{ib}. For example, at $T_{fi} = 120\ °C$, the value of U_l, decreases from 12.28 to 6.34 and 1.65 W/m^2 K respectively for the above two cases.

Effect of Concentration Ratio: In Fig. 7.23, the impact of decreasing the absorber tube's size while increasing the concentration ratio is shown. It is noticeable that efficiency is rising. This result is evident from Eq. (7.29). The optical efficiency value varies very little when the concentration ratio rises without the intercept factor reducing. However, the losses from the absorber tube which are inversely proportional to C decrease, and hence, the collection efficiency increases.

7.6 Compound Parabolic Collector (CPC)

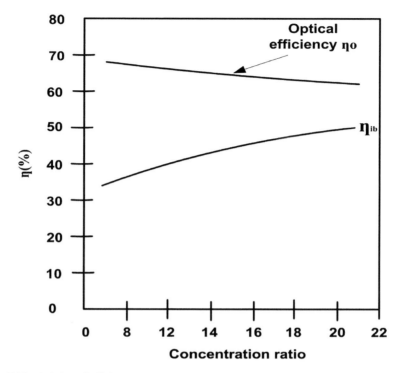

Fig. 7.23 Variation of efficiency with concentration ration—data of Example 7.5

7.6 Compound Parabolic Collector (CPC)

Like the flat-plate collector with plane reflectors, the compound parabolic concentrating collector is also a non-imaging device. It has a large acceptance angle and requires only intermittent tracking. The usefulness of the geometry of the compound parabolic collector for solar energy collection was noted by Winston (1974) and it has been the subject of considerable attention.

7.6.1 Geometry

The geometry of an ideal two-dimensional CPC is shown in Fig. 7.24. The concentrator consists of two segments AB and DC which are parts of two parabolas 1 and 2. AD is the aperture of width W, and BC is the absorber surface of width b. The axes of two parabolas are oriented to each other at an angle in such a manner that the point C is the focus of parabola 1 and point B is the focus of parabola 2. Tangents drawn to the parabolas at points A and D are parallel to the axis of the CPC.

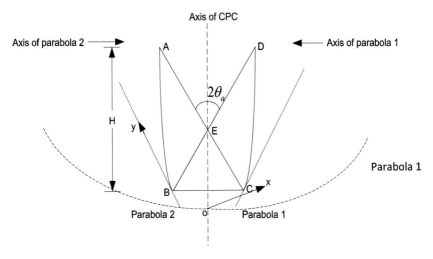

Fig. 7.24 Geometry of a compound parabolic concentrating collector

The acceptance angle of the CPC is the angle AED ($2\theta_a$), made by the lines obtained by joining each focus to the opposite aperture edge. The concentration ratio is given by $C = (W/b)$. It can be shown that $(W/b) = (1/\sin\theta_a)$ and this concentration ratio is the maximum possible for the acceptance angle $2\theta_a$.

Using the x–y coordinate system shown in Fig. 7.24, with origin O at the vertex of parabola 2, it is easy to show that the equation for parabola 2 is

$$y = \frac{x^2}{2b(1+\sin\theta_a)}, \tag{7.51}$$

where the focal length OB = $(1+\sin\theta_a)b/2$. The coordinates of the end point of the segment CD are as follows:

Point C: $x = b\cos\theta_a \quad y = \dfrac{b}{2(1-\sin\theta_a)},$

Point D: $x = (b+W)\cos\theta_a \quad y = \dfrac{b}{2}(1-\sin\theta_a)\left(1+\dfrac{1}{\sin\theta_a}\right)^2.$

The height-to-aperture ratio of the concentrator is given by

$$\frac{H}{W} = \frac{1}{2}\left(1+\frac{1}{\sin\theta_a}\right)\cos\theta_a = \frac{1}{2}(1+C)\left(1-\frac{1}{C^2}\right)^{1/2}. \tag{7.52}$$

The surface area of the concentrator is obtained by integrating along the parabolic arc. Rabl (1976b) has shown that the ratio of the surface area of the concentrator to the area of the aperture is given by the expression

7.6 Compound Parabolic Collector (CPC)

$$\frac{A_{con}}{WL} = \sin\theta_a(1 + \sin\theta_a)$$
$$\left[\frac{\cos\theta_a}{\sin^2\theta_a} + \ln\left\{\frac{(1 + \sin\theta_a)(1 + \cos\theta_a)}{\sin\theta_a[\cos\theta_a + (2 + 2\sin\theta_a]^{1/2}}\right\} - \frac{\sqrt{2}\cos\theta_a}{(1 + \sin\theta_a)^{3/2}}\right]. \quad (7.53)$$

For values of concentration ratio greater than 3, it can be shown that the following simple expression (which predicts values to accuracy better than 5%) may be used in place of Eq. (7.53)

$$\frac{A_{con}}{WL} = 1 + C. \quad (7.54)$$

Rabl (1976b) has also shown that the average number of reflections m undergone by all radiation falling within the acceptances angle, before reaching the absorber surface, is given by expression

$$m = \frac{1}{2\sin\theta_a}\left(\frac{A_{con}}{WL}\right) - \frac{(1 - \sin\theta_a)(1 + 2\sin\theta_a)}{2\sin^2\theta_a}, \quad (7.55)$$

where the value of (A_{con}/WL) is to be calculated from Eq. (7.53). Thus, the effective reflectivity of the concentrator surface is given by

$$\rho_e = \rho^m, \quad (7.56)$$

where

ρ_e effective reflectivity, and
ρ reflectivity value for a single reflection.

Example 7.6 A CPC, 1 m long, has an acceptance angle of 20°. The absorber surface of the collector is flat and has a width of 10 cm. Calculate the concentration ratio, the aperture, the height, and the surface area of the concentrator.

Solution Concentration ratio $C = 1/\sin 10° = 5.76$.
   Therefore, aperture $W = 5.76 \times 10 = 57.6$ cm.
   From Eqs. (7.52) and (7.54),

$$\frac{H}{W} = \frac{1}{2}\left(1 + \frac{1}{\sin 10°}\right)\cos 10° = 3.328, \text{ and } \frac{A_{con}}{WL} = 1 + 5.76 = 6.76.$$

Therefore, height $H = 3.328 \times 57.6 = 191.7$ cm.
Surface area of concentrator,

$$A_{con} = 6.76 \times 0.576 \times 1 = 3.90 \, \text{m}^2.$$

The results of Example 7.6 show that compared to a cylindrical parabolic collector, a compound parabolic collector (CPC) is very deep and requires large concentrator area for given aperture. Fortunately, however, it has been show that a large portion of the top of a CPC can be removed with negligible loss in performance. Thus, in practice, a CPC is generally truncated (reduced in height) by about 50% in order to reduce cost. A detailed study on the effects of truncation has also been carried out by Rabl (1976b).

7.6.2 Tracking Requirements

A two-dimensional CPC is usually oriented with its length parallel to the horizontal E-W direction and aperture plane sloping towards the south. As stated earlier, only intermittent tracking is required, the frequency of adjustment depending upon the concentration ratio. Thus, for example, for a concentration ratio of 10, the acceptance angle is 11.5° and a tracking adjustment may be needed once every few days in order to ensure collection for 7 or 8 h every day. On the other hand, for a lower concentration ratio of 5, the acceptance angle is 23.1° and a tracking adjustment may be needed only once in a month or two months.

The apparent plane of motion of the sun does not coincide with the E-W plane passing through the axis of the CPC. Hence the sun "rises" and "falls" with respect to the CPC, an extreme position being attained at solar noon. It is thus necessary to calculate the solar "swing" for the time period of the day for which collection is to be done. Referring to Fig. 7.25, assume that OG is a vertical stick whose shadow in horizontal plane is GE. EFGH is a rectangle in the horizontal plane, with EF and GH being E-W lines. Thus, angle OEG is the solar altitude angle α_a. The projection of this angle in a vertical north–south plane, i.e. angle OFG, may be called the *solar elevation angle*. We will denote it by the symbol α_v. The change in the angle α_v over the specified time period is the solar swing.

From Fig. 7.25,

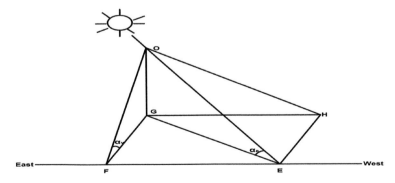

Fig. 7.25 Calculation of solar swing

7.6 Compound Parabolic Collector (CPC)

$$\tan \alpha_v = \frac{OG}{FG} = \frac{\left(\frac{OG}{OE}\right)}{\left(\frac{EH}{OE}\right)} = \frac{\sin \alpha_v}{\cos OEH}.$$

Substituting expression for $\sin \alpha_a$ and $\cos OEH$ from Eqs. (2.6) and (2.8) respectively, we have

$$\tan \alpha_v = \frac{\sin \phi \sin \delta + \cos \phi \cos \delta \cos \omega}{\sin \phi \cos \delta \cos \omega - \cos \phi \sin \delta}. \tag{7.57}$$

At noon, when $\omega = 0$, we have,

$$\tan \alpha_v = \frac{\cos(\phi - \delta)}{\sin(\phi - \delta)} = \cot(\phi - \delta).$$

$$\text{Therefore, } (\alpha_v)_{\omega=0} = \frac{\pi}{2} - (\phi - \delta). \tag{7.58}$$

The solar swing angle over a time period corresponding to an hour angle $+\omega_t$ to $-\omega_t$ is change in α_v from the time corresponding to the angle $\pm\omega_t$ to solar noon. Thus, the magnitude of solar swing angle

$$= \left| \left[\frac{\pi}{2} - (\phi - \delta) \right] - \tan^{-1} \left[\frac{\sin \phi \sin \delta + \cos \phi \cos \delta \cos \omega_t}{\sin \phi \cos \delta \cos \omega_t - \cos \phi \sin \delta} \right] \right|. \tag{7.59}$$

Equation (7.59) has been expressed in a neater form by Rabl (1976a). He defines a solar elevation angle α_v' measured with reference to the equatorial plane rather than the horizontal plane. α_v' is related to α_v by the equation

$$\alpha_v' = \alpha_v - \left(\frac{\pi}{2} - \phi \right). \tag{7.60}$$

Thus, $\tan \alpha_v' = -\cot(\alpha_v + \phi) = -\left[\frac{\cos \phi - \tan \alpha_v \sin \phi}{\tan \alpha_v \cos \phi + \sin \phi} \right]. = -\left[\frac{\cos \phi - \tan \alpha_v \sin \phi}{\tan \alpha_v \cos \phi + \sin \phi} \right].$
Substituting the expression for an $\tan \alpha_v$ from Eq. (7.57), we get

$$\tan \alpha_v' = \tan \delta / \cos \omega \text{ or } \alpha_v' = \tan^{-1}(\tan \delta / \cos \omega). \tag{7.61}$$

Thus, the magnitude of the solar swing angle

$$= |(\alpha_v')_{\omega=0} - (\alpha_v')_{\omega=\omega_t}| = \left| \delta - \tan^{-1} \left(\frac{\tan \delta}{\cos \omega_t} \right) \right|. \tag{7.62}$$

Equations (7.59) and (7.62) give the same values for the solar swing angle. However, Eq. (7.62) has the advantage of being simpler and of showing explicitly that the solar swing angle is independent of the latitude.

396   7   Solar Collectors and Thermal Conversion

Example 7.7 A CPC is located in Mumbai (19.12° N) and is to be used for 8 h of collection on December 21 without making a tracking adjustment during the day. Calculate the minimum acceptance angle required for the collector, its concentration ratio and its orientation.

Solution First, we calculate the solar swing angle on December 21 from 0800 to 1200 h (LAT) can be obtained by substituting $\delta = -23.45°$ and $\omega_t = 60°$ in Eq. (7.62),

$$\text{Solar swing angle} = \left| -23.45° - \tan^{-1}\left[\tan(-23.45°)/\cos 60°\right]\right|$$
$$= |-23.45° - (-40.94°)| = 17.5°.$$

The minimum acceptance angle required for the collector is obviously equal to the solar swing angle. Thus,

$$2\theta_a = 17.5°, \text{ and concentration ratio } C = (1/\sin 8.75°) = 6.57.$$

The slope of the collector aperture plane would have to be adjusted such that the sun's rays enter parallel to the axis of the upper parabolic segment at 0800 h. This will ensure that the sun's rays at noon enter parallel to the axis of the lower parabolic segment. Now from Eq. (7.58),

$$(\alpha_v)_{\omega=0} = \frac{\pi}{2} - (19.12° + 23.45°) = 47.43°.$$

Therefore, angle made by the axis of the lower parabolic segment with the horizontal $= 47.43°$.

Angle made by axis of CPC with the horizontal

$$= 47.43° - \theta_a = 38.68°.$$

It should be noted that the solar swing angle is maximum on the solstice days, viz. June 21 and December 21, and equal to zero on the equinox days, viz. March 21 and September 21. Thus a collector having an acceptance angle of 17.5° would give 8 or more hours of collection on all the days of the year without requiring a diurnal tracking adjustment.

Equation (7.62) can also be used for calculating the collection time per day for a CPC with a given acceptance angle or for calculating the number of adjustments required in a year for a CPC with a given acceptance angle and a required minimum collection time per day. This calculation has been done by Rabl (1976a). The procedure adopted was as follows: On June 21, the collector is assumed to be oriented with its axis pointing at an angle θ_a above the solar noon altitude angle. It is left in this position until the day when the collection time falls below the minimum specified value. On this day, it is adjusted with its axis again pointing at an angle θ_a above the

7.6 Compound Parabolic Collector (CPC)

solar noon altitude angle. The procedure is repeated over a time span of one year in order to obtain the number of adjustment per year.

7.6.3 Performance Analysis

Consider a compound parabolic collector having an aperture W, length L, and acceptance angle $2\theta_a$ (Fig. 7.26). The absorber surface has a width b. The heat collected at the absorber surface is transferred to a fluid flowing through N tubes, having an outer diameter D_0 and an inner diameter D_i, attached to the bottom side.

The fluid enters at a temperature T_{fi}, leaves at a temperature T_{fo}, and has a mass flow rate $\dot{m}$. The aperture, which is covered with a transparent sheet, is assumed to be sloping south at such an angle that the beam radiation incident on it is within the acceptance angle of the collector. It is also assumed that the length of the absorber surface is a little more than the length L so that it intercepts all the reflected radiation.

We first derive an expression for the flux S absorbed at the absorber surface. Because of its large acceptance angle, a CPC accepts both beam and diffuse radiation. The beam radiation flux falling on the aperture plane is $I_b r_b$, while the diffuse radiation flux within the acceptance angle is given by (I_d/C). Thus, the total effective flux entering the aperture plane is $[I_b r_b + (I_d/C)]$, and

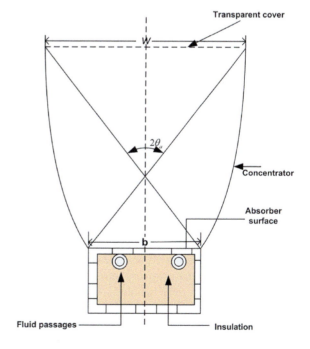

Fig. 7.26 Compound parabolic collector with flat absorber surface

$$S = \left[I_b r_b + \frac{I_d}{C} \right] \tau \rho_e \alpha, \tag{7.63}$$

where

τ transmissivity of the cover,
ρ_e effective reflectivity of the concentrator surface for all radiation,
α absorptivity of the absorber surface.

The values of τ, ρ_e, and α are assumed to be the same for beam and diffuse radiation. It is to be noted that the flux S is based on the area of the aperture. In order to obtain an expression for the useful heat gain rate, we take an energy balance on an elementary slice dx of the absorber surface at a distance x from the inlet. This yields the equation

$$dq_u = \left[S - \frac{U_l}{C}(T_p - T_a) \right] W dx. \tag{7.64}$$

Proceeding along lines similar to adopted for cylindrical parabolic collector, we obtain

$$q_u = F_R W L \left[S - \frac{U_l}{C}(T_{fi} - T_a) \right], \tag{7.65}$$

where

$$F_R = \frac{\dot{m} C_p}{b U_l L} \left[1 - \exp \left\{ \frac{-F' b U_l L}{\dot{m} C_p} \right\} \right], \tag{7.66}$$

$$\text{and } \frac{1}{F'} = U_l \left[\frac{1}{U_l} + \frac{b}{N \pi D_i h_f} \right]. \tag{7.67}$$

The instantaneous collection efficiency is then given by Eq. (7.31).

The main difficulty associated with the use of Eq. (7.65) is that the overall loss coefficient U_l cannot be calculated with the same degree of accuracy as in be case of a flat-plate collector or a cylindrical parabolic collector. Adequate convective heat transfer correlations are not available for the purpose and the radiation heat exchange calculation is also more complicated. Based on some approximate correlations, Rabl (1976b) has the estimated the value of U_l, for different values of the absorber plate temperature, plate emissivity and the concentration ratio. These values are given in Table 7.6. It is seen that they vary from 4 to 19.4 W/m^2 K. values of U_l, for situations differing from those given in Table 7.6 may be obtained by interpolation.

Apart from the flat one-sided absorber surface shown in Fig. 7.26, CPCs can also be designed for other absorber surfaces. Two shapes, (a) tubular and (b) tubular with longitudinal fins, are shown in Fig. 7.27. Unlike the flat absorber surface, these shapes have the advantage of not having a back side through which heat can be lost.

7.6 Compound Parabolic Collector (CPC)

Table 7.6 Overall loss coefficient in a compound parabolic concentrator (*Units* W/m² K)

T_{pm} (°C)	C = 1.6		4.0		8.0	
	$\varepsilon_p = 0.1$	0.9	0.1	0.9	0.1	0.9
110	4.0	8.2	5.6	10.5	6.9	11.8
210	5.0	11.8	7.8	16.0	10.4	19.4

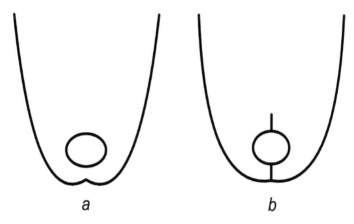

Fig. 7.27 Other absorber shapes used in CPCs: **a** Tabular and **b** tubular with longitudinal fins

The CPCs discussed so far have been symmetrical. Asymmetrical CPCs have also been developed for meeting seasonally varying outputs. Figure 7.28 shows a sketch of an asymmetric CPC. The concentrator consists of a single parabola CD with focus at A and apex at C and a circular arc BC, while the absorber is the underside of the flat surface AB. The entire radiation incident on the aperture AD of the concentrator is ultimately reflected to the underside of AB.

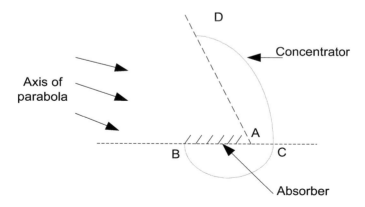

Fig. 7.28 Schematic diagram of a an asymmetric CPC

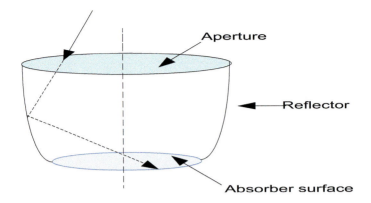

Fig. 7.29 Schematic diagram of a 3-D CPC

In addition to 2-D CPCs, 3-D CPCs have also been developed. Figure 7.29 shows a sketch of a 3-D CPC. It is a surface formed by rotating a 2-D CPC. 3-D CPCs have been used in few applications. An example is their use as secondary concentrators for augmenting the concentration of solar flux in a central receiver collector (see Sect. 7.8.3). The aperture of the 3-D CPC is placed at the focal plane of the primary concentrator, while the absorber part coincides with the aperture of the solar receiver.

CPCs are also being used for various optical applications. Recently they have been used for photo-catalytic disinfection of water. In this case, the reactor is made up of a non-tracking CPC with absorber tubes through which the water to be disinfected flows (Blanco-Galvez et al. 2007).

Example 7.8 A CPC is mounted on a horizontal E-W axis and oriented with its aperture plane sloping at an angle of 40°. The concentration ratio of the collector is 6.5, the width of its absorber plated is 6 cm, and its length is 2 m. The collector is used for heating a fluid $(C_p = 2.35 \text{ kJ/kg K})$ which enters at a temperature of 130 °C. Calculate the exit temperature of the fluid and the instantaneous collection efficiency for the following situation:

Location of collector	New Delhi (28.58° N)
Date	November 5
Time	1100 h (LAT)
I_g	0.735 kW/m^2
I_d	0.162 kW/m^2
Number of tube	2
Outer diameter of tube	18 mm
Inner diameter of tube	14 mm
Glass cover transmissivity	0.89

(continued)

7.6 Compound Parabolic Collector (CPC)

(continued)

Location of collector	New Delhi (28.58° N)
Concentrator reflectivity	0.87
Absorber surface absorptivity	0.94
Overall loss coefficient	10.5 W/m^2 K
Heat transfer coefficient on inside of absorber tube	230 W/m^2 K
Fluid mass flow rate	1.25 kg/min
Ambient temperature	21 °C

Solution

Geometry of Collector

Half acceptance angle $\theta_a = \sin^{-1}(1/6.5) = 8.85°$.

Therefore, acceptance angle $= 17.70°$.

Aperture width $W = 0.06 \times 6.5 = 0.39$ m.

Substituting the value of θ_a in Eq. (7.53), we get

$$\frac{A_{con}}{WL} = 0.1538 \times 1.1538(41.7459 + 1.7832 - 1.1275) = 7.524.$$

Therefore, from Eq. (7.55).

Average number of reflections $m = 1.12$.

Flux S

The flux S is calculated from Eq. (7.63). Before doing so, we must check that the beam radiation is within the acceptance angel of the collector.

On November 5, $n = 309$ and $\delta = -16.55°$. Therefore, from Eq. (7.57), the solar elevation angle is given by

$$\tan \alpha_v = \frac{\sin 28.58° \sin(-16.55°) + \cos 28.58° \cos(-16.55°) \cos 15°}{\sin 28.58° \cos(-16.55°) \cos 15° - \cos 28.58° \sin(-16.55°)} = 0.9766,$$

$$\alpha_v = 44.32°.$$

Since the slope of the aperture plane is 40°, the axis of the CPC makes an angle of 50° with the horizontal. Thus, beam radiation having a solar elevation angle between the limits of (50° ± 8.85°), i.e. between 58.85° and 41.15°, would be accepted by the collector. In the present problem, the beam radiation is within these limits.

Now, from Eq. (2.36 from Chap. 2), tilt factor

$$r_b = \frac{\begin{array}{c}\sin(-16.55°)\sin(28.58° - 40°)\\+\cos(-16.55°)\cos 15° \cos(28.58° - 40°)\end{array}}{\sin 28.58° \sin(-16.55°) + \cos 28.58° \cos(-16.55°)\cos 15°} = 1.4243.$$

Therefore, $S = (573 \times 1.4243 + \frac{162}{6.5})0.89 \times 0.87^{1.12} \times 0.94 = 602.3\,\text{W/m}^2 .0.89 \times 0.87^{1.12} \times 0.94 = 602.3\,\text{W/m}^2$.

Useful Heat Gain Rate

From Eqs. (7.65)–(7.67), $\frac{1}{F'} = 10.5\left[\frac{1}{10.5} + \frac{0.06}{2\times\pi\times0.014\times230}\right] = 1.0311$,

$$F' = 0.9698$$

$$\frac{\dot{m}C_p}{bU_lL} = \frac{1.25}{60} \times \frac{2350}{0.06 \times 10.5 \times 2} = 38.856,$$

$$F_R = 38.856\left[1 - \exp(-0.9698/38.856)\right] = 0.9578,$$

Useful heat gain rate $q_u = 0.9578 \times 0.39 \times 2 \times \left[602.3 - \frac{10.5}{6.5}(130 - 21)\right] = 318.4\,\text{W}$.

Exit Fluid Temperature and Collection Efficiency

The exit fluid temperature is calculated from the equation

$$(1.25/60) \times 2350 \times (T_{fo} - 130) = 318.43,$$
$$T_{fo} = 136.50\,°\text{C}.$$

From Eq. (7.31)

$$\eta_i = \frac{318.4}{\left[573 \times 1.4243 + \frac{162(1+\cos 40°)}{2}\right]0.39 \times 2} = 0.4256.$$

7.7 Paraboloid Disc Collector

Science Applications International Corp. and STM Power, Inc., USA, has developed the Sun Dish system with a peak output of 22.9 kW. The paraboloid concentrator of the system is made up of 16 rounds, stretched-membrane mirror facets, each 3.2 m diameter, which is mounted on a truss structure. The total reflecting area is 117.2 m² with a reflectivity of 0.95. The truss is attached to an azimuth/elevation drive on top of a pedestal. A distinguishing feature of the concentrator is that the facets are attached to the truss in a staggered manner so that there are gaps between the facets.

7.7 Paraboloid Disc Collector 403

This helps to reduce the wind load. The focal length of the concentrator is 12 m. The receiver is a cavity with an aperture having a diameter of 38 cm. The cavity contains the heater head of a Stirling engine in the shape of a truncated cone on which the insolation falls directly. The intercept factor for the radiation reflected through the receiver aperture is 0.90 and the concentration ratio is 2500. A four-cylinder Stirling engine is used. Burners are provided in the receiver to use other fuel for running the engine when solar energy is not available (Mancini et al. 2003b).

The "Euro Dish" system has been developed under a joint-venture project between the European Community and German/Spanish industries and research institutions. The concentrator consists of a glass-fibre composite shell onto which glass mirrors (reflectivity $= 0.94$) are bonded with an adhesive. It is 8.5 m in diameter and the total glass area is 60 m^2. The concentrator dish is supported by a space frame ring truss, and the whole concentrator is suspended on a space frame turntable rolling on six wheels. A cavity-type receiver with an aperture diameter of 15 cm is kept at the focus (focal length $= 4.5$ m) and is directly attached to the cylinder heads of a Stirling engine. The engine is coupled to an induction generator which has produced a peak output of 8.5 kW.

The SES system has a peak output of 25.3 kW. The concentrator is approximately paraboloid and is made from 82 reflecting glass facets with a total area of 91.0 m^2. The reflectivity of the glass is 0.91. The diameter of the aperture of the receiver is 20 cm, and the collector has an intercept factor of 0.97. Like other systems, the temperature in the receiver is around 700 °C. The prime mover is a four-cylinder Stirling engine operating at 1800 rpm.

The receiver is an important component of a paraboloid dish system. It is difficult to design because it has to receive a high heat flux (of the order of 10^6 W/m^2 and at about 700 °C), absorb it, and transfer the energy to the working fluid of the engine. A number of new ideas are therefore being tried out. Three suggestions are as follows:

1. Heat pipes using liquid metals like sodium are being used to transfer the heat from the receiver to the engine head.
2. Hybrid receivers which can absorb solar energy as well as energy from some other source like a fossil fuel or bio-gas are also being developed. The development of such receivers would enable power to be developed in the system at all hours.
3. The use of volumetric receivers is also being experimented with. These receivers are similar to those described in Sect. 7.8 for use in central receiver power systems.

Finally, it should be mentioned that the paraboloid dish collector can be effectively used for providing process heat in industries which require heat at temperatures up to 250 °C (Kedare et al. 2006).

7.8 Central Receiver Collector

The principle of working of a central receiver power plant had been described in Sect. 7.1.1. The idea of building such a plant was first suggested by Baum et al. (1957). Based on their calculations, they indicated the possibility of erecting an installation in the sunny regions of the USSR, to produce 11–13 t of steam per hour at 30 atm and 400 °C. The optical system was calculated to consist of 1293 mirrors of dimensions 3 × 5 m. These heliostats were proposed to be mounted on carriages which moved on rails in arcs around the tower.

A number of small pilot plants were built by Francia (1968) in Italy in the period 1965–1967. In one of these, he collected 50 kW of energy. After a break of a few years, the design of central receiver collector systems again attracted attention in the eighties and a number of plants ranging in capacity from 0.5 to 10 MWe were built. These have been listed in Table 7.2 along with some technical specifications. The two major components making up a central receiver collector are the heliostats and receiver. These will now be described.

7.8.1 Heliostats

The heliostats form an array of circular arcs around the central tower. They intercept, reflect, and concentrate the solar radiation on to the receiver. The array is served by a tracking control system which ensures that each heliostat focuses beam radiation towards the receiver during collection. A two-axis tracking mode involving adjustments to the surface azimuth angle and the slope of the heliostat is usually adopted. In addition, when solar radiation is not being collected, the control system orients the heliostats in a safe direction so that the receiver is not damaged.

As stated earlier, the 10 MWe plant at Barstow was the largest of the pilot plants built. It was operated for six years, from 1982 to 1988 as Solar One and again after modifications from 1996 to 1999 as Solar Two. Solar One had a field of 1818 heliostats positioned all around a central tower of height 80 m. Each heliostat was an assembly of 12 slightly concave glass mirrors mounted on a support structure and geared drive that could be controlled for azimuth as well as elevation. The total reflective area of each heliostat was 39.3 m^2. A rear-view sketch is shown in Fig. 7.30. The 12 mirror panels in each heliostat were 1 × 3 m in size and were made from 3-mm-low iron float glass. When clean, the heliostats had an average reflectivity of 0.903. However, dirt accumulated due to exposure to the environment reduced the average reflectivity to 0.82. Solar Two had an additional 108 heliostats each having area of 95 m^2.

Degradation of the mirrored surface of the heliostats and low availability of the heliostats are major issues in the development of central receiver systems. Thus, an increase in the reflectivity and in the reliability of the heliostat system is important. In addition, reduction in cost is necessary because the cost of the heliostats and their control systems forms a significant part (about 40–50%) of the initial investment. In

7.8 Central Receiver Collector

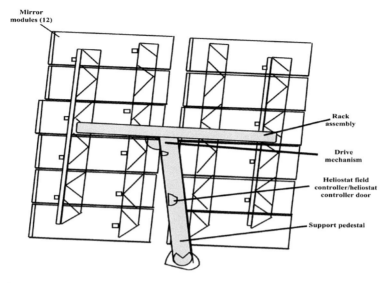

Fig. 7.30 Heliostat (Bartel 1984)

order to achieve these goals, many new concepts have been tried. For example, larger size glass-mirror heliostats having areas of 150 m^2 and reflectivity values up to 0.94 have been built. Also, a new type of heliostat using a stretched membrane has been developed. In this heliostat, the reflector is a silvered polymer film laminated to a thin metal foil which is stretched over a large-diameter metal ring. The reflectivity of this surface has been measured to be 0.92. Because of its simplicity and reduced weight, a stretched-membrane heliostat would be about 30% less costly than a glass-mirror design.

Chen et al. (2005) have suggested the concept of a non-imaging, focusing heliostat. Generally, there is a spread of the image at the receiver even if there is a slight concavity in the heliostat mirrors. As a result, the size of the receiver has to be large. In order to overcome this problem, Chen et al. have proposed a new design. The heliostat consists of a master mirror and surrounding slave mirrors, forming a two-dimensional array. A new tracking method is proposed such that all the solar images of the slave mirrors are superposed onto that formed by the master mirror at the receiver. The movements are divided into two parts. The master mirror moves with the moving frame for primary tracking, while the slave mirrors are given further local movements for realizing the desired image size. Chen et al. have built a prototype having an area of 12.96 m^2 and tested it extensively. The tracking mechanism showed much less variation in the image spread, leading to a more uniform flux, compared to a traditional heliostat (Chen et al. 2004).

7.8.2 Receiver

The receiver is the most complex part of the collection system. The main factor influencing its design is its ability to accept the large and variable heat flux which results from the concentration of the solar radiation by the heliostats. This flux has to be transferred to the receiver fluid. The value of the heat flux can range from 100 to 1000 kW/m^2, and this results in high temperatures, high thermal gradients, and high stresses in the receiver. The value depends on the concentration ratio and varies with the season and the day. It also varies over the surface of the receiver. For these reasons, attention has to be given to the absorber shape, the heat transfer fluid, the arrangement of tubes to carry the fluid, and the materials used for construction.

There are two types of receiver designs: the external type and the cavity type. The external receiver (Fig. 7.31) is usually cylindrical in shape. The solar flux is directed onto the outer surface of the cylinder consisting of a number of panels and is absorbed by the receiver fluid flowing through closely spaced tubes fixed on the inner side. On the other hand, in a cavity receiver, the solar flux enters through one or more small apertures in an insulated enclosure. The cavity contains a suitable tube configuration through which the receiver fluid flows. The geometry of the cavity is such that it maximizes the absorption of the entering radiation, minimizes heat losses by convection and radiation to the ambient, and at the same time accommodates the heat exchanger that transfers the radiant energy to the receiver fluid. Both types of receivers have their advantages and disadvantages. The external type has a very wide acceptance angle, while the cavity type has a small acceptance angle. On the other hand, the cavity-type traps the solar flux more effectively and consequently has a higher efficiency than the external type.

Solar One had an external type, tubular receiver in which water was heated directly and converted to superheated steam. The receiver was a cylinder, 7 m in diameter and 13.5 m in height, made up of 24 vertical panels painted black. Tubes made from Incoloy 800 (0.6 cm I.D., 1.25 cm O.D.) were fixed on the inside. The receiver was located on a tower 80 m high and produced steam at 102 bar and 510 °C. The receiver had an annual efficiency of 0.69, which was rather low. In order to achieve a higher efficiency and reduce the failure rate, the receiver for Solar Two was re-designed to use molten nitrate salt as the heat transfer fluid instead of water/steam. With a molten salt, the receiver could operate with higher incident solar fluxes. Consequently, the size of the receiver was reduced resulting in smaller thermal losses. In addition, the molten salt was essentially at atmospheric pressure, thereby permitting the use of thinner walled tubes in the receiver. An annual efficiency of 0.88 was attained.

A cavity-type receiver was used in the Themis power plant in France. The receiver was almost cubic having dimensions of 4 m $\times$ 4 m $\times$ 3.5 m (Fig. 7.32). The inner walls had tubes (diameter: 18 mm, thickness: 1.5 mm) in which the heat transfer fluid was circulated. The aperture of the receiver was in the focal plane of the heliostat field. Hitec salt (40% NaNO$_2$, 7% NaNO$_3$, and 53% KNO$_3$ by weight) was used as the working fluid. The typical value of the outlet temperature of the fluid was 430 °C for an inlet of 250 °C (Bonduelle and Cazin-Bourguignon 1986).

7.8 Central Receiver Collector

Fig. 7.31 External receiver

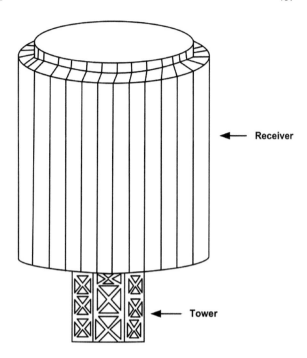

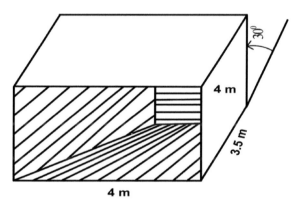

Fig. 7.32 Cavity receiver (Bonduelle and Cazin-Bourguignon 1986)

In the nineties, suggestions were made that air could be used as the heat transfer fluid in the receiver. As a result, cavity type, volumetric receivers in which tubes are not used for circulating the fluid were developed. Highly porous materials like metallic wire meshes, ceramics, or honeycomb structures are used to absorb the concentrated radiation in these receivers. Because of the porous nature, the radiation penetrates and is absorbed deep within its interior. A fan sucks the air through the absorbent pores and hot gas is obtained. Since the absorption of radiation takes place over a certain depth and not at the surface, thermal losses are reduced. On the other hand, the heat transfer area is more compared to a tubular surface. As a result, a higher quantity of heat can be transferred, resulting in a reduction of receiver size. Further, the absorption of solar radiation as well as convective heat transfer to the working fluid occurs at the same surface; this reduces the thermal stress on the material. Another advantage of using volumetric receivers is that they have provided opportunities to integrate the solar power plant with gas turbines for higher plant efficiency. This result in reduced heliostat area and a reduction in the investment cost (Romero et al. 2002).

Volumetric receivers can be open-loop or closed-loop type. A number of such receivers have been developed and tested. The open-loop volumetric receiver in the TSA plant is one example. This receiver delivered hot air at a temperature of 700–750 °C. A closed-loop receiver was developed in the REFOS project (Buck et al. 2002). This yielded hot air at 15 bar and 800 °C, a 3-D CPC being used as a secondary concentrator to increase the concentration of radiation. Fend et al. (2004) have shown that ceramic foams and ceramic fabrics are promising material from the point of view of large surface area per unit volume of material and low pressure drop. Consequently, research and development activities are in progress on open volumetric receivers with ceramic absorbers for realizing high temperatures (Hoffschmidt et al. 2003).

Finally we describe two recent suggestions concerning the design of central receiver systems. The first is the dual receiver concept (Buck et al. 2006), and the second the multi-tower solar array concept (Schramek and Mills 2004). The dual receiver concept has been proposed in order to avoid the difficulties encountered in converting feed water from the sub-cooled liquid state to superheated vapour in one tubular receiver. Figure 7.33 shows a schematic view of a dual receiver. It consists of a tubular receiver (1) located in front of an open volumetric receiver (2). The heating is now done in three stages. First the feed water is preheated from the sub-cooled liquid state to the saturated liquid state in a heat exchanger (3) that is fed by air heated in the volumetric receiver. In the second stage, the water evaporates as it flows through the tubular receiver (1) and is converted into saturated steam. Finally the saturated steam flows through another heat exchanger (4) where it is superheated. The heat required for this purpose is also supplied by air heated in the volumetric receiver. The advantages of the dual receiver concept are (i) a lower temperature in the volumetric receiver resulting in a higher efficiency and (ii) a lower auxiliary energy requirement for air circulation.

The multi-tower solar energy concept has been proposed in order to make efficient use of solar radiation falling on a given ground area. In this concept, several tower-mounted receivers with partly overlapping heliostat fields are used (Fig. 7.34). This allows utilization of radiation which would remain unused in a single tower system due to mutual blocking of the heliostats. Schramek and Mills (2004) have shown that specially shaped hexagonal reflectors can yield ground coverage of up to 100%.

7.8 Central Receiver Collector

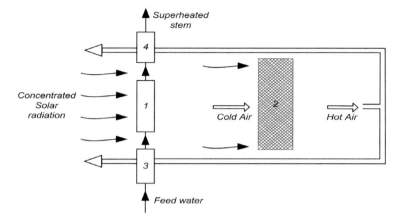

Fig. 7.33 Schematic diagram of a dual receiver. (1) Tubular receiver. (2) Volumetric receiver. (3) Heat exchanger. (4) Heat exchanger

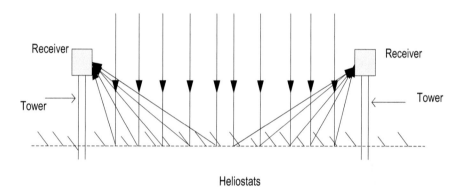

Fig. 7.34 Multi-tower concept (Schramek and Mills 2004)

7.8.3 Analysis

Consider a central receiver collector consisting of N mirrors* (*The mirrors could be circular in which case $A_m = \pi w^2/4$, or square in which case $A_m = w^2$) each of area A_m and span w covering a ground area A_g around the tower. The mirrors have to be laid out in such a manner that incident or reflected radiation associated with one heliostat is not blocked by a neighbouring heliostat. As a result, they have to be spaced apart and cover only a fraction of the ground area. If the fraction of the ground area covered is ψ, then

$$N A_m = \psi A_g. \tag{7.68}$$

In most central receiver collector designs, the value of ψ is around 0.4.

Let the receiver's absorber surface (on which the radiation is focused) have an area A_p and be located on top of a tower of height H. Taking an energy balance on the absorber, we obtain the following expression for the useful heat gain rate,

$$q_u = I_b \left[\sum_{j=1}^{N} r_{bj} \right] \rho \tau \alpha A_m - U_1 A_p (T_{pm} - T_a). \tag{7.69}$$

Defining an average tilt factor $(r_b)_{av} = \dfrac{1}{N} \sum_{j=1}^{N} r_{bj}.$ (7.70)

From using Eq. (7.68), we have

$$\begin{aligned} q_u &= I_b (r_b)_{av} \psi A_g \rho \tau \alpha - U_1 A_p (T_{pm} - T_a) \\ &= \psi A_g \left[I_b (r_b)_{av} \rho \tau \alpha - \frac{U_1}{C} (T_{pm} - T_a) \right], \end{aligned} \tag{7.71}$$

where $C = (N A_m / A_p) = (\psi A_g / A_p)$.

In order to obtain expressions for the size of the external absorber and the concentration ratio, we assume that the mirror field is circular with the tower at the centre. The rim angle, i.e. the angle made by the line joining the absorber and the outermost mirror with the vertical is taken to be ϕ_r (Fig. 7.35). Thus, the distance between the outermost mirror, and the absorber is $(H/\cos \phi_r)$. If the mirrors are flat, it follows that the size of the image at the absorber is given by

$$L_i = \frac{H}{\cos \phi_r} (\theta_s + \theta_e) + w, \tag{7.72}$$

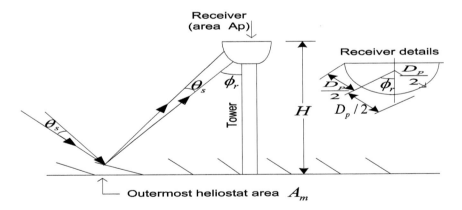

Fig. 7.35 Analysis of central receiver collector

7.8 Central Receiver Collector

where θ_s is the angle subtended by the sun at the earth and θ_e is the total angular error associated with the reflection due to factors like mirror surface imperfection and mirror orientation. If the mirrors are suitably dished, then the spread of the image due to mirror span could be eliminated, and the size L_i is obtained by putting $w = 0$ in Eq. (7.72).

Vant-Hull and Hildebrandt (1976) and Sukhatme and Nayak (2012) have suggested that an appropriate shape for the absorber could be a spherical segment with a conical section as shown in Fig. 7.35. The area of such a shape is given by

$$A_p = \frac{\pi}{2} D_p^2 \left(1 + \sin \phi_r - \frac{\cos \phi_r}{2} \right), \tag{7.73}$$

where D_p is the diameter of the sphere as well as the apparent height of the absorber when viewed from the outermost mirror. Equating D_p with the image size given in Eq. (7.72), we have

$$A_p = \frac{\pi}{2} \left\{ \frac{H}{\cos \phi_r} (\theta_s + \theta_e) + w \right\}^2 \left(1 + \sin \phi_r - \frac{\cos \phi_r}{2} \right). \tag{7.74}$$

Thus, the concentration ratio for this shape is given by

$$
\begin{aligned}
C &= \frac{N A_m}{\frac{\pi}{2} \left\{ \frac{H}{\cos \phi_r} (\theta_s + \theta_e) + w \right\}^2 \left(1 + \sin \phi_r - \frac{\cos \phi_r}{2} \right)} \\
&= \frac{\psi \pi H^2 \tan^2 \phi_r}{\frac{\pi}{2} \left\{ \frac{H}{\cos \phi_r} (\theta_s + \theta_e) + w \right\}^2 \left(1 + \sin \phi_r - \frac{\cos \phi_r}{2} \right)}. \tag{7.75}
\end{aligned}
$$

For the suitably dished mirrors, in which the spread of the image due to mirror span is eliminated, the concentration ratio is obtained by putting $w = 0$ in Eq. (7.75). Expression similar to Eq. (7.75) can be obtained for other absorber shapes.

Example 7.9 In a central receiver collector, the height of the tower is 150 m, the rim angle is 50°, and the diameter of the mirror is 4.5 m. Find the size of the image formed by the outermost mirror at the receiver, the area of the absorber (if it is as shown in Fig. 7.35) and the corresponding concentration ratio. Assume that mirrors are (i) flat and (ii) dished. Take $\psi = 0.38$ and $\theta_e = 0.002$ radians.

Solutions

(i) From Eqs. (7.72) and (7.73),

$$L_i = \frac{150}{\cos 50°} \left(\frac{32\pi}{60 \times 180} + 0.002 \right) + 4.5 = 7.14 \, \text{m},$$

$$A_p = \frac{\pi}{2} \times 7.14^2 \left(1 + \sin 50° - \frac{\cos 50°}{2} \right) = 115.67 \, \text{m}^2.$$

412   7 Solar Collectors and Thermal Conversion

Thus, $C = \frac{0.38 \times \pi \times 150^2 \times \tan^2 50^\circ}{115.67} = 330. C = \frac{0.38 \times \pi \times 150^2 \times \tan^2 50^\circ}{115.67} = 330.$

(ii) If the mirrors are dished,

$$L_i = \frac{150}{\cos 50^\circ}\left(\frac{32\pi}{60 \times 180} + 0.002\right) = 2.64\,\text{m},$$

$$A_p = \frac{\pi}{2} \times 2.64^2\left(1 + \sin 50^\circ - \frac{\cos 50^\circ}{2}\right) = 15.82\,\text{m}^2.$$

Thus, $C = \frac{0.38 \times \pi \times 150^2 \times \tan^2 50^\circ}{15.82} = 2412. C = \frac{0.38 \times \pi \times 150^2 \times \tan^2 50^\circ}{15.82} = 2412.$

Review Questions

(1) What do you mean by the solar collector concentration ratio?
(2) Sketch and discuss the flat-plate solar collector.
(3) Discuss the some benefits of flat-plate solar collector.
(4) What is a CPC collector's estimated value for the concentration ratio?
(5) Identify 3 collectors that need one-axis solar tracking.
(6) Which temperature range can a paraboloidal dish collector reach?
(7) What concentration ratios can be achieved with a central tower receiver collector?
(8) Describe how a solar collector works. How may collector coating be applied to enhance a collector's functionality?
(9) Show the structural features of a flat-plate collector and identify several solar thermal collector types. What are the key benefits?
(10) List concentrating collectors' benefits and drawbacks in comparison with flat-plate collectors.

Problems

(1) Obtain the expression

$$\psi = (\pi + \beta + 2\phi - 2\delta)/3$$

for the correct inclination of a south-facing specular reflector fixed on the top edge of a flat-plate collector and having the same dimensions as the flat-plate collector.

(2) A cylindrical parabolic collector is operated in Kolkata (22.65° N, 88.45° E) in tracking mode II. Calculate the variation in the slope of the aperture plane from 0800 to 1600 h (LAT) on June 21.

Ans: Values of are symmetric about solar noon and $\gamma = 180^\circ$

LAT (h)	0800	0900	1000	1100	1200
β (deg)	18.29	8.88	3.96	1.53	0.80

7.8 Central Receiver Collector 413

(3) A cylindrical parabolic collector is located in Pune (18.53° N) and operates in tracking mode II on May 1 Calculate the values of the slope of the aperture plane from 0600 to 1200 h (LAT) at hourly intervals and the corresponding angle of incidence. Calculate also the time at sunrise.

Ans: Sunrise at 0540 h (LAT)

LAT (h)	0600	0700	0800	0900	1000	1100	1200
γ (deg)	180	180	180	180	0	0	0
β (deg)	71.47	27.26	9.49	2.09	1.45	3.13	3.63
θ (deg)	75.10	68.98	56.81	43.10	28.89	14.48	0

(4) A cylindrical parabolic concentrating collector is operating in tracking mode III. Calculate the slope of the aperture plane and the tilt factor for beam radiation for the following conditions:

Location	21°0.06′ N,79°03′ E
Date	November 13
Time	1500 h (LAT)

Ans: 52.7°, 1.652

(5) Calculate the overall loss coefficient for an evacuated glass tube cylindrical parabolic focusing collector with the following data:

Absorber tube: outer diameter	6.5 cm
Inner diameter	6.0 cm
Glass cover: outer diameter	15.8 cm
Inner diameter	15.0 cm
Aperture	1.90 m
Length of concentrator	3.50 m
Emissivity of absorber tube surface	0.22
Emissivity/absorptivity of glass	0.88
Average temperature of absorber tube	200 °C
Ambient temperature	20 °C
Wind velocity	1.5 m/s

Ans: 2.83 W/m^2 K

(6) A cylindrical parabolic focusing collector is used for heating a thermic fluid (c_p = 2.2 kJ/kg K) which enters with a temperature of 160 °C. The concentrator has an aperture of 1.8 m and a length of 3 m. The absorber tube has an inner diameter of 2.8 cm and outer diameter of 3.2 cm and has a concentric glass cover around it. Given that

414 7 Solar Collectors and Thermal Conversion

Specular reflectivity of concentrator surface	0.82
Intercept factor	0.91
$(\tau\alpha)_b$	0.8
Beam radiation incident normally on aperture plane	556 W/m^2
Diffuse radiation incident on aperture plane	152 W/m^2
Overall loss coefficient	9.5 W/m^2 K
Convective heat transfer coefficient on inside of absorber tube	325 W/m^2 K
Ambient temperature	27 °C
Mass flow rate of fluid	360 kg/h

Calculate the useful heat gain rate, the exit temperature of the fluid and the instantaneous efficiency.

Ans: 1368 W, 166.2 °C, 35.8%

(7) A cylindrical parabolic concentrating collector is operating under tracking mode II. It is used for heating a thermic fluid ($C_p = 2.3$ kJ/kg K), which enters the receiver at a temperature of 125 °C with a flow rate of 330 kg/h. The following data are available:

Location	21°06′ N,79°03′ E
Date	October 21
Time	1400 h (LAT)
Concentrator width	2 m
Concentrator length	6 m
Absorber outer diameter	8.1 cm
Absorber inner diameter	7.5 cm
Collector overall loss coefficient	6.2 W/m^2 K
Convective heat transfer coefficient on the inside surface of the absorber tube	205 W/m^2 K
Intercept factor	0.94
Reflectivity of the concentrating surface	0.86
(ra), of the absorber and cover assembly	0.81
Ambient temperature	23 °C
Global solar radiation on horizontal surface	681 W/m^2
Diffuse solar radiation on horizontal surface	136 W/m^2

Calculate the exit temperature of the fluid, the useful heat gain rate and the instantaneous efficiency of the concentrating collector.

Ans: 144.3 °C, 4060 W, 51.1%

7.8 Central Receiver Collector   415

(8) Calculate the performance of the cylindrical parabolic collector of Example 7.2 for the situation when (i) the absorber tube surface is selective ($\varepsilon_p = 0.15$), (ii) the annulus is evacuated, and (iii) the fluid inlet temperature is 180 °C. All other data remain the same.

Ans: $U_l = 2.59$ W/m^2 K, $T_{fo} = 187.8°C$, $\eta_i = 45.1\%$, $\eta_{ib} = 60.4\%$, $\Delta p = 0.71$ cm of water

(9) Study the effect of having two glass covers instead of one on the performance of the cylindrical parabolic collector of Example 7.5. Assume that the outer glass cover has an outer diameter of 8.0 cm and an inner diameter of 7.2 cm, while the dimensions of the inner glass cover are unchanged. All other data remain the same. Calculate the value of η_{ib}, for inlet temperatures of 120, 150, and 180 °C and plot the variation. It should be noted that the curve obtained with two glass covers will intersect the curve with one glass cover (Fig. 7.21, curve 1).

Ans: 44.2, 40.2,35.7%

(10) Study the effect of varying the annular gap between the absorber tube and the glass cover on the performance of the cylindrical parabolic collector of Example 7.5. Keep all data the same and calculate the efficiency for two situations (i) $D_{co} = 5.5$ cm. $D_{ci} = 4.8$ cm, and (ii) $D_{co} = 7.1$ cm, $D_{ci} = 6.4$ cm. Compare the results with those obtained in Example 7.5. Is there an optimum annular gap?

Ans: (i) $\eta_i = 25.6\%$, $\eta_{ib} = 34.2\%$ (ii) $\eta_i = 28.9\%$, $\eta_{ib} = 38.7\%$

(11) Derive an expression for the useful heat gain rate of a parabolic dish collector having a cylindrical absorber at its focus. The diameter and length of the absorber on the outside are D_o and L_o, and on the inside, D_i and L_i. Assume that the heat transfer fluid changes its phase as it passes through the absorber and consequently remains at a constant temperature T_{fi}.

(12) An untruncated CPC has a half-acceptance angle of 36° and an absorber width of 10 cm. Calculate the concentration ratio, the aperture, the height, and the surface area of the concentrator per metre length.

Ans: 1.70, 17.0 cm, 18.6 cm, 0.382 m^2

(13) A CPC has an acceptance angle of 7.5°. Find the maximum collection period possible on April 15 without making a tracking adjustment.

Ans: 7.6 h

(14) Solve Example 7.9 by assuming that the absorber has the shape of a hemispherical bowl.

Ans: 321, 2351

(15) A central receiver collector system consists of 1800 heliostats, each 6.5 × 6 m in size. The height of the central tower is 80 m, and the rim angle is 81°. Find the size of the image formed by the outermost arc of heliostats at the receiver.

416   7 Solar Collectors and Thermal Conversion

The receiver is a vertical cylinder whose height is two times the diameter. Calculate the size of the receiver and the concentration ratio. Assume that the angular error associated with the heliostats does not exceed 0.0055 radians and that the heliostats are suitably dished.

Ans: 7.57 m. Receiver diameter $=$ 7.57 m, Receiver height $=$ 15.14 m, Concentration ratio $=$ 195

Objective Type Questions

(1) A flat-plate collector's concentration ratio value is

    (a) 10
    (b) 1000
    (c) 100
    (d) 1

(2) Aperture area in solar collector defines

    (a) receiver area
    (b) system area
    (c) area occupied by the system after installation
    (d) receiver cross-sectional area

(3) A flat-plate collector's heat-removal factor falls within the following range:

    (a) 0–1
    (b) 0.5–0.6
    (c) 0–0.1
    (d) 0.9–0.95

(4) The concentration type solar collector

    (a) boosts the solar radiation's density before capturing it
    (b) initially radiation is absorbed then the concentration increases
    (c) density dilution of radiation before absorption happens
    (d) first radiation density increases then reflection takes place.

(5) Generally solar thermal collector

    (a) absorbs radiation and dissipate to the atmosphere
    (b) collects the solar power and reflect it back
    (c) collect and transfer the solar energy into thermal energy and send it to the next stage of the system
    (d) collect and transform the solar energy into electrical energy

(6) In a central type solar collector the heliostats

    (a) have double axis tracking facility
    (b) are fixed

(c) have t-axis tracking structure

(d) are modified as per season

(7) Selective surface has the characteristics of

(a) selective absorption of short wavelength radiation

(b) selective absorption of long wavelength radiation

(c) lower absorptivity value for incoming solar radiation and higher emissivity value of outgoing reradiation

(d) higher absorptivity value for incoming solar radiation and lower emissivity value of outgoing reradiation

(8) Booster mirrors in flat-plate collector

(a) decreases the reflection to the atmosphere

(b) increases the reflection to the atmosphere

(c) increases the beam radiation component on the absorber

(d) increases the diffused radiation component on the absorber

(9) A CPC concentrator needs

(a) single axis tracking

(b) double axis tracking

(c) seasonal modification

(d) none

(10) The transparent cover in a flat-plate collector used for

(a) To minimize transmission of the incident radiation into the box

(b) To maximize transmission of the incident radiation into the box

(c) To ensure partial transmission of the incident sunlight into the box

(d) To entirely reflect the incident sunlight back

Answers

1. d	2. a	3. c	4. a	5. c
6. a	7. d	8. c	9. b	10. b

References

Bartel J (1984) 10-MWe solar thermal central receiver pilot plant. J Solar Energy Eng Trans ASME 106:50

Baum VA, Aparasi RR, Garf BA (1957) High power solar installations. Solar Energy 1:6

Blanco-Galvez J, Fernandez-Ibanez P, Malato-Rodriguez S (2007) Solar photocatalytic detoxification and disinfection of water: recent overview. J Solar Energy Eng Trans ASME 129:4

Bonduelle B, Cazin-Bourguignon AM (1986) Themis receiver: thermal losses and performance. In: Becker M (ed) Proceedings of the 3rd international workshop on solar thermal central receiver systems, Vol I: Design, construction and operation of CRS-plants, June 23–27, Konstanz, Germany, pp 273–282

Buck R, Braeuning T, Denk T, Pfaender M, Schwarzboezi P, Tellez F (2002) Solar-hybrid gas turbine-based power tower systems (REFOS). J Sol Energy Eng Trans ASME 124:2

Buck R, Barth C, Eck M, Steinmann W (2006) Dual-receiver concept of solar towers. Sol Energy 80:1249

Chen YT, Kribus A, Lim BH, Lim CS, Chong KK, Karni J, Buck R, Pfahl A, Bligh TP (2004) Comparison of two sun tracking methods in the application of a heliostat field. J Solar Energy Eng Trans ASME 126:638

Chen YT, Chong KK, Lim CS, Lim BH, Tan BK, Lu YF (2005) Report on the second prototype of non-imaging focusing heliostat and its application in food processing. Sol Energy 79:280

Churchill SW, Bernstein M (1977) A correlating equation for forced convection from gases and liquids to a circular cylinder in cross flow. J Heat Transf Trans ASME 99:300

Date AW (2000) Numerical prediction of Laminar flow and heat transfer in a tube with twisted-tape insert: effects of property variations and Buoyancy. J Enhanc Heat Transf 7:217–229. https://doi.org/10.1615/JENHHEATTRANSF.V7.I4.10

Energy Information Administration (1980) Annual report to congress, Volume Two: Data. DOE/EIA-0173 (80)/2. U.s. Department of Energy

EnviroMission Solar Tower. http://pesn.com/Radio/FreeEnergyNow/shows/2006/10/28/9700213EnviroMissionsolartower. Accessed on 27 Nov 2006

Fend T, Hoffschmidt B, Pitz-Paal R, Reutter O, Rietbrock P (2004) Porous materials as open volumetric solar receivers: experimental determination of thermophysical and heat transfer properties. Energy 29:823

Francia G (1968) Pilot plants of solar steam generating stations. Sol Energy 12:51

Gehlisch K, Heikal H, Mobarak A, Simon M (1982) Large parabolic dish collectors with small gas-turbine, stirling engine or photovoltaic power conver- sion systems. Proc Intersoc Energy Convers Eng Conf Los Angeles, USA 5:2186

Grassie SL, Sheridan NL (1977) The use of planar reflectors for increasing the energy yield of flat-plate collectors. Sol Energy 19:663

Hilpert R (1933) Warmeabgabe von geheizen Drahten und Rohren. Forsch Gebiete Ingenieurw 4:220

Hoffschmidt B, Tellez FM, Valverde A, Fernandez J, Fernandez V (2003) Performance evaluation of the 200-kWth HiTRec-II open volumetric air receiver. J Sol Energy Eng Trans ASME 125:87

Hong SW, Bergles AE (1976) Aug of of Hea Hong & by means of twisted tape inserts. J Heat Transf Trans ASME 98:251

Kedare SB, Nayak JK, Paranjape AD (2006) Development, installation and evaluation of large scale concentrating solar collector for medium temperature industrial thermal applications. Final Report of Project No. 15/ST/2002

Kelkar KM (1982) Performance analysis of a cylindrical parabolic collector, B. Tech. Thesis, Department of Mechanical Engineering, IIT Bombay

Kreider JF, Kreith F (1981) Solar energy handbook. McGraw-Hill Book Company, New York

Kreith F (1976) Principles of heat transfer. Harper and Row, New York

Kreith F, Kreider JF (1978) Principles of solar engineering. McGraw-Hill Book Company, New York

Mancini T, Heller P, Butler B, Osborn B, Schiel W, Goldberg V, Buck R, Diver R, Andraka C, Moreno J (2003a) Dish-Stirling systems: An overview of development and status. J Solar Energy Eng Trans ASME 125:135

Mancini T, Heller P, Butler B, Osborn B, Schiel W, Goldberg V, Buck R, Diver R, Andraka C, Moreno J (2003) Dish-stirling systems: an overview of development and status. J Solar Energy Eng Trans ASME 125:135

References

McDaniels DK, Lowndes DH, Mathew H, Reynolds J, Gray R (1975) Enhanced solar energy collection using reflector-solar thermal collector combinations. Sol Energy 17:277

Mullick SC, Nanda SK (1989) An improved technique for composting to heat loss factor of a tubular absorber. Sol Energy 42:1

Rabl A (1976a) Comparison of solar concentrators. Sol Energy 18:93

Rabl A (1976b) Optical and thermal properties of compound parabolic consensus tors. Sol Energy 18:497

Rabl A (1985) Active solar collectors and their applications. Oxford University Press, New York

Raithby GD, Hollands KGT (1975) A general method of obtaining approximate solutions to laminar and turbulent free convective problems. Adv Heat Transf 11:265

Romero M, Buck R, Pacheco JE (2002) An update on solar central receiver systems, projects and technologies. J Solar Energy Eng Trans ASME 124:98

Schlaich J, Bergermann R, Schiel W, Weinrebe G (2005) Design of commercial solar updraft tower systems-utilisation of solar induced convective flows for power generation. J Heat Trans Trans ASME 127:117

Schramek P, Mills DR (2004) Heliostats for maximum ground coverage. Energy 29:701

Sukhatme SP, Nayak JK (2012) Solar energy, principles of thermal collection and storage. McGraw-Hill Book Co., New Delhi

Vant-Hull LL, Hildebrandt AF (1976) Solar thermal power system based on optical transmission. Sol Energy 18:31

Winston R (1974) Principles of solar concentrators of a novel design. Sol Energy 16:89

Chapter 8
Solar Pond

Abstract In this chapter fundamentals, architecture and performance traits of solar ponds have been thoroughly discussed. Effects of temperature dispersion in solar ponds as well as, a comprehensive assessment of the efficiency and viability of solar ponds as solar energy collecting systems have been discussed.

Keywords Solar energy conversion · Solar pond · Convective zone · Transmissivity · Temperature distribution

8.1 Introduction

More affordable methods of solar energy collection and storage must be created to lower the price of large solar thermal systems. In this respect, the focus has been brought to the potential for exploiting vast areas of shallow water for solar radiation absorption and storage in place of flat-plate collectors and hot water storage tanks. Experience suggests that, due to the natural convection current flow that is put into motion as soon as energy is captured at the bottom, the water in such a pond often warms up by only a few degrees. One would get a significant increase in the temperature only if the convection could be controlled. An artificially constructed pond in which significant temperature rises are caused to occur in the lower regions by preventing convection is called a "*solar pond*" (El-Sebaii et al. 2011).

Convection is often avoided by dissolving a salt in the water and maintaining a concentration gradient. The more precise phrase "salt-gradient solar pond" is used to describe these ponds. We will mainly talk about the "salt-gradient solar pond" in this chapter because it has advanced well. Other concepts like the gel solar pond and the equilibrium solar pond will only be described briefly.

Many salt-gradient solar ponds have been constructed in many countries over the past half a century, ranging in size from a few hundred to a few thousand square metres of surface area, largely for testing and demonstration purposes (El-Sebaii et al. 2011). They would likely be cost-effective for applications requiring low-temperature processing heat up to 70 °C or 80 °C, according to the indications. However, issues

© The Author(s), under exclusive license to Springer Nature Singapore Pte Ltd. 2024
K. Namrata et al., *Wind and Solar Energy Systems*, Energy Systems in Electrical
Engineering, https://doi.org/10.1007/978-981-99-9710-7_8

with their long-term maintenance and operation have been observed. This needs to be satisfactorily resolved if solar ponds are to be used extensively and developed commercially.

Early in the 1960s, Tabor and his colleagues constructed the first solar pond in Israel. Temperatures near the maximum value of 100 °C were achieved at the bottom, thereby demonstrating the working concept, however because of several practical issues, the project was abandoned. Since then, several solar ponds have been constructed all over the world, and the thermal energy contained inside has been utilized to produce both process heat and electricity. The 250,000 m^2 solar pond at Beit Ha'arvah in Israel is one of the biggest ones yet to be constructed. An organic fluid Ranking cycle (Einav 2004) was employed to produce electrical power from the heat stored in this pond. Currently, a 3240 m^2 solar pond with a depth of 3.2 m has been operated at El Paso, Texas (USA) since 1985. The major research focused on the desalination and brine management technologies (Lu et al. 2004). Another experimental pond is located at Pyramid Hill, Australia, and is being used to supply heat to dry air which is used in a salt production process (RMIT University 2007).

India was the first Asian country which initiate the utilization of the solar pond concept and built the massive solar pond having an area of 6000 m^2 at Bhuj, Gujarat, for several research purposes. The pond was being utilized to serve the dairies in nearby locations by supplying processed heat energy.

8.2 Working Principle of Solar Pond

The idea for a solar pond came from the findings that the lower portions of certain lakes that are found naturally do experience a large increment of temperature (of the order of 40–50 °C). As a result of the natural gradient in salt content, which is present in these lakes, causes the bottom-level water to remain denser even when the water at the surface is warm. Convection does not take place as a result, and conduction is the only way heated water loses heat. The amount of salt in these lakes is usually stabilized due to the availability of salt deposits at the lakes' bottoms, which result in congregations that are almost at saturation in the lower parts, and by fresh streams that pass at the surface.

Referring to Fig. 8.1 will help to understand how a solar pond operates. Consider a body of water with a depth L that has salts dissolved in it. We consider that there is a concentration gradient from top to bottom and that the concentration at the top C_1 is lower than the concentration at the base C_2. For the two concentrations, the density changes with temperature as indicated. Let T_1 and ρ_1 be the temperature and density of the top layer represent the depth of water, T_2 represents the temperature, and ρ_2 represents the density of the bottom layer. The curve AB is drawn to indicate the difference in density as one descends below in the pond and similar points are found for the intermediate layers. It is clear that no convection will happen as long as curve AB has a positive slope (Einav 2004).

8.2 Working Principle of Solar Pond

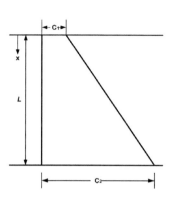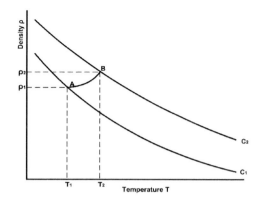

Fig. 8.1 Solar pond working principle

The mathematical criterion that the lower layers continue to be denser than the upper layers can be defined as

$$\frac{d\rho}{dx} > 0. \tag{8.1}$$

$\because \rho = \rho(C, T)$; consequently, the need for stability is

$$\left.\begin{array}{l} \left[\dfrac{\partial \rho}{\partial C}\right]_T \left[\dfrac{\partial C}{\partial x}\right] + \left[\dfrac{\partial \rho}{\partial T}\right]_C \left[\dfrac{dT}{dx}\right] > 0 \text{ or} \\[2mm] \dfrac{dC}{dx} > -\left\{\left(\dfrac{\partial \rho}{\partial T}\right)_C \left(\dfrac{dT}{dx}\right) \bigg/ \left(\dfrac{\partial \rho}{\partial C}\right)_T\right\} \end{array}\right\}. \tag{8.2}$$

From a slightly more sophisticated analysis which considers the effect of small perturbations, it can be shown that

$$\frac{dC}{dx} > -\left\{\left(\frac{v+\alpha}{v+D}\right)\right\}\left\{\left(\frac{\partial \rho}{\partial T}\right)_C \left(\frac{dT}{dx}\right) \bigg/ \left(\frac{\partial \rho}{\partial C}\right)_T\right\}, \tag{8.3}$$

where D, v, and α denote the salt diffusivity in water, kinematic viscosity, and thermal diffusivity, respectively.

The value of the term $(v + \alpha)/(v + D)$ for saltwater solutions under the circumstances present in solar ponds is around 1.15. As a result, the stability requirement provided by Eq. (8.3) is a bit more explicit than the stability criterion provided by Eq. (8.2). Calculating the minimal concentration necessary to maintain a certain thermal gradient at a specific level in a solar pond may be done using Eqs. (8.2) or (8.3). In real fact, some certain safety margin is encouraged, and the concentration gradient is kept at a value that is roughly double what Eq. (8.3) suggests.

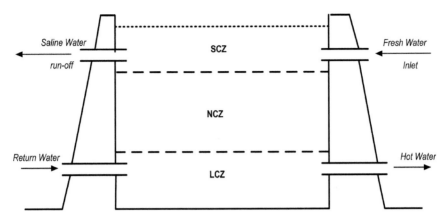

Fig. 8.2 Solar pond schematic graph

8.3 Description of Solar Pond

In Fig. 8.2, a solar pond has been shown schematically. As mentioned previously, it combines the functions of heat collection with long-term storage and can provide sufficient heat for the entire year. It often has a thick, durable plastic liner put down at the bottom and is 2–3 m deep. Low density polyethylene (LDPE), high density polyethylene (HDPE), woven polyester yarn (XB-5), and hypalon reinforced with nylon mesh are the materials required to make the liner. The salts such as sodium chloride (NaCl), magnesium chloride ($MgCl_2$), or sodium nitrate (Na_2No_3), are dissolved in the water with concentrations ranging from 20 to 30% at the lower region to nearly zero at the surface (Karakilcik et al. 2006a). If left to its own, the upward salt diffusion will eventually cause the gradient in salt concentration to disappear. At the surface of the pond, fresh water is added to sustain it, while somewhat salty water is run-off. Concentrated brine is also added to the pond's bottom at the same time. About 50 g/m^2 per day of salt is needed for this purpose, which is a significant amount when seen on a yearly basis. Hence the salt is generally recycled by evaporation of salty water at the top and discharge in the evaporating pond adjacent to it.

Hot water is continually extracted from the bottom, circulated via a heat exchanger, and then restored to the bottom to recover the energy contained below. As an alternative, water is used to remove heat from a heat exchanger coil that is immersed at the bottom. The solar pond can be categorized based on the fluid mixing at the top and bottom levels and movement in three regions:

1. Surface Convective Zone (SCZ): It typically has a low thickness, between 10 and 20 cm. It has a reasonably consistent temperature that is near the atmospheric temperature and a small, homogeneous concentration that is almost zero.
2. Non-convective Zone (NCZ): It fills more than half of the pond's depth and is substantially thicker. Temperature and substance concentration increases with depth in this region. Its primary function is to act as an insulating layer, which

8.4 Performance Analysis

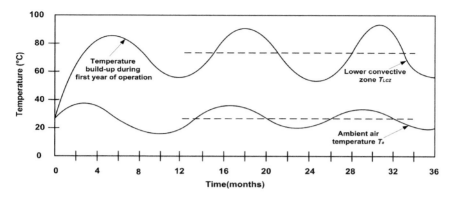

Fig. 8.3 Typical annual cyclic variation of daily mean temperature in a solar pond

lowers upward heat losses. This region also acts as a portion of the thermal storage and is where the heat is collected in small quantities.
3. Lower Convective Zone (LCZ): Its thickness is similar to that of the non-convective zone. In this region, the temperature and concentration are both relatively constant. It functions as both the primary heat-collection and thermal-storage medium. The storage region or bottom zones are two terms often used to describe the lowest convective zone.

A zone of lower convection temperatures in a well-designed, large operating pond in India may see seasonal changes from a peak of 85–95 °C in the summer to a low of 50–60 °C in the cold season. It can be seen in Fig. 8.3, which also depicts the fluctuation in average ambient temperatures. It will be noted that there is a phase difference of a month or two between the two curves. The annual collection efficiency each year varies between 15 and 25%.

These readings are lower than what a flat-plate collector would provide. Nonetheless, solar ponds are more cost-effective since they are much less expensive per square metre than liquid flat-plate collector systems when the area is more than a thousand square metres.

8.4 Performance Analysis

The first step to assess the solar pond's performance is to ascertain how the radiation that strikes the solar pond is reflected, absorbed, and propagated through the water (Karakilcik et al. 2006b).

8.4.1 Transmissivity Based on Reflection–Refraction at the Air–Water Interface

By taking into account reflection–refraction and absorption independently, the transmissivity of a collector's cover system may be determined with enough precision and is provided by the product form

$$\tau = \tau_r \tau_a, \tag{8.4}$$

where τ_r = Transmissivity determined only by reflection and refraction and τ_a = Transmissivity calculated using only absorption.

When a beam of light travelling through transparent medium 1 of intensity I_{bn} encounters the interface separating it from transparent medium 2, it is reflected and refracted (Fig. 8.4). The reflected beam has a lower intensity of I_r and is directed in a way that makes the incident angle and reflection equal. The directions of the incident and refracted rays, on the other hand, are connected by Snell's law, which stipulates that

$$\frac{\sin \theta_1}{\sin \theta_2} = \frac{n_2}{n_1}, \tag{8.5}$$

where θ_1, θ_2, n_1, and n_2 denote the incident angle, refraction angle, and refractive index of two media, respectively. The reflectivity $\rho \left(= \frac{I_r}{I_{bn}} \right)$ is related to the incidence and refraction angle by the following equations:

$$\rho = \frac{1}{2}(\rho_{\mathrm{I}} + \rho_{\mathrm{II}}),$$

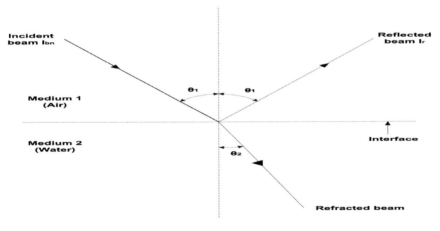

Fig. 8.4 Reflection and refraction at the interface of two media

8.4 Performance Analysis   427

Table 8.1 Transmissivity at the water surface based on reflection and refraction

Angle of incidence θ_1 (degree)	Angle of refraction θ_2 (degree)	ρ_I Equation 8.6	ρ_{II} Equation 8.7	$\rho = \frac{1}{2}(\rho_I + \rho_{II})$	$\tau_r = (1 - \rho)$
0	0	0.020	0.020	0.020	0.980
15	11.32	0.022	0.018	0.020	0.980
30	22.08	0.030	0.012	0.021	0.979
45	32.12	0.052	0.003	0.027	0.973
60	40.63	0.114	0.004	0.059	0.941
75	46.57	0.312	0.111	0.211	0.789
90	48.75	1	1	1	0

$$\rho_I = \frac{\sin^2(\theta_2 - \theta_1)}{\sin^2(\theta_2 + \theta_1)}, \tag{8.6}$$

$$\rho_{II} = \frac{\tan^2(\theta_2 - \theta_1)}{\tan^2(\theta_2 + \theta_1)}. \tag{8.7}$$

ρ_I and ρ_{II} are the reflectivities of the two components of polarization. For the special case of normal incidence ($\theta_1 = 0°$), it can be represented that

$$\rho = \rho_I = \rho_{II} = \left(\frac{n_1 - n_2}{n_1 + n_2}\right)^2. \tag{8.8}$$

The transmissivity τ_r is given by an expression similar to that for ρ. Thus

$$\tau_r = \frac{1}{2}(\tau_{rI} + \tau_{rII}), \tag{8.9}$$

where τ_{rI} and τ_{rII} are the transmissivities of the two components of polarization. We assume that the refractive index of water in comparison with air is 1.33. Table 8.1 lists the values τ_r obtained at various incidence angles. For incidence angles between $0°$ and $60°$, it can be shown that the loss from reflection is minimal and varies from 2 to 6%. The loss is considerable at higher angles. Large angles of incidence, which are linked to low radiation levels, are often not of interest from the perspective of energy collecting in the pond.

8.4.2 Transmissivity Based on Absorption

Rabl and Nielsen (1975) have shown the transmissivity based on absorption for the water surface in the following form:

428   8 Solar Pond

$$\tau_a = \sum_{j=1}^{4} A_j e^{-K_j x},\qquad (8.10)$$

where $x =$ water depth and $K =$ extinction coefficient. The constants A_j and K_j are found to have the following values after fitting the available experimental data.

$$A_1 = 0.237,\ K_1 = 0.032\ \text{m}^{-1}\ \text{for}\ 0.2 < \lambda < 0.6\,\mu\text{m}$$
$$A_2 = 0.193,\ K_2 = 0.45\ \text{m}^{-1}\ \text{for}\ 0.6 < \lambda < 0.75\,\mu\text{m}$$
$$A_3 = 0.167,\ K_3 = 3\ \text{m}^{-1}\ \text{for}\ 0.75 < \lambda < 0.9\,\mu\text{m}$$
$$A_4 = 0.179,\ K_4 = 35\ \text{m}^{-1}\ \text{for}\ 0.9 < \lambda < 1.2\,\mu\text{m}.$$

It is estimated that Eq. (8.10) is accurate to within 3%.

The four different values of A in Eq. (8.10) correspond to the wavelength range 0.2–1.2 µm and add up to 0.776. Therefore, only 77.6% of the radiation is addressed. The remaining 22.4%, or radiation with wave lengths greater than 1.2 µm, is absorbed within the first 1–2 cm very near the surface. Thus, Eq. (8.10) is valid for all depths except the first 1 or 2 cm.

Bryant and Colbeck (1977) proposed an alternate and straightforward equation for computing τ_a. It is as follows:

$$\tau_a = 0.36 - 0.08 \ln x,\qquad (8.11)$$

where $x =$ depth of water in metres. Equation (8.11) is as accurate as Eq. (8.10) for $x > 0.01$ m. It is advised to substitute $x/\cos\theta_2$, where θ_2 shows the refraction angle, for x in both equations if the radiation is not striking normally.

Example 8.1 A 1.5 m deep solar pond is built in Pondicherry (11°56′ N). The following values of global and diffuse radiation are measured by a horizontal pyranometer placed beside the pond on April 20 at 1300 h (LAT):

$$I_g = 0.964\ \text{kW/m}^2,\ \ I_d = 0.210\ \text{kW/m}^2.$$

Calculate the variation of the solar radiation flux as it penetrates through the pond.

Solution $n = 110$ (for April 20),

$$\delta = 23.45 \sin\left\{ \frac{360}{365}(284 + 110) \right\} = 11.23°.$$

From Eq. (2.6), the incident angle of the beam radiation can be calculated as

$$\cos\theta_1 = \sin 11.93° \sin 11.23° + \cos 11.93° \cos 11.23° \cos(-15°) = 0.9672,$$
$$\theta_1 = 14.71°.$$

8.4 Performance Analysis

Hence from Eq. (8.5) the refraction angle $\theta_2 = \sin^{-1}\left\{\frac{\sin 14.71°}{1.33}\right\} = 11.01°.\theta_2 = \sin^{-1}\left\{\frac{\sin 14.71°}{1.33}\right\} = 11.01°$.

From Table 8.1, $\rho_b = 0.020$.

For diffuse radiation, here we are considering the incident angle to be 60°. Hence, from Table 8.1, refraction angle $= 40.63°$ and $\rho_d = 0.059$.

Flux reflected from the water surface

$$= I_b\rho_b + I_d\rho_d = (0.964 - 0.210) \times 0.020 + (0.210 \times 0.059) = 0.027\,\text{kW/m}^2.$$

Therefore, flux entering the water medium $= 0.964 - 0.027 = 0.937\,\text{kW/m}^2 = 0.964 - 0.027 = 0.937\,\text{kW/m}^2$.

Equation (8.11) is used for calculating the transmissivity based on absorption at depths of 0.01, 0.1, 0.5, 1.0, and 1.5 m.

At $x = 0.01$ m,

τ_a for beam radiation $(\tau_{ab}) = 0.36 - 0.08\ln(0.01/\cos 11.01°) = 0.7269$,

τ_b for diffuse radiation $(\tau_{ad}) = 0.36 - 0.08\ln(0.01/\cos 40.63°) = 0.7063$.

Hence, solar flux (I) at a depth of 0.01 m

$$= I_b\tau_{rb}\tau_{ab} + I_d\tau_{rd}\tau_{ad}$$
$$= (0.964 - 0.210)(1 - 0.020) \times 0.7269 + 0.210 \times (1 - 0.059) \times 0.7063$$
$$= 0.677\,\text{kW/m}^2.$$

In the same way, we get the following solar flux values at different depths:

Depth (m)	0.1	0.5	1.0	1.5 m
Solar flux (I) (kW/m^2)	0.541	0.384	0.332	0.301 kW/m^2

Figure 8.5 illustrates the solar radiation flux variation with depth. It is clear to notice how much energy is absorbed close to the surface. Assuming that the surface convective zone is 10 cm thick in this instance, it can be shown that 0.396 kW/m^2, or 41% of the incoming energy, is absorbed there. One of the primary causes of a solar pond's poor collecting efficiency is that nearly all of the energy is dissipated to the environment. Additionally, it can be observed in Fig. 8.5 that the flux reaching the pond's bottom is 0.301 kW/m^2 or 31% of the incident energy.

Finally, it is important to note that Eqs. (8.10) and (8.11), which calculate the values of τ_a, are applicable to water. There are certain values available for the salt solutions used in solar ponds. These are lower than the values for water alone.

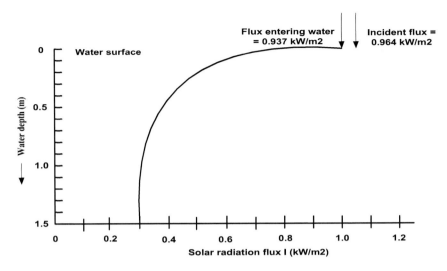

Fig. 8.5 Solar radiation absorption and reflection of a solar pond

8.4.3 Temperature Distribution and Collection Efficiency

Since a solar pond has three zones, calculating the temperature distribution in the pond requires some effort. For an accurate solution, one must solve the proper differential equation for each zone, employ matching conditions at the zone interfaces, and meet the pond's surface boundary conditions at its top and bottom. Because of how complicated it is, people usually make some changes to make it easier to understand. Here, we will provide a formulation that results in a single set of equations. This formula makes the assumption that the upper convective zone and the lower convective zone are fully mixed layers with homogeneous temperatures that only vary over time (Karakilcik et al. 2006b; Bryant and Colbeck 1977).

The differential equation for the non-convective zone is the heat conduction equation, assuming that the lateral dimensions of the pond are large relative to its depth L (such that the temperature changes exclusively in the vertical direction) can be expressed as

$$\rho C_p \frac{\partial T_{II}}{\partial t} = k \frac{\partial^2 T_{II}}{\partial x^2} - \frac{dI}{dx}, \tag{8.12}$$

with $I = I_b \tau_{rb} \tau_{ab} + I_d \tau_{rd} \tau_{ad}$.

The solar radiation absorbed in the pond is represented by the expression $(-dI/dx)$. Energy balances are used to determine the differential equations that must be met for the top and bottom layers of the pond (Fig. 8.6).

For the Surface Convective Zone (SCZ): Rate of change energy change in the SCZ having thickness of l_1.

8.4 Performance Analysis

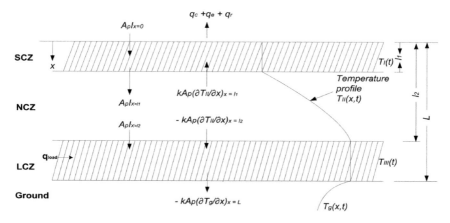

Fig. 8.6 Flow of energy in and out of the SCZ and the LCZ

= (Rate at which heat is conducted in from the non-convective zone) + (Solar radiation absorbed in the thickness l_1) − (Rate at which heat is lost by convection, evaporation, and radiation from the top surface).

Thus,

$$\rho l_1 C_p \left(\frac{dT_\mathrm{I}}{dt}\right)_{x=l_1} = k\left(\frac{\partial T_\mathrm{II}}{\partial x}\right)_{x=l_1} + [(I)_{x=0} - (I)_{x=l_1}] - \frac{1}{A_p}(q_c + q_e + q_r). \tag{8.13}$$

For the Lower Convective Zone (LCZ): Rate of energy variation in the LCZ having a thickness of $(L_1 - l_2)$

= (Rate at which heat is conducted in from the NCZ) + (Solar radiation absorbed in the thickness l_2) − (Rate at which heat is conducted out to the ground underneath) − (Effective heat extraction rate).

Thus,

$$\rho(L - l_2)C_p\left(\frac{dT_\mathrm{III}}{\partial t}\right)_{x=l_2} = -k\left(\frac{\partial T_\mathrm{II}}{\partial x}\right)_{x=l_2} + (I)_{x=l_2} - \left[-k_g\left(\frac{\partial T_g}{\partial x}\right)_{x=L}\right] - \frac{q_\mathrm{load}}{A_p}. \tag{8.14}$$

Numerous researchers have discovered solutions to the set of equations from (8.12)–(8.14) or similar sets (derived by using different simplifications). They have been addressed for the situation when there is no useful heat extraction from the pond ($q_\mathrm{load} = 0$) and the case where there is heat extraction. Here is a discussion of these solutions.

By using a superposition approach, Weinberger provided the first analytical solution of the differential Eq. (8.12), taking into account the effects of radiation absorption at the top, inside the body of water, and at the bottom individually. In order to further simplify the issue, he ignored the convective zone thicknesses and made the boundary assumption that the temperature of the pond surface is the same as the ambient air temperature. Thus, correlations were not required to determine the losses from the surface q_c, q_e, and q_r. Weinberger provided the first estimates of the temperature increase that may be anticipated in a solar pond using a sine–cosine series for the solar radiation falling on the pond surface and a sine series for the ambient temperature variance. He estimated a temperature spike of almost 100 °C for an Israeli pond one metre deep under typical circumstances when there is no heat extraction.

While the pond is allowed to heat up initially, the energy needed for utilization is then extracted, preventing the bottom layer's temperature from rising further. Weinberger (Weinberger 1964) discovered an important outcome that for a given mean extraction temperature there is a certain pond depth at which the rate of heat extraction is maximum. He also demonstrated that the solar radiation that reaches the pond's bottom is equal to the maximum rate of energy removal. Results for a site in Israel with a mean ambient temperature of 26 °C and an average yearly daily global radiation of 21,140 kJ/m^2-day are presented in Fig. 8.7. The ideal mean extraction temperature and the associated yearly collection efficiency may be found for a particular pond depth. Weinberger's study overestimates these values, according to further analyses. In areas with comparable yearly values of air temperature and global radiation, Fig. 8.7 may be used to roughly estimate solar pond behaviour.

A mathematical solution for the yearly temperature change in a solar pond used for room heating has also been provided by Rabl and Nielsen (1975). They took into

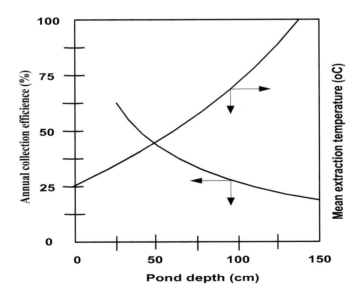

Fig. 8.7 Annual collection efficiency and optimum extraction temperature as a function of pond depth (Weinberger 1964)

8.4 Performance Analysis

433

account a pond that was several metres deep and had a lower convective zone with a limited thickness but a thin surface convective zone. Neglecting daily changes, the lower convective zone's temperature is calculated using the formula

$$T_{III} = \hat{T}_{III} + \tilde{T}_{III} \cos(wt - \delta),$$

where $\hat{T}_{III}$ = Lower convective zone annual average temperature, $\tilde{T}_{III}$ = Amplitude, ω = frequency, and δ = phase lag with insolation variation. By resolving the heat conduction Eq. (8.12) in the steady state, one may quickly and simply get the yearly average temperature $\hat{T}_{III}$, which is independent of time. The solution obtained is as follows:

$$\hat{T}_{III} - \hat{T}_a = \frac{\tau_r \hat{H}_g}{k} \sum_{j=1}^{4} \frac{A_j}{K'_j} \left(1 - e^{-K'_j l_2} \right) - \frac{l_2}{k} \frac{\hat{q}_{load}}{A_p}. \tag{8.15}$$

In the above equation $\hat{T}_a$ = annual average ambient temperature, $\hat{H}_g$ = annual average global radiation, $K'_j = (K_j / \cos \theta_2)$, where θ_2 is a refraction angle that corresponds to an effective incident angle. This is assumed to represent the angle of incidence on the equinox day at 1400 LAT at the place under examination and $l_2 =$ Pond's depth at the lower area of NCZ, $\hat{q}_{load}$ = average heat extraction rate per year. The other symbols have been explained in earlier sections.

Equation (8.15) is a clear and effective method for calculating average performance or evaluating the dimensions of a solar pond for a given requirement.

Calculating $\hat{T}_{III} \cos(wt - \delta)$, the time-dependent component of T_{III}, is more challenging.

With the help of their approach, Rabl and Nielsen (1975) have performed detailed assessments for several locations in the USA with significantly varying climates. They discovered that the solar pond works effectively in every location and can provide sufficient heat even in areas close to the Arctic Circle. At latitudes of about 40°, they discover that the pond should be approximately comparable in surface area and volume to the space it is to heat. Calculations also indicate that for a particular location, heat load and annual mean extraction temperature, there is an optimum value of l_2 corresponding to which the pond area is a minimum.

Example 8.2 Estimate the area of a solar pond required for supplying 5×10^9 kJ of energy per year at an annual mean temperature of 70 °C for an industrial process heat application. The pond is located in Nagpur (21°6′ N, 79°5′ E) where the annual average daily global radiation is 19600 kJ/m^2-day and the annual average ambient temperature is 26.3 °C. Assume that the depth of the pond at the bottom of the non-convective zone is 0.95 m.

Solution First, we figure out the effective incidence and refraction angle which are collected at 1400 h (LAT) on the equinox day ($\delta = 0°$).

$$\cos \theta_1 = \cos \phi \cos \omega = \cos 21.1° \cos(-30°) = 0.8080,$$
$$\theta_1 = 36.10°.$$

Hence, Refraction angle $\theta_2 = \sin^{-1}\left(\frac{\sin 36.10°}{1.33}\right) = 26.30°$, $\cos \theta_2 = 0.8965$, $\cos \theta_2 = 0.8965$,

$$\therefore \quad K'_j = K_j / 0.8965.$$

According to Table 8.1, the τ_r value, which corresponds to an incident angle of 36.10°, is 0.976. We'll choose a water thermal conductivity value that roughly corresponds to the mean temperature over the whole pond. Using 50 °C as the base, we get $k = 0.648$ W/m K. $k = 0.648$ W/m K.

Now, annual average daily global radiation

$$\hat{H}_g = 19{,}600 \, \text{kJ/m}^2\text{day} = \frac{19{,}600 \times 1000}{3600 \times 24} = 226.9 \, \text{W/m}^2.$$

Annual average heat extraction rate

$$\hat{q}_{\text{load}} = 5 \times 10^9 \, \text{kJ/year} = \frac{5 \times 10^9 \times 10^3}{365 \times 24 \times 3600} = 158549 \, \text{W}.$$

Substituting into Eq. (8.15), we have

$$70 - 26.3 = \frac{0.976}{0.648} \times 226.9 \sum_{j=1}^{4} \frac{0.8965 A_j}{K_j} \left(1 - e^{-0.95 K_j / 0.8965}\right) - \frac{0.95}{0.648} \times \frac{158549}{A_p}.$$

Using the K_j and A_j values provided in Eq. (8.10), we get

$$70 - 26.3 = 141.85 - \frac{232{,}433}{A_p} \quad \text{or } A_p = 2368 \, \text{m}^2.$$

Annual collection efficiency $= \frac{158{,}549}{226.9 \times 2368} \times 100 = 29.5\%. \times 100 = 29.5\%.$

Additionally, Kooi (1979) examined the solar pond as a steady-state device and came up with solutions expressed in terms of time-independent yearly average values. In a steady state, Eq. (8.12) for the non-convective zone shrinks to

$$k \frac{d^2 \hat{T}_{\text{II}}}{dx^2} = \frac{d}{dx}\left(\hat{H}_g \, \tau_r \tau_a\right) = \hat{H}_g \, \tau_r \frac{d\tau_a}{dx}, \tag{8.16}$$

8.4 Performance Analysis

where the effective incident angle is used to get the values of τ_r and τ_a. Integrating twice, we obtain

$$k \hat{T}_{II} = \hat{H}_g \tau_r \int_{l_1}^{x} \tau_a dx + c_1 x + c_2, \tag{8.17}$$

where c_1 and c_2 are constants of integration.

Substituting the conditions

$$\text{at } x = l_1, \ \hat{T}_{II} = \hat{T}_I \ \text{(a constant)},$$

$$\text{at } x = l_2, \ \hat{T}_{II} = \hat{T}_{III} \ \text{(a constant)}.$$

The formulas for the distribution of temperature gradients and temperature distribution in the non-convective zone may be found by solving for c_1 and c_2. With the use of these formulations and the knowledge that, in a steady-state condition, the rate of energy flow at the interface $x = l_2$ into the lower convective zone is equal to the average yearly heat extraction rate from the pond, we obtain

$$\hat{q}_{load} = A_p \left[\left\{ \frac{\hat{H}_g \tau_r}{(l_2 - l_1)} \int_{l_1}^{l_2} \tau_a dx \right\} - \frac{k}{(l_2 - l_1)} \left\{ \hat{T}_{III} - \hat{T}_I \right\} \right]. \tag{8.18}$$

The solar pond's annual average efficiency is calculated by dividing both sides of Eq. (8.18) by $\hat{H}_g A_p$. We have,

$$\hat{\eta} = \left\{ \frac{\tau_r}{(l_2 - l_1)} \int_{l_1}^{l_2} \tau_a dx \right\} - \frac{k}{(l_2 - l_1)} \left\{ \frac{\hat{T}_{III} - \hat{T}_I}{\hat{H}_g} \right\}. \tag{8.19}$$

Using Rabl and Nielsen's Eq. (8.10) for τ_a and substituting K'_j for K_j we have

$$\int_{l_1}^{l_2} \tau_a dx = \sum_{j=1}^{4} \frac{A_j}{K'_j} \left(e^{-K'_j l_1} - e^{-K'_j l_2} \right). \tag{8.20}$$

Thus Eqs. (8.18) and (8.19) will be

$$\hat{q}_{load} = A_p \left[\left\{ \frac{\hat{H}_g \tau_r}{(l_2 - l_1)} \sum_{j=1}^{4} \frac{A_j}{K'_j} \left(e^{-K'_j l_1} - e^{-K'_j l_2} \right) \right\} - \frac{k}{(l_2 - l_1)} \left\{ \hat{T}_{III} - \hat{T}_I \right\} \right], \tag{8.21}$$

and

$$\hat{\eta} = \left\{ \frac{\tau_r}{(l_2 - l_1)} \sum_{j=1}^{4} \frac{A_j}{K'_j} \left(e^{-K'_j l_1} - e^{-K'_j l_2} \right) \right\} - \frac{k}{(l_2 - l_1)} \left\{ \frac{\hat{T}_{III} - \hat{T}_1}{H_g} \right\}. \quad (8.22)$$

Equation (8.21) has a structure that is similar to Eq. (8.15). Like Eq. (8.15), it may be used to determine the average performance of a solar pond or to determine the area of a pond for a particular need. It should be observed that Eq. (8.21) simplifies to Eq. (8.15) if the thickness of the surface convective zone is ignored ($l_1 = 0$) and it is assumed that there is no temperature difference between the surface of the pond and the ambient, i.e. $\hat{T}_1 = \hat{T}_a$.

Example 8.3 Use Eq. (8.21) based on Kooi's analysis for calculating the area of a solar pond working under the same conditions as in Example 8.2. Take the depth of the surface convective zone to be 0.10 m and assume that the temperature of this zone is equal to the ambient air temperature.

Solution In Eq. (8.21) by substitution, we find

$$158{,}549 = A_p \left[\left\{ \frac{226.9 \times 0.976}{(0.95 - 0.10)} \sum_{j=1}^{4} \frac{0.8965 A_j}{K_j} \left(e^{-K_j \times 0.10/0.8965} - e^{-K_j \times 0.95/0.8965} \right) \right\} - \frac{0.648}{(0.95 - 0.10)} (70 - 26.3) \right].$$

Using the K_j and A_j values provided in Eq. (8.10), we have

$$158{,}549 = A_p[226.9 \times 0.412 - 0.76 \times 43.7]. \quad (8.23)$$

$$\therefore \quad A_p = 2639 \text{ m}^2.$$

This value is somewhat higher than the 2368 m^2 value that Rabl and Nielsen's algorithm provided. Since the surface convective zone's limited thickness has been taken into account, the estimate is probably more accurate.

In order to generate somewhat accurate estimates of pond performance, analytical approaches like those by Weinberger or Rabl and Nielsen must make a number of simplifying assumptions. It becomes necessary to solve the fundamental equations numerically using finite difference techniques in order to get more accurate solutions. The numerical techniques give the greater freedom to incorporate appropriate initial and boundary conditions and permit a more realistic representation of climatic conditions as well as load variations. A number of such investigations have been carried out (Mansour et al. 2006; Angeli et al. 2006; Jubran et al. 2004; Sukhatme and Nayak 2009).

8.4 Performance Analysis

Review Questions

(1) What do you mean by "salt-gradient pond" and what is the present status of solar ponds all over the world?
(2) Briefly explain the working principle of the solar pond.
(3) Draw the schematic graph of a solar pond and explain.
(4) Discuss the three zones of solar pond.
(5) Discuss the transmissivity at the water surface based on reflection and refraction.

Problems

8.1 Sodium chloride is used as the salt in a solar pond. Estimate the minimum concentration (kg of salt per kg of water) required at the bottom if the concentration at the top is 0.02 and a temperature difference of 65 °C is to be maintained. Assume that the concentration and temperature profiles are straight lines and take the average values of $(\partial \rho / \partial T)$ and $(\partial \rho / \partial C)$ to be $- 0.5$ kg/m^3 °C and 650 kg/m^3, respectively.

Ans: 0.07

8.2 A solar pond, 2.5 m deep, is proposed to be built in Bhavnagar (21°45′ N, 72°11′ E). Calculate the solar flux which would be received at the bottom of the pond on March 15 for which the following radiation data is available:

Time (h LAT)	0730	0930	1130	1330	1530	1730
I_g (kW/m^2)	0.308	0.721	0.921	0.840	0.496	0.070
I_d (kW/m^2)	0.111	0.157	0.183	0.165	0.140	0.040

Ans: 0.072, 0.189, 0.248, 0.225, 0.128, 0.014 kW/m^2

8.3 A solar pond located in New Delhi (28°35′ N, 77°12′ E) is to be used for supplying the heat energy input for a low-temperature Rankine cycle power unit using R-11 as the working fluid. The unit develops an output of 20 kW, has an energy conversion efficiency of 9.0%, and requires its energy input at an annual mean temperature of 85 °C. Use the Rabl and Nielsen analysis to calculate the pond area for values of the depth l_2 varying from 1 to 3 m. Show that the area is a minimum for a particular value of l_2. The following data is given for New Delhi:
Annual average daily global radiation = 19 690 kJ/m^2 day.
Annual average ambient temperature = 25.1 °C.

Ans:

l_2 (m)	1	1.5	2	2.5	3
A_p (m^2)	3862	3536	3480	3500	3551

Minimum area at $l_2 = 2$ m.

438   8   Solar Pond

8.4 Use the data of problem 8.3 and the analysis due to Kooi, to calculate the area of
a solar pond for a situation in which $l_1 = 0.25$ m, $l_2 = 1.75$ m, and $L = 2.50$ m.
Assume that the SCZ's temperature is the same as the ambient air temperature.
Calculate also the quantities equivalent to the transmissivity-absorptivity term
and the loss term for this pond.

Ans: $Ap = 3823$ m^2, 0.3695, 0.435 W/m^2 K

Objective Type Questions

(1) Generally, solar ponds are utilized for

   (a) Solar energy reflection
   (b) Solar energy collection and reflection
   (c) Solar energy diversion
   (d) None of these

(2) Solar pond has

   (a) Fresh water
   (b) Salty water
   (c) Sea water
   (d) None of these

(3) The solar pond has a combination of

   (a) Solar heat collection and energy storage
   (b) Solar energy collection and energy storage
   (c) Solar heat storage and energy collection
   (d) None

(4) The salt concentration in salty water

   (a) Remains the same throughout the pond
   (b) Increases from top layer to bottom layer
   (c) Decreases from top layer to bottom layer
   (d) None

(5) Which layer or region stores the energy in the solar pond

   (a) Non-convective layer
   (b) Lower convective layer
   (c) Upper convective layer
   (d) Both a and b

(6) The upper convective layer

   (a) Has homogenous low salinity
   (b) Formed due to heat transfer and wind
   (c) Not useful for energy collection
   (d) All

References 439

(7) Which layer in the solar pond has a diffusive transport of salt and heat

   (a) Non-convective layer
   (b) Lower convective layer
   (c) Upper convective layer
   (d) None

(8) The non-convective layer generally refers

   (a) Double diffusion
   (b) Single diffusion
   (c) Triple diffusion
   (d) None

(9) Which types of salts are dissolved in solar ponds?

   (a) Sodium chloride
   (b) Magnesium chloride
   (c) Sodium nitrate
   (d) All

(10) What are the salt concentrations at the bottom layer?

   (a) 5–10%
   (b) 10–15%
   (c) 15–20%
   (d) 20–30%

Answers

1. b	2. b	3. c	4. b	5. b
6. d	7. a	8. a	9. d	10. d

References

Angeli C, Leonardi E, Maciocco L (2006) A computational study of salt diffusion and heat extraction in solar pond plants. Sol Energy 80:1498–1508. https://doi.org/10.1016/J.SOLENER.2005.10.015

Bryant HC, Colbeck I (1977) Solar pond for London [For space heating of houses]. Sol Energy (United States) 19(3):321–322. https://doi.org/10.1016/0038-092X(77)90079-2

Einav A (2004) Solar energy research and development achievements in Israel and their practical significance. J Sol Energy Eng 126:921–928. https://doi.org/10.1115/1.1758246

El-Sebaii AA, Ramadan MRI, Aboul-Enein S, Khallaf AM (2011) History of the solar ponds: a review study. Renew Sustain Energy Rev 15(6):3319–3325. https://doi.org/10.1016/J.RSER.2011.04.008

Innovative Technology to Collect Solar Energy Energy for Heating Purposes and Reduce Greenhouse Gas Emissions, Mechanical and Automative Engineering, RMIT University. Australia [online] http://www.rmit.edu.au; Accessed on 21 May 2007

Jubran BA, Al-Abdali H, Al-Hiddabi S, Al-Hinai H, Zurigat Y (2004) Numerical modelling of convective layers in solar ponds. Sol Energy 77:339–345. https://doi.org/10.1016/J.SOLENER.2004.04.004

Karakilcik M, Dincer I, Rosen MA (2006a) Performance investigation of a solar pond. Appl Therm Eng 26(7):727–735. https://doi.org/10.1016/J.APPLTHERMALENG.2005.09.003

Karakilcik M, Kiymaç K, Dincer I (2006b) Experimental and theoretical temperature distributions in a solar pond. Int J Heat Mass Transf 49(5–6):825–835. https://doi.org/10.1016/J.IJHEAT MASSTRANSFER.2005.09.026

Kooi CF (1979) The steady state salt gradient solar pond. Sol Energy 23:37–45. https://doi.org/10.1016/0038-092X(79)90041-0

Lu H, Swift AHP, Hein HD, Walton JC (2004) Advancements in salinity gradient solar pond technology based on sixteen years of operational experience. J Sol Energy Eng 126:759–767. https://doi.org/10.1115/1.1667977

Mansour RB, Nguyen CT, Galanis N (2006) Transient heat and mass transfer and long-term stability of a salt-gradient solar pond. Mech Res Commun 33:233–249. https://doi.org/10.1016/J.MEC HRESCOM.2005.06.005

Rabl A, Nielsen CE (1975) Solar ponds for space heating. Sol Energy 17:1–12. https://doi.org/10.1016/0038-092X(75)90011-0

Sukhatme SP, Nayak JK (2009) Solar energy principles of thermal collection and storage, 3rd edn. TMH

Weinberger H (1964) The physics of the solar pond. Sol Energy 8:45–56. https://doi.org/10.1016/0038-092X(64)90046-5